GERMAN PRIVATE AND COMMERCIAL LAW:
AN INTRODUCTION

GERMAN PRIVATE AND COMMERCIAL LAW: AN INTRODUCTION

BY

NORBERT HORN
Professor of Law at the University of Bielefeld and Director of the Centre for Interdisciplinary Research

HEIN KÖTZ
Professor of Law at the University of Hamburg and Director of the Max Planck Institute for Foreign and International Private Law

AND

HANS G. LESER
Professor of Law at the University of Marburg and Director of the Institute of Comparative Law

TRANSLATED BY
TONY WEIR
Fellow of Trinity College, Cambridge

CLARENDON PRESS · OXFORD
1982

Oxford University Press, Walton Street, Oxford OX2 6DP
London Glasgow New York Toronto
Delhi Bombay Calcutta Madras Karachi
Kuala Lumpur Singapore Hong Kong Tokyo
Nairobi Dar es Salaam Cape Town
Melbourne Auckland
and associate companies in
Beirut Berlin Ibadan Mexico City

Published in the United States by
Oxford University Press, New York

© Norbert Horn, Hein Kötz, and
Hans G. Leser 1982

All rights reserved. No part of this publication may be reproduced,
stored in a retrieval system, or transmitted, in any form or by any means,
electronic, mechanical, photocopying, recording, or otherwise, without
the prior permission of Oxford University Press

British Library Cataloguing in Publication Data

Horn, Norbert
  German private and commercial law.
  1. Civil law — Germany (Federal Republic, 1949–)
  I. Title II. Kötz, Hein III. Leser, Hans G.
  344.306

ISBN 0-19-825382-6
ISBN 0-19-825383-4 Pbk

Lanchester Library

PO 3536

Typeset by Oxford Verbatim Limited
Printed in Great Britain
at the University Press, Oxford
by Eric Buckley
Printer to the University

Foreword

A survey of modern German private law is a subject of great attraction to the lawyer in the common law jurisdictions. Modern German law is still founded, to a large extent, on the great codes of 1897, which came into operation on 1 January 1900. This codification has sometimes been criticized as Professorenrecht, because undue attention has been paid to the controversy of the Germanistic school, led by von Gierke, and the Romanistic school of Windscheid. That the criticism is unjustified clearly emerges from a study of this volume. The codes have provided a firm foundation on which the German courts have been able to develop, in a pragmatic fashion, a modern legal system that satisfies the needs of a complex modern industial society. In that respect the German codes can be compared with the great Victorian codifications of commercial law in the United Kingdom. These commercial codes, though operating in a much more restricted field, are still, in essence, with us. Their finest examples are the Bills of Exchange Act 1882, the Sale of Goods Act 1893 (now 1979), the Partnership Act 1890 and the Marine Insurance Act 1906. The modern lawyer can but admire the intellectual achievement of his forerunners at the end of the ninteenth century.

The aim of this slim volume is to introduce the lawyer working in one of the common law jurisdictions to the study of German law. The work should be of interest to the practitioner, the academic lawyer, and the student of comparative law. It will enable the practitioner to understand the methodology and traditional way of thinking of his German colleague. It will indicate to the academic lawyer how a code law system — incomprehensible if one only reads the codes — is made to operate in practice, and it will give the comparative lawyer a key to an understanding of the science of comparative law, which is perhaps the most stimulating and interesting branch of modern law, so necessary to appreciate the growing interdependence of international commercial and personal relations.

The outstanding features of this work are utmost clarity of

presentation and its practical approach. Three distinguished and leading German scholars, Professors Norbert Horn, Hein Kötz, and Hans Leser, have combined to introduce the common lawyer to German law, as it is today. They are all experts in comparative law. They all have studied the common law at English or American institutions and approach their exposition of German law with a view to making clear what the common lawyer should know about it in order to appreciate the differences between his own legal system and German law. A warm tribute should be paid to Mr Tony Weir, the distinguished Cambridge scholar, who has translated the work into impeccable English with consummate skill.

This work owes its origin to the practical experience of teaching the elements of German private law to English lawyers. The International Summer Courses of Law, sponsored by the British Institute of International and Comparative Law and the London Office of the German Academic Exchange Service, have since 1973 annually arranged a course in Modern German Law for English Lawyers. This course, which is held in London, has attracted considerable interest, as the growing number of participants shows. The three authors of this book have taught at these courses regularly, and this book is the result of their accumulated teaching experience. They have wisely introduced their treatment of private law by a brief description of the German constitutional situation and a survey of the organization of the German courts.

The work constitutes an invaluable enrichment of the relatively scanty literature on comparative law in the English language. Following the publication of *An Introduction to Comparative Law* by Professors K. Zweigert and H. Kötz, likewise translated into English by Mr Tony Weir, it will become an indispensable tool for the teaching of comparative law at the law schools of the English-speaking countries and will acquire a position similar to that held by the renowned classic in French law, *Amos and Walton's Introduction to French Law*. The function of comparative law in the setting of our small modern world need not be emphasized. Every lawyer, whether in practice or in law teaching, knows that today this approach is a necessity for the understanding of his own legal system and for the proper performance of the tasks with which he will be faced.

Foreword vii

The thanks of the legal profession in the common law countries are due to the three German scholars who have undertaken the task of presenting this work and to their English translator. They have widened the horizon of the common lawyer.

Clive M. Schmitthoff

Authors' Preface

This book is based mostly on teaching experience the authors have gained since 1973 in summer courses on Modern German law held at the London School of Economics and sponsored by the British Institute of International and Comparative Law and the German Academic Exchange Service (DAAD). The authors hope that the book will serve as a useful first guide to the same sort of readers as participated in their courses over the years: young students of comparative law as well as experienced scholars in this field, practising lawyers as well as members of the civil service dealing with legal matters involving German law. The authors wish to thank the DAAD for promoting the project of this book and, in particular, the head of the London Branch of DAAD 1976–81, Franz Eschbach, who encouraged them to write it as well as Tony Weir who, in a most enjoyable collaboration, put their texts into readable English. The book was produced with the financial assistance of the Stifterverband für die deutsche Wissenschaft and the Deutscher Akademischer Austauschdienst.

The authors are grateful to the North-Holland Publishing Company for permission to paraphrase material which first appeared in chapters 17–20 of *Introduction to Comparative Law* by Konrad Zweigert and Hein Kötz (vol. 2).

The authors

Contents

Foreword		v
Authors' Preface		viii

1 Historical Introduction (*Horn*)

I	THE BUNDESREPUBLIK, THE FEDERAL REPUBLIC OF GERMANY	1
II	TRADE AND INDUSTRY IN GERMANY	4
III	GERMANY AS A CIVIL LAW COUNTRY	8

2 Constitutional Law (*Kötz*)

I	INTRODUCTION	14
II	THE SYSTEM OF GOVERNMENT	16
III	THE FEDERAL CONSTITUTIONAL COURT	20
IV	BASIC RIGHTS	21

3 The Administration of Justice (*Kötz*)

I	THE SYSTEM OF COURTS	27
	1 Ordinary Jurisdiction	27
	2 Labour Courts	33
	3 Administrative Courts	33
	4 Social Courts	34
	5 Tax Courts	34
II	JUDGES, ATTORNEYS, AND NOTARIES	35
	1 Legal Education	35
	2 Judges	37
	3 Attorneys	41
	4 Notaries	44
III	ELEMENTS OF CIVIL PROCEDURE	45
IV	EXPENSES OF LITIGATION AND LEGAL AID	48

4 Sources, Legal Literature, and the Code (*Leser*)

I	SOURCES; THE MAIN DIVISIONS OF LAW AND THEIR CODIFICATIONS	51
	1 The Divisions of Law	51
	2 Areas of Law, Main Codes, and Enactments	52
II	FINDING THE LAW	56
	1 Approaching German Legal Literature	56

	III	LEGISLATION, CODIFICATION, AND INTERPRETATION	58
		1 The Supremacy of Legislation	58
		2 Codification and Interpretation	60
		3 Creation of Law by Judges?	63
	IV	THE CIVIL CODE AND ITS STRUCTURE	64
		1 The Birth of the BGB	64
		2 Characteristics of the BGB	65
		3 Structure	66
		4 Fundamental Concepts	68

5 Contract: Capacity, Formation, and Freedom of Contract (*Leser*)

	I	PERSONS	71
		1 Persons with Rights	71
		2 The Legal Capacity of Human Beings	71
		3 Capacity to Act	73
	II	DECLARATION OF WILL AND FORMATION OF CONTRACT	74
		1 Declaration of Will, Juristic Act, and Interpretation	74
		2 Formation of Contract – Meeting of the Minds	76
		3 The Legal Effect of Silence in Negotiations	77
		4 The Factual Contract	79
	III	DEFECT OF INTENTION AND AVOIDANCE OF CONTRACT	80
		1 Deceit and Duress	80
		2 Mistake	82
		3 Rescission	83
	IV	FREEDOM OF CONTRACT AND GENERAL CONDITIONS OF BUSINESS	84
		1 Freedom of Contract	84
		2 Limits to Contractual Freedom	85
		3 General Conditions of Business	87

6 Breach of Contract   (*Leser*)

	I	THE CONTRACT AS A BLUEPRINT FOR PERFORMANCE	90
		1 The Contract as a Blueprint	90
		2 The Completion of the Programme: Time and Place of Performance	91
		3 Performance	93

II	IRREGULARITIES IN PERFORMANCE	93
	1 Basic Principles	93
	2 The Contribution of the Pandectists and of Roman Law	94
	3 Impossibility and Delay as the Only Categories of Irregularity	95
III	IMPOSSIBILITY OF PERFORMANCE	96
	1 The Forms of Irregularity	96
	2 Effect on Unilateral Obligations	98
	3 Effect on Bilateral Contracts	99
	4 Summing Up	101
IV	DELAY	101
	1 Time and Responsibility	101
	2 Consequences of Delay	103
	3 Fixing Time and Freeing Oneself from the Contract	103
	4 Anticipatory Refusal to Perform	104
V	POSITIVE BREACH OF CONTRACT AND *CULPA IN CONTRAHENDO*	105
	1 Impossibility and Delay Cease to be the Only Categories	105
	2 Positive Breach of Contract as a Residual Category	107
	3 Culpa in Contrahendo	108
VI	REMEDIES FOR BREACH OF CONTRACT	109
	1 Specific Performance and Rescission	109
	2 Damages for Nonperformance	110
	3 Rescinding the Contract	111
VII	THE FAULT PRINCIPLE AND LIABILITY FOR CONTRACTUAL ASSISTANTS	112
	1 The Fault Principle	112
	2 Liability for Others	114

7 Sale (*Leser*)

I	FUNDAMENTALS	116
	1 Historical	116
	2 Sale as a Special Type of Contract	116
	3 The Duties in a Contract of Sale	118
II	THE DUTIES OF THE SELLER: GENERAL REMEDIES OF THE BUYER	120
	1 What can be sold	120

xii *Contents*

		2 *Performance and Transfer of Risk*	120
		3 *Remedies for Non-performance by the Seller*	122
		4 *Legal Defects*	124
	III	THE DUTIES OF THE SELLER: SPECIFIC REMEDIES OF THE BUYER	125
		1 *Defects*	125
		2 *Wandlung and Minderung*	126
		3 *Guaranteed Attributes*	127
		4 *Remedies for Defects and General Provisions*	129
		5 *Prescription*	131
	IV	THE DUTIES OF THE BUYER: REMEDIES OF THE SELLER	131
		1 *The Duty to Pay the Price*	131
		2 *Accepting Delivery*	132
	V	VARIANT FORMS OF SALE	133
		1 *Instalment Contracts*	133
		2 *The Financed Instalment Sale*	133
8	The Principle of Good Faith: §242 BGB (*Leser*)		
	I	FUNDAMENTALS	135
		1 *Origins*	135
		2 *Development*	136
		3 *Function*	137
		4 *Does it Provide a System of New Remedies?*	138
	II	SPECIFICATION OF CONTRACTUAL RIGHTS AND DUTIES	138
		1 *Construing and Completing Contracts*	138
		2 *Judicial Reconstruction of Contract: the Revalorization Decisions of 1923*	140
		3 *The Collapse of the Foundation of the Transaction*	141
	III	CONTROLS ON THE EXERCISE OF RIGHTS	143
		1 *Control of General Conditions of Business*	143
		2 *Formal Requirements and §242 BGB*	143
		3 *Estoppel, or Sleeping on One's Rights*	144
9	The Law of Tort (*Kötz*)		
	I	THE MAIN HEADS OF TORTIOUS LIABILITY	147
		1 *§823 par. 1 BGB*	147
		2 *§823 par. 2 BGB*	155
		3 *§826 BGB*	156
	II	VICARIOUS LIABILITY	157

	III	STRICT LIABILITY	160
		1 Instances	160
		2 Characteristics	163
	IV	PROTECTION OF HONOUR, REPUTATION, AND PRIVACY	165
10	The Law of Property *(Leser)*	
	I	FUNDAMENTALS	169
		1 Property Law in the BGB	169
		2 The Basic Principles of Property Law	169
		3 Possession and Ownership	171
	II	OWNERSHIP	172
		1 Freedom and Duty: the Two Poles of Ownership	172
		2 The Protection of Ownership against Dispossession	174
		3 Protection against Other Invasions	174
	III	MOVEABLES	175
		1 Acquisition by Agreement and Delivery	175
		2 Good Faith Acquisition	176
		3 Other Modes of Acquisition	176
	IV	LAND	177
		1 Basic Changes since the BGB	177
		2 Acquisition of Ownership in Land	179
		3 The Grundbuch	180
		4 The Grundbuch in Operation	181
	V	SECURITY RIGHTS	182
		1 Security in the Law of Obligations	182
		2 Security in the Law of Property	184
		3 Ownership as Security	185
		4 Security Assignments	187
		5 Conflicts between Security Rights	188
11	Family Law and the Law of Succession	
	I	FAMILY LAW *(Kötz)*	189
	II	THE LAW OF SUCCESSION *(Leser)*	195
		1 Principles	195
		2 Liability of the Heirs, and the Devolution of the Estate	196
		3 Wills	197
		4 Statutory Intestate Succession	198
		5 Pflichtteil or Legitim	199

12 Conflict of Laws and Nationality *(Horn)*

    I   NATIONALITY   200
   II   PRIVATE INTERNATIONAL LAW   202
  III   JURISDICTION AND PROBLEMS OF THE LAW OF PROCEDURE   208

13 Commercial Law *(Horn)*

    I   THE MERCHANT AND HIS LAW: THE COMMERCIAL CODE (HGB)   211
         1 Survey   211
         2 Scope of Application: Who is a Merchant?   211
         3 The Philosophy of the Commercial Code Exemplified: Some Rules on Commercial Acts   215
         4 Commercial Code and Civil Code   217
         5 Other Sources of Commercial Law   218
   II   THE BUSINESS NAME (FIRMA) AND THE COMMERCIAL REGISTER   219
         1 The Firma   219
         2 Changes of Ownership   221
         3 Legal Protection of the Business and the Firma   222
         4 The Commercial Register   223
  III   AGENCY IN CIVIL AND COMMERCIAL LAW   225
         1 The German Concept of Agency   225
         2 Commercial Power of Attorney: Prokura and Other Forms   227
         3 The Commercial Agent (Handelsvertreter)   229
         4 The Kommissionär   232
         5 Agency and Sales Organization   234
   IV   CREDIT AND SECURITY   235

14 Partnerships and Companies: Business Organization *(Horn)*

    I   COMPARATIVE SURVEY   239
         1 The Models in the BGB and their Variants in Business Law   239
         2 Ownership, Liability, and Legal Personality   241
         3 Lifting the Corporate Veil   242
         4 Management and Representation   242
   II   THE GENERAL COMMERCIAL PARTNERSHIP (OHG)   243
         1 Formation   243
         2 The Partners: Rights and Duties   244

	3 Representation and Liability	244
	4 Dissolution	246
	5 Changes in Membership	247
III	THE LIMITED PARTNERSHIP (KG)	248
	1 Legal Structure	248
	2 The KG in Business Life	248
	3 The GmbH & Co. KG	250
	SILENT PARTNERSHIP	250
V	THE PRIVATE LIMITED COMPANY (GmbH)	251
	1 Definition and Formation	251
	2 Organization	252
	3 Shareholders' Rights and Duties	254
	4 Dissolution; Exclusion of Shareholders	256
VI	THE STOCK COMPANY — PUBLIC LIMITED COMPANY (AG)	257
	1 Definition and Formation	257
	2 Organization: Management and Control	258
	3 Shares; Shareholders' Rights and Duties	264
	4 Accounting, Appropriation of Profits, and Disclosure	266
	5 Financing the AG	268
	6 Legal Protection of Minority Shareholders	270
	7 Dissolution and Liquidation; Merger	271
VII	COMBINED ENTERPRISES (KONZERNE)	272
	1 The Approach of German Company Law	272
	2 Three Types of Combine	273
	3 Law of Combines not Involving an AG: the ITT case	276
VIII	CO-DETERMINATION OF EMPLOYEES	276
	1 Co-determination Legislation in its Political and Legal Setting	276
	2 The Legal Mechanics of Co-determination	278
	3 Issues and Prospects of Co-determination	279
IX	TAXATION	279
	1 Individual and Company Income Tax	279
	2 Capital Transfer Tax	281
	3 Business Tax (Gewerbesteuer)	281

15 The Law of Competition *(Horn)*

I THE LAW AGAINST UNFAIR COMPETITION — 282
  1 The Unfair Competition Act (UWG): Purpose, Scope, and Sanctions — 282
  2 What is Unfair? — 284
  3 Bonuses and Rebates — 287
  4 The Protection of Trade-marks and Business Marks — 288
  5 International Protection of Industrial Property — 289

II THE LAW AGAINST RESTRAINTS ON COMPETITION — 289
  1 Survey: Scope and Sanctions of the Law — 289
  2 Cartels and Concerted Conduct — 291
  3 Vertical Agreements and Price Recommendations — 295
  4 Abuse of a Dominant Position in the Market; Discriminatory Practices — 297
  5 Control of Mergers — 303
  6 Procedures — 307
  7 The GWB and the Cartel Law of the EEC — 308

16 Labour Law *(Kötz)*

I INTRODUCTION — 310
II COLLECTIVE AGREEMENTS AND INDUSTRIAL ACTION — 312
  1 Trade Unions (Gewerkschaften) — 312
  2 The Parties to a Collective Agreement — 314
  3 Collective Agreements — 315
  4 Industrial Conflict — 316
III WORKS COUNCIL — 318
IV THE CONTRACT OF EMPLOYMENT — 320
  1 Continuation of Pay during Business Interruptions — 320
  2 Industrial Accidents — 321
  3 Dismissal — 322

Appendix I — Abbreviations — 326
Appendix II — Laws and Publications on German Law in English — 332
Index of German terms — 337
Index — 340

1

Historical Introduction

One cannot really understand the law of a country without knowing something of its legal history; and its legal history must be seen in the context of its general history, that is, of the political, economic, and social conditions in which the law developed. The student of German law in particular needs (1) some knowledge of the emergence of the Bundesrepublik as a federal democracy committed to the rule of law, (2) some idea of Germany's growth as a mercantile and industrial power, and (3) a sense of those factors in the history of ideas which have put their mark on German law and contributed to its characteristic mode of legal thinking.

1 THE BUNDESREPUBLIK, THE FEDERAL REPUBLIC OF GERMANY

The Bundesrepublik emerged as a political entity after the Second World War (1939–45). At the end of that war a line of demarcation was drawn which bisected Germany and divided Europe into two political blocs, the countries of the North Atlantic Treaty Organization on the one side, and the members of the Warsaw Pact on the other. It was on that part of the German Empire which was occupied at the end of the war by the Western Allies, viz. the United States of America, England, and France, that the Bundesrepublik was formed. Its constitution, or Basic Law (Grundgesetz, GG), came into force on 23 May 1949 after being debated in draft and adopted by a representative pre-parliamentary body, the Parlamentarischer Rat. Elections for the first Parliament of the Bundesrepublik were held on 14 August 1949, and the first federal government was formed by Konrad Adenauer, whom the Bundestag had chosen to be the first Federal Chancellor. The Bundesrepublik obtained sovereignty as a state on 5 May 1955 pursuant to treaties with the Western Allies, and sees itself as the legal successor of the German Empire and as the political homeland of all Germans, whether they live in it or not. Thus a German from Leipzig in the German Democratic Republic is not treated as a foreigner or settler if he comes to the Bundesrepublik: he is a citizen and may claim a passport as such. The Bundesrepublik nevertheless recognizes the separate political existence of the German Democratic Republic, which was formed in the part of Germany occupied by the troops of the Soviet Union.

Germany's geography, situated in the centre of Europe, explains much of its history: its internal politics have been of concern to its neighbours rather than just a matter for Germans to settle among themselves. Thus the Peace Treaty of Munster and Osnabruck of 1648, signed by the European powers which had joined in the Thirty Years War on German territory, regulated constitutional issues in the old German Empire as well as questions of international law. Wars and other difficulties between Germany and its westerly neighbour, France, explain not only the ending of the Holy Roman Empire of the German People in 1806, but also the re-establishment of the German Empire, under Prussian leadership but excluding Austria, in 1871: indeed, it was at Versailles that the King of Prussia was proclaimed Emperor of Germany on 18 January of that year. The present boundaries of the Bundesrepublik are the result of a world war, and if the two parts of Germany are to be reunited, as most of its people would like, it will require an international settlement.

The Bundesrepublik is a federal state, composed of eleven Länder at present, including West Berlin. After the Second World War democracy grew from the roots, first in the cities and communities, then through the formation of the several Länder with their own constitutions, parliaments and governments, until finally the Bundesrepublik itself was created. The federal structure of the Bundesrepublik has a long tradition behind it. The modern territorial state in continental Europe emerged from a process in the sixteenth, seventeenth, and eighteenth centuries whereby the kings and princes managed to consolidate their political power as against the nobles and the estates and so build up a central administration. The Emperor could do this in Austria, on his own territory, but he could not do it in the Empire as a whole because of its constitution and the prevailing political conditions. Accordingly, it was in the individual territories of the electors, princes, and cities in Germany that this development took place, and in this respect Germany differs markedly from France and Spain, which became centrally organized states in the Age of Absolutism. Even the Empire that was recreated in 1871 under the leadership of Bismarck's Prussia consisted of a federation of princedoms and cities, and it was not until the First World War (1914–18) and the ensuing revolution had put paid to the primacy of the princes that the Reichstag, or central Parliament, and the central government managed to gain power at the expense of the individual federated territories in the Republic under the new constitution of 1919. Under Hitler's regime of National Socialism (1933–45), the federal structure of the Empire was abandoned.

The Bundesrepublik is a democracy. 'All the power of the state proceeds from the people. It is exercised by the people in elections which are free, equal, direct and secret' (art. 20 I GG). The Bundesrepublik has often been called a young democracy. It is true that, after the totalitarianism of the National Socialist regime and the calamity of the Second World War, the daunting task of rebuilding public life on democratic principles in the newly formed Bundesrepublik was successfully carried out on the whole, but the democratic tradition in Germany actually goes back to the nineteenth century. It found powerful expression in the revolution of 1848–49. A general election was held in May 1848, and the German National Assembly met in St Paul's Church in Frankfurt to consider a constitution for the German Empire. The revolution foundered, admittedly, but constitutions were introduced in the individual German Länder (in Prussia on 31 January 1850), and their parliaments gradually acquired the political rights of a modern chamber. The constitutional contest between Bismarck and the Prussian House of Representatives over Parliament's right to control the budget (1862–6) is one dramatic chapter in the story. In 1871, when the German Empire was re-established, there were national political parties; and the national Parliament, the Reichstag, was democratically elected although the imperial government under the Chancellor was still not directly dependent on its confidence. The Constitution of the Weimar Republic of 1919 was the first to embody democratic principles to the full, but the

4 Historical Introduction

prevailing conflict of political philosophies and the differences of opinion on the causes and consequences of losing the First World War made it difficult for them to take proper root in the political consciousness of the people. Democratic government disappeared in 1933 when Hitler and the National Socialists seized power and set up the totalitarian dictatorship which took Germany into the Second World War in 1939.

The Bundesrepublik is committed to the rule of law. 'The legislature is bound by the Constitution, and the executive and the judiciary are bound by all law' (art. 20 III GG). In Germany, as elsewhere, the rule of law is a nineteenth-century phenomenon. While Germany was by no means in the forefront of its development, it would be an error to suppose that Germany was slow or backward in accepting the rule of law because the princes retained power so long or because the Empire of 1871 was a monarchy under the Kaiser. The state had increasingly come to accept that it was constrained by law, enacted or decided, to respect private ownership, freedom of enterprise and contract, and other vested rights. The ordinary civil courts protected ownership against encroachment, and special administrative courts to deal with legal problems between the citizen and the authorities were set up in the Länder, starting with Baden in 1863. Thus Germany followed the rule of law in the nineteenth century to much the same extent as other European countries with comparable social and political conditions. After the bitter taste of totalitarianism between 1933 and 1945 the Bundesrepublik has striven to promote the rule of law: this is one reason for its unusually comprehensive system of specialized courts, which will be discussed elsewhere (Chapter 3 below).

II TRADE AND INDUSTRY IN GERMANY

The Bundesrepublik has a highly developed industrial economy; nearly 40 per cent of its manpower is engaged in industry (45 per cent if one includes artisans), and it has a leading place in world trade. England led the world in the process of industrialization, and France was first on the continent of Europe. Germany was a relative latecomer. Not until the 1860s did it begin to catch up on England and France, but by the end of the century Germany surpassed both of them in industrial expansion and rate of growth.

The process of industrialization began, as it did elsewhere, with the construction of railways, which occurred in the 1840s and led to the growth of mining and steelworks in the Ruhr and Silesia. The same period saw the rise of the textile industry, especially in the Rhineland. In the later nineteenth century, as industry expanded, Germany obtained a commanding position in certain fields, such as chemicals, pharmaceuticals, and electrical goods. Holdings in industry started to become heavily concentrated from the 1880s onwards as a result of cartelization on the one hand and the formation of mammoth enterprises on the other. Cartelization was to remain a prominent feature of the German economy until 1945.

In recent German economic history the major events have been the collapse after the Second World War and the astonishing reconstruction thereafter, owing to the highly disciplined work ethic of the population, the material assistance provided by the United States through the Marshall Plan, the introduction of market principles by Ludwig Ehrhard along with the currency reform of 1948, and the great stability of relations between management and labour in industry.

Industrialization in Germany, like everywhere else, was bottomed on the political postulates of liberalism: economic activity must be free from state intervention and control. At the same time, the protection of freedom and ownership by private law laid the basis for the development of a market economy. The restrictive customs regulations within the country were dismantled by the German Customs Union of 1843, which also created a common external tariff to protect the infant German economy against foreign competition.

Commercial law was the first law to be unified in Germany: in 1848 the various Länder enacted a Uniform Law on Bills of Exchange (Allgemeine Deutsche Wechselordnung), and in 1861 a General Commercial Code for Germany (Allgemeines Deutsches Handelsgesetzbuch, ADHGB). This Code contains the private law specially applicable to persons and transactions in commerce, like the French Code de commerce, as well as the law of partnerships and companies. A change in company law in 1870 conferred freedom of incorporation, and freedom of enterprise was recognized in the Trade Ordinance of 1869 (§1). The triumph of economic liberalism was now manifest.

As time went by, however, Parliament found it necessary to derogate from these liberal principles and to make increasing use of its power to regulate the economy in order, *inter alia*, to protect the public and the environment from noxious activities, to promote safety in factories by inspections, and to provide employees with some social security. The legislature also intervened in company law. Since the amending law of 1884, freedom of contract in this area has been severely curtailed in order to protect shareholders and company creditors from fraudulent flotations and irresponsible management.

During the two world wars the market economy was replaced by a planned state economy: administrative devices such as authorizations, permits, and ration cards replaced the market as the means of distributing the raw materials and consumer goods that were in such short supply. Thanks to the efforts of Ludwig Ehrhard, later to become Economics Minister and Federal Chancellor, this state planned economy was largely set aside even before the Bundesrepublik was formed, and a market economy was reintroduced along with the currency reform. Competition is now safeguarded and maintained by the Law against Restraints on Competition of 1957 (Gesetz gegen Wettbewerbsbeschränkungen, GWB). This enactment, which is partly based on American experience, has prevented the cartelization that used to exist in the German economy until 1945, but it has not provided a complete solution to the problem of creeping concentration by means of takeovers and multiples. The Bundesrepublik is a member of the European Community (European Coal and Steel Community, 1951; European Economic Community (Treaty of Rome), 1957; Euratom, 1957), with its common market for industrial products, its intricately controlled market for agricultural products, and its special laws to protect competition within the market.

In Germany, as in other countries, the legislator was reluctant to grapple with, and slow to solve, the social problems posed by the worker in nineteenth-century industry. Opposition to trade unionism is evident enough in the prohibition of trade unions by the Prussian Koalitionsverbot of 1845, and in the provisions of the German Trade Ordinance of 1869, whereby claims based on collective agreements were rendered unenforceable (§§152–3). The first progress was made in the area of protecting employees against exploitation and danger, by prohibiting child labour, limiting hours of work, especially for minors, and having official inspections of the safety of workplaces. As to the social security of the worker, the German Empire led the world by making it compulsory for employers to insure their employees against sickness (1883), accident (1884), and incapacity (1889), and to provide retirement and survivors' pensions (1883–9, 1911). This legislation was triggered by an address from the Emperor in 1881.

The revolution of 1918, which effected the transition from monarchy to republic, met many of the demands of the workers' movement for an appropriate labour law. The Collective Bargaining Ordinance of 1918 regulated collective bargains and accorded them legal validity; the Works Councils Law of 1920 provided for works councils in individual factories and also at the regional level to represent the interests of workers; the Labour Court Law of 1926 created a special jurisdiction for questions of labour law. Since then it has become the practice in Germany to go to court, if need be, to resolve almost all the disputes that may arise from the individual work relationship. The worker was also given some protection, in the form of a money payment (Entschädigungsanspruch), against unfair dismissal. Indeed, it is really due to the Weimar Republic that Germany has its present labour law and is ready to submit industrial relations to control by the courts. Experience with compulsory state mediation of industrial conflict, on the other hand, was rather unfortunate, so no such institution exists in the Bundesrepublik today.

After the Second World War it was decided that there should be only one trade union per industry to represent all the workers in that industry who wanted to join, whatever their trade or politics. These industry-based trade unions (Industriegewerkschaften) are all members of an organization called the Deutscher Gewerkschaftsbund (DGB), the only union outside it being the Deutsche Angestelltengewerkschaft (DAG), a union of managerial employees. Works councils were set up again everywhere and operate under the rules laid down by the Labour Management Relations Acts of 1952 and 1972

(Betriebsverfassungsgesetz). The employee's protection against unfair dismissal was strengthened, and his ability to participate in the running of the factory through the works council was reinforced.

Workers were also given some influence in the management of businesses. Under the law of 1952 one-third of the members of a company's supervisory board must be representatives of the workers. Miners had been granted parity of representation on supervisory boards in 1951, when also a seat on the board of management (Vorstand) was earmarked for a so-called Labour Manager. The Law of Co-determination of 1976 (Mitbestimmungsgesetz) now provides for parity of votes on the supervisory boards of all larger enterprises, though the owners' representatives still have the final say because of the president's casting vote.

Industrial relations in the Bundesrepublik have been very stable. That this is so is doubtless due in part to the fact that the country's economic development has been generally successful and that earnings have consequently risen; but it is also due to the fact that the trade unions are industrial unions and that the labour law is sophisticated enough to absorb most industrial disputes and yet leave scope for individual responsibility and co-operation on both sides of industry. Finally, one should mention that the social security of workers has been further improved, by means of an advanced, comprehensive, and subtle system of public insurance against sickness and for retirement pensions.

III GERMANY AS A CIVIL LAW COUNTRY

Germany belongs to the realm of the civil law: in other words, its private law and its whole style of legal thinking have been greatly influenced by the law of ancient Rome. In the centuries after 1100 when the Corpus Juris Civilis (AD 529–35) of the Emperor Justinian began to be studied in Bologna, the legal learning of the Glossators (1100–1250) and the Commentators (1250–1500) spread throughout Europe. The same period also saw the development of the canon law, the law of the Church, itself strongly marked by the ways of Roman Law.

The practice of going to university to obtain one's basic

training in law and to acquire the style of lawyerly discourse has been followed ever since then in Europe, though not in England until recently. In the course of the thirteenth, fourteenth, and fifteenth centuries the Roman law of the medieval scholars came to be recognized in the German Empire as a secondary source of law which could be prayed in aid of the local laws, and when the Reichskammergerichtsordnung of 1495 laid down that the imperial court, or Reichskammergericht, half of whose members were to be jurists, was to base its decisions, in the absence of special statutes and customs, on 'imperial and common law' ('nach des Reiches und Gemeinen Rechten'), it meant the Roman law as received. The contemporary idea was that Roman law was valid in Germany because it was the law of the Emperor, the supposed successor of the Roman Caesars. Since this theory was naturally inapplicable in kingdoms and principates with other sovereigns, it was thought until fairly recently that it was only in Germany that Roman law was received at all. As a matter of fact, however, Roman law was received throughout continental Europe (and even in Scotland, though not in Scandinavia or Russia), thanks to the part played by the universities, the practice of courts, especially the church courts, and the presence of learned jurists in the secular courts. Arthur Duck (1580–1648), an English civilian, i.e. trained in Roman law, summed up the situation quite aptly: 'Roman law was recognized in Europe not by reason of imperial power, but by the imperious power of reason (*non ratione imperii, sed rationis imperio*).'

The other source of German law is Germanic law, the unwritten legal customs of the regions and localities. Some of these laws were collected and written down in the thirteenth century, such as the important Sachsenspiegel (1221–7) and the Schwabenspiegel (1275). Subsequent collections of local laws, such as city laws (Magdeburg, 1444), were more imbued with learned Roman and canon law, and in the fifteenth and sixteenth centuries many city laws were deliberately reworked so as to harmonize with the received Roman law ('reformed city laws'; e.g., Nuremberg, 1479, Frankfurt, 1509).

In the sixteenth, seventeenth, and eighteenth centuries, legal scholars on the Continent tried to adapt the Roman legal tradition to the requirements of modern practice while remaining faithful to it. The leaders of this *usus modernus pandectarum*, as it was called, were Ulrich Zasius (1461–1535) and Benedict Carpzow (1595–1666). During this period Roman lawyers in Germany and elsewhere were producing lengthy syntheses of modern Roman law in the name of the Law of Nature or the Law of Reason. The aim of these authors, rather in the classical and Christian tradition of natural philosophy, was to present the material in a logical and orderly progression of concepts and principles. The Dutchman Hugo Grotius (1583–1645), one of the great European authorities on international law (*De iure belli ac pacis*, 1625), was the leader of the Natural Law movement; among the Germans, mention should be made of Samuel Pufendorf (1632–94) (*De iure naturae ac gentium libri octo*,1672), who taught constitutional and international law; Christian Thomasius (1655–1728), famous for his campaign against witch-hunting and rather critical of traditional Roman law; and his pupil, Christian Wolff (1679–1754). Now scholars were concerned not only with clarifying the concepts of Roman law, as revived and refined by the medieval jurists, but also with presenting the law in a logical and systematic manner and in making legal argument turn on deduction from its axioms. These elements, which are relatively absent from Anglo-Saxon legal thought, are still evident, though perhaps not so prominent, in German and continental legal thinking today.

The Law of Reason prepared the way for the codification movement of the eighteenth century, which embraced the ideas of the Enlightenment, and strove for a single law, a comprehensive and comprehensible codification, which would apply to every citizen and every aspect of his life. Bavaria enacted a civil code in 1756 (Codex Maximilianeus Bavaricus Civilis), and in 1794 there appeared the General Law of the Prussian States (Allgemeines Landrecht der Preussischen Staaten, ALR).

Many Germans wanted a national code of private law such as the French had obtained with their Code civil in 1804, but the Congress of Vienna in 1815 at the end of the Napoleonic Wars left Germany in a state of political disunity, and the time was not yet ripe. Having no code to attend to, the private law scholars in Germany continued to work on the Roman sources, to great effect. These Pandectists, as they were called, gained international acclaim. Their programme was worked out in the

Historical School of Law, whose most distinguished representative, Friedrich Carl von Savigny (1779–1861), rejected the idea of codification and stressed that law is a social phenomenon which develops dynamically in intimate relation with the experience and ideas of the people. While 'historical' in this context really means 'empirical', much attention, under the inspiration of the Romantic Movement, was devoted to legal history, including the history of Roman law. Legal history still plays an important part, though now somewhat reduced, in the training of German lawyers at university. As well as studying the past, the Pandectists also applied themselves to the problems of the present. Unconstrained by a code, they continued to apply the conceptual and systematic methods inherited from the school of Natural Law and the Law of Reason, much like John Austin, the English lawyer, in his analytical jurisprudence. The Pandectists were criticized for their conceptualistic and systematic way of thinking by Rudolf von Jhering (1818–92), who made an important contribution to our understanding of law by describing it as a means of defending the interests of the individual (Kampf ums Recht, 1872) and emphasizing that rules could be understood only in the light of their purpose and the interests they were designed to protect (Zweck im Recht, 1877–84).

Only after political unification under the Empire did Germany obtain a comprehensive code of private law — about a century late, so to speak. The Bürgerliches Gesetzbuch (BGB) came into force on 1 January 1900. There had previously been legislation, as has already been mentioned (p. 5 above), in the area of commercial and company law; and general imperial laws (Reichsjustizgesetze) on the constitution of courts, civil procedure, criminal procedure, and bankruptcy had been enacted in 1877, effective 1 January 1879.

The codification movement, which failed to prosper in England despite Bentham's efforts, expresses another characteristic feature of continental legal thought, namely the view that, ideally, the rules should all be contained in a major enactment which the judges are dutifully and obediently to apply. If a court is doing no more than applying the law when it makes a decision, the decision itself can have no force as law beyond the very case in which it is rendered. This view squares with the constitutional theory propounded by Montesquieu, and generally accepted on the Continent, that the legislative, judicial, and executive powers should be kept separate. The German lawyer and the common lawyer thus have quite different conceptions of the role of the judge.

Subsequent experience with the code and with enacted law generally has shown that, even if it keeps on changing and adding to private law, the legislature can no more regulate everything in this sphere than it can in other areas of lawmaking. It is therefore admitted nowadays that the judge must fill in the gaps in the law and develop it as he applies it. To understand the law as it actually operates, the German lawyer today needs to be almost as familiar with the wealth of judicial decisions as the common lawyer. The methods of interpreting and applying legal texts have become more flexible, thanks to legal theorists, especially the school of Interessenjurisprudenz (notably Philipp Heck, 1858–1943), which proceeded on von Jhering's view that one can understand a law only when one knows what particular interests it was designed to promote. Specialists in legal methods now talk in terms of 'teleological' construction, of interpretation linked to legislative purpose. More recently, emphasis has been put on the importance of grasping the social and economic context in which the law is to be applied, though no agreed method of proceeding has yet emerged.

In the light of this historical sketch, we may try to summarize the principal points of difference between continental and Anglo-American legal thinking. First, the legacy of Roman law has conduced to a greater clarity in the language used in laws and contracts on the Continent, whereas the common lawyer's approach to a problem is often more practical. Second, the kind of systematization evident in the works produced by the School of Natural Law and reflected in part in the codes is more attractive to lawyers on the Continent. It is true that reasoning by deduction from axioms is less compelling nowadays than it was in the time of the Pandectists, but the legal vocabulary, the textbooks, and the laws themselves all betoken a concern to conceptualize problems. Actually, the bulk of decisional law is causing common lawyers to systematize, abstract and digest; in this respect the American Restatements, for example, have much in common with the writings of the School of Natural Law. Third, legislation has more significance as a source of law on the Continent, as the great codes powerfully testify. Here, too, however, the different systems may be moving together, for the importance of decisional law is constantly increasing on the Continent, while in the United States and England there is growing use of legislation and recognition of its importance. One can thus discern a gradual approximation of the viewpoints of the different systems on the role of legislation.

2

Constitutional Law

I INTRODUCTION

During the London Conference at the end of 1947, the four Occupying Powers finally realized that it was going to be impossible to construct a common framework for political life in all four of the Zones of Occupation in Germany. The three Western powers therefore invited the Chief Ministers of the Länder in their zones to proceed to work out a constitution for West Germany alone. The Chief Ministers were quite aware that such a constitution might well confirm the division of Germany and defer its reunification for a long time, but they accepted the invitation nevertheless. Great pains were, however, taken to make it appear that the constitution was only a provisional one. Rather than use the German word for constitution and call it a *Verfassung*, the less solemn, more neutral, word *Grundgesetz*, or 'Basic Law', was chosen. Furthermore, the purpose of the Basic Law was stated in the Preamble to be 'to give a new order to political life for a transitional period' only, and the Preamble proceeds to call upon the entire German people 'to achieve in free self-determination the unity and freedom of Germany'.

The political conditions of the time also affected the way the Basic Law was created and brought into force: instead of being discussed and adopted by a national constituent assembly directly elected by the people, it was drafted by a body whose members were selected by the parliaments of the Länder then in existence, and it was to come into force on being ratified by the parliaments of two-thirds of the Länder. All the Länder except Bavaria did in fact ratify it, and the Basic Law came into force on 23 May 1949.

The draftsmen of the Basic Law followed the Weimar Constitution of 1919 on many points, but they also deviated from it in certain critical respects, mainly where experience had shown that the rules of the Weimar Constitution were unsatisfactory or unsafe. In particular, the powers of the head of state under the Basic Law are much restricted as compared with the Weimar Constitution; the position of the government, and of the Chancellor within it, is greatly strengthened; and the role of the Länder in the federation is enlarged.

As to the last point, the Occupying Powers, whose concurrence was needed before the Basic Law could come into effect, insisted that the independence of the Länder be entrenched. Indeed, the federal structure of the Bundesrepublik is one of the matters which cannot be changed even by constitutional amendment (art. 79 par. 3): it would take a revolution. Since the German people had been used to various forms of local autonomy throughout its history, this insistence by the Occupying Powers was by no means unwelcome, but it is perhaps a little surprising that the Länder should have remained much as they were when the Bundesrepublik was founded, since, apart from Bavaria and the two Hanseatic cities of Hamburg and Bremen, they were created by the Occupying Powers more or less by chance and without much regard for historical and cultural links. This has not, however, greatly worried the German people, doubtless because so much historical tradition and regional feeling was lost in the catastrophic maelstrom of 1945, and because of the influx of millions of refugees, first from the regions east of the Oder and Neisse, and later from the German Democratic Republic.

There are ten Länder on the territory of the Bundesrepublik today (their population in millions, as of 1978, is given in brackets): Baden-Württemberg (9.1), Bavaria (10.8), Bremen (0.7), Hamburg (1.7), Hesse (5.5), Lower Saxony (7.2), North Rhine-Westphalia (17.0), Rhineland-Palatinate (3.6), Saarland (1.1), and Schleswig-Holstein (2.6). To these must be added West Berlin (1.9), which has a special status. When the Basic Law was being approved, the Western Occupying Powers made certain reservations with regard to West Berlin. It was not yet to be incorporated fully into the Bundesrepublik, and the executive and legislative branches of the Federation were to have no direct political effect upon it. This is why, for example, federal laws do not come into force automatically in Berlin as they do in other Länder, but only after a resolution of the Parliament of the Land Berlin. This is also why the representatives of Berlin in the Bundestag do not have full voting powers.

II THE SYSTEM OF GOVERNMENT

The provisions of the Basic Law fall into two groups. One group contains the 'basic rights', which regulate the relationship between the citizen and the state, while the other provisions relate to the governmental institutions of the Bundesrepublik and the functions they perform. We shall take the latter group first.

The most important governmental institution in the Bundesrepublik is the *Bundestag*. It represents the people, and its members are elected by the people for four years at a time. The Basic Law states that elections must be 'general, direct, free, equal and secret', but leaves all the details of electoral procedure to be determined by federal enactment (art. 38). Thus the decision whether to adopt the majority system or some form of proportional representation was left by the constitution to the federal legislature. So far it has opted for a slightly modified form of proportional representation, which ensures that Parliament accurately reflects the political forces in the country. The danger of such a system is that it can give rise to many small splinter groups, but this is countered by a provision of the federal electoral law which ensures that no party obtains any seat in Parliament at all unless it has attracted at least 5 per cent of the votes cast.[1]

Of the powers which the Basic Law explicitly allocates to the Bundestag, the most important are its power to legislate and its power to nominate the Chancellor; but almost as important, though unmentioned as such in the Basic Law, is its role in the formation of public opinion. The Bundestag is the forum in which the central controversies of politics are debated, the viewpoints of the different parties formulated, and the Opposition's powers of control exercised.

About the function and powers of the *federal government* the Basic Law does not say very much. According to art. 62, the federal government consists of the Federal Chancellor and the federal ministers. The Chancellor's position is a strong one. He determines the 'principles of policy', and the policies he prescribes must be followed by the ministers in the conduct of their departments, although they are otherwise independent (art.

---

[1] The Bundesverfassungsgericht has held that the '5 per cent clause' is not in conflict with art. 38 GG, which requires elections to be 'equal'; see *BVerfGE* 1, 208, 256.

65). The federal ministers are chosen by the Chancellor, and the President is bound to execute his proposals for their appointment or dismissal. Thus only the Chancellor himself is selected by the Bundestag. The Bundestag may call upon the Chancellor to dismiss a particular minister with whose performance it is dissatisfied, but whether the Chancellor follows such a resolution or not depends entirely on his view of what is politically expedient. It is only the Chancellor himself whom the Bundestag can require to withdraw, but a simple vote of no confidence is not sufficient for the purpose: the Bundestag must *at the same time* nominate his successor (art. 67). The purpose of this provision is to prevent the instability of an interregnum such as occurred several times under the Weimar Constitution, when there was a majority in Parliament for dismissing the head of government, but not for choosing his replacement. The position of the Federal Chancellor is thus particularly powerful and influential: people have even called the Bundesrepublik a 'Chancellor-democracy', and it is undeniable that it has certain points of resemblance with the American presidential system.

In the Bundesrepublik the head of state, the *Federal President*, has but little scope for exercising political influence. His main function is to accord formal validation to decisions taken by others: he promulgates the laws that have been voted in Parliament, appoints the Federal Chancellor chosen by the Bundestag and the ministers proposed by the Chancellor, and represents the Bundesrepublik in the sphere of international law, for example in the accreditation of diplomats. The Federal President symbolizes the unity of the state and can thus exercise considerable influence on public opinion; but this depends more on his personal authority than on his constitutional powers.

The most important attribute of the Bundestag is its power to *legislate*. This power is subject to two constraints. First, the Bundestag can enact laws only on matters which the Basic Law allocates to the legislative competence of the Federation rather than to that of the Länder; and second, such legislation needs the co-operation of the *Bundesrat*.

The constitution starts out from the proposition that legislation is a matter for the Länder: federal laws can only be enacted where express power has been granted to the Federation in the

Basic Law (art. 70). The areas in which *exclusive* power to legislate is granted to the Federation include foreign affairs, national defence, nationality, the currency, and the law relating to patents, trade-marks, and copyrights. In areas where the Federation has *concurrent* legislative power, the Länder may legislate, but only if the Federation has failed to exercise its power; in fact, the Federation has exercised its power so fully by now in these areas that there is hardly any room left for the Länder to legislate. Such areas include the whole of private and criminal law and procedure, labour law, the law of social security, and the vast field of law concerning the economy and agriculture. Finally, there are areas in which the Federation only has the power to enact *skeleton laws*. For example, the Federation may legislate on the status of civil servants within the Länder, the protection of the countryside, and the structure of universities, which are run exclusively by the Länder. Any such law, however, may contain only skeleton provisions, and must leave other matters to be filled in by legislators in the Länder. The matters for which the Länder retain exclusive competence today are in reality only collateral and residual, such as the law relating to local administration and cultural institutions, notably education.

The Länder do, however, take part in federal legislation through the Bundesrat. It is composed of members of the governments of the Länder, and they vote as instructed by the Land that sends them. A Land has a minimum of three votes, four if it has more than two million inhabitants, and five if it has more than six million.

If the Bundesrat disapproves of a bill which has been voted by the Bundestag, it can enter an objection (Einspruch). The Bundestag must then consider the bill anew, and only if the majority in the Bundestag in favour of the bill is as great as the majority in the Bundesrat against it may the law come into force. Often, however, the Bundesrat has a more effective veto, for the Basic Law expressly provides that the concurrence of the Bundesrat is required for any law that significantly affects the interests of the Länder, as many laws in fact do. In all these cases agreement between the Bundestag and the Bundesrat may be facilitated by the Vermittlungsausschuss or Committee of Mediation for which the Basic Law provides: it consists of equal numbers of members of the two bodies, and works out compromise proposals to lay before them (art. 77).

The Bundesrat's involvement in federal legislation is of great importance in two practical respects. First, it permits the input of the Länder's administrative experience and technical expertise. In principle, it is for the authorities in the Länder to implement federal laws (art. 83), and it might be very difficult or impossible for them to do this if bills promoted by the zealous reformers in the Bundestag could not be modified on technical administrative grounds on their way through the Bundesrat. Second, the role played by the Bundesrat in federal lawmaking can be of critical political importance, especially when the parties that have a majority in the Bundestag, and have therefore chosen the Federal Chancellor and formed the government, are in a minority in the Bundesrat. In such a situation, which is by no means rare, the Bundesrat can obviously obstruct the legislative policies of the federal government. It is questionable whether it is right for a Land to use its voting power in the Bundesrat for general political purposes, to frustrate the legislative policies of the federal government and the majority in the Bundestag, rather than to protect any special interest of its own. However this may be, the part played by the Bundesrat in the process of federal lawmaking is certainly great — somewhat greater indeed than the draftsmen of the Basic Law had in mind.

If the place of the Länder in the Bundesrat has enabled them to increase their influence on federal policy-making, the past twenty years have seen a marked decrease in their ability to take independent measures on their own territory. It is not that the Basic Law has been amended to this effect; it is just that technological, economic, and social changes have made it much more sensible to do the planning and set the goals at the federal level rather than in the individual Länder. In the event, the Länder have developed various forms of mutual co-operation, often involving the Federation as well; these permit the assimilation of conditions in the different Länder and the harmonization of administrative practice, but they also tend to reduce the Länder's individual freedom of action. Furthermore, the Federation has powers whose exercise tends to weaken the powers of the Länder. For example, when the Basic Law was amended in 1959, the Federation was granted the power to give the Länder and local communities financial assistance with very costly projects (art. 91a). It may also now help the Länder with improving their regional economies and with constructing and extending their universities, matters that fall within the competence of the Länder. In these cases the Federation bears half of the costs involved (art. 91a). It will be evident that the offer of half the cost of a project which the Federation fancies is bound to curtail the practical options of the Land in question.

The independence and self-sufficiency of the Länder depend, more than anything else, on the manner in which tax revenues are divided. The Basic Law states specifically which taxes go the Federation, to the Länder, and to the communities respectively. Half of the proceeds of income tax and corporation tax go to the Federation and half to the Länder; value-added tax is divided in proportions fixed by federal law, adjusted at stated intervals to take account of the respective needs of the Federation and the Länder: such legislation naturally requires the concurrence of the Bundesrat. Allowance is also made for the great disparities of wealth between the various Länder, by means of an arrangement whereby the relatively 'poor' Länder, such as Schleswig-Holstein and Lower Saxony, receive contributions from the Federation and the relatively 'rich' Länder, such as Baden-Württemberg, North Rhine-Westphalia, and Hamburg.

III THE FEDERAL CONSTITUTIONAL COURT

The resolution of disputes between the organs of state under the constitution of many countries is a matter for compromise and agreement. Other constitutions, including the Basic Law, provide for an independent court. In Germany this is the Bundesverfassungsgericht, or Federal Constitutional Court, and the Basic Law gives it a particularly important position (arts. 92–4). It decides disputes between the Federation and the Länder, or between different Länder, and also resolves any difficulties regarding their respective rights and duties under the Basic Law that may arise between the Bundestag, the Bundesrat, the federal government, the Federal President, and certain other institutions with constitutional roles.[2]

[2]Detailed regulation is to be found in art. 93 GG and in the Law on the Federal Constitutional Court of 1951. On the Vorlageverfahren and the Verfassungsbeschwerde, see p. 22, below.

The Bundesverfassungsgericht meets in Karlsruhe and sits in two senates, each of which has eight judges. All the judges must possess the qualification for judicial service (see p. 36 below), and at least three in each senate must have served on the highest courts in the Federation. Judges, who must retire at the age of sixty-eight, are appointed for twelve years. The minimum age for appointment is forty, but since the usual age is fifty or more, service on the court is normally the final stage in the judge's career, which helps to ensure his independence. Half of the judges are chosen by a committee of the Bundestag consisting of twelve representatives selected by proportional representation, and the other half by the Bundesrat. A two-thirds majority is required in each case, in order to prevent a party with a simple majority in the Bundestag or Bundesrat from pushing through the election of its own candidate.

IV BASIC RIGHTS

The first chapter of the Basic Law is entitled 'Basic Rights' ('Grundrechte'). These are in line with the general human and civil rights now recognized in all constitutional states in the West, and covered by the European Convention for the Protection of Human Rights. Here one finds the basic principles that everyone has the right to life, corporeal integrity, and the unhampered development of his personality (art. 2); that everyone is equal before the law (art. 3); that freedom of belief, conscience, religion, and ideology is inviolable (art. 4); and that everyone has the right to express himself freely in speech, writing, or pictures (art. 5). Here, too, are the basic right of assembly (art. 8), the inviolability of the home (art. 13), and the right of petition (art. 17). In addition, all Germans are granted the right to form associations and societies, especially in the field of industrial relations (art. 9), to make a free choice of their career, place of work, and place of training (art. 12), and to refuse on grounds of conscience to serve in the armed forces (art. 4). The state affords special protection to marriage and the family (art. 6), and it guarantees the right of property and inheritance by providing that there shall be no expropriation except for the common good and against a fair compensation (art. 14). Some other basic rights are conferred elsewhere in the Basic Law, such as the right to a hearing in the courts, the freedom from punishment under retroactive laws, and the freedom from double punishment for a single act (art. 103). There are also provisions on the citizen's rights in relation to arrest and detention (art. 104).

According to art. 1 par. 3 of the Basic Law, these basic rights constitute directly applicable law binding on the legislature, the executive, and the judiciary. However great the majority in parliament, therefore, the legislature may not enact any law that conflicts with these basic rights; nor may the courts affect them by the form of their proceedings or the substance of their decisions; and the executive is bound to respect these basic rights both in imposing burdens on individuals and in distributing benefits.

It follows that, whenever the outcome of any civil, criminal, or administrative proceeding depends on its validity, the court may inquire whether a law or other act of the public power is unconstitutional and void as infringing a basic right. Usually it will be an interested party that draws the court's attention to the question, but the court must inquire into it *ex officio* if need be. As to the constitutionality of a *law*, a distinction must be drawn: if the court can dispose of any objections to the constitutionality of the law, it must apply the law and make its decision accordingly; but if it is convinced that the objections are sound, it cannot decide that the law is unconstitutional, but must stay the proceedings and submit the question to the Bundesverfassungsgericht (art. 100).[3] The name given to this submission procedure is 'Vorlageverfahren'; but there is another, more important, method by which the Bundesverfassungsgericht can be seised of the question whether an official act is constitutional or not: it is open to a citizen to complain directly to the Bundesverfassungsgericht that one of his basic rights has been infringed by an act of the public power. The name given to this remedy is 'Verfassungsbeschwerde' (compare art. 93 par. 1 no. 4a).

Several conditions must be satisfied before a Verfassungsbeschwerde is admissible. In particular, the complainant must have exhausted all the other remedies, including litigation, that

---

[3]This only applies to laws which came into force after the Basic Law: see *BVerfGE* 11, 126. It is open to any court to hold unconstitutional and void any law adopted before the Basic Law and any regulation made by the executive pursuant to delegated legislative power.

are open to him in respect of the harmful official act. Furthermore, a Verfassungsbeschwerde is only admissible if the complainant has *locus standi*: he must be personally affected by the act in question, rather than just a member of the public. If the act in question is a decision of an *administrative authority*, only the person to whom the decision was addressed may complain of it. If the act in question is a *law*, the citizen must in principle wait until a court or administrative authority has applied the law to his disadvantage; only when this has been done, and he has exhausted his legal remedies against the administrative or judicial decision, may the citizen raise a Verfassungsbeschwerde. This, however, does not apply where the mere enactment of the law has 'directly and immediately' affected the complainant in his basic rights.

It was so held in a case in which the complainant claimed that the Law of 1957 on the Equal Rights of Man and Woman was unconstitutional in so far as it provided that, in cases of a difference of opinion between spouses, the husband was entitled to the final decision. The Bundesverfassungsgericht held that the Verfassungsbeschwerde was admissible because the complainant was 'directly and immediately' affected by the law; furthermore, it held that the complaint was justified, because the Basic Law declares that men and women are equal in law (art. 3 par. 2) and there was no adequate justification for giving the husband a superior right in this way.[4]

A Verfassungsbeschwerde may likewise be raised against a *court decision*, for decisions of the courts are acts of the public power as well. Of course, the complainant must allege that the judgment in question misconstrued the Basic Law, but this requirement does not entirely remove the danger that a litigant who is dissatisfied by the judgment of the ultimate civil, criminal, or administrative court may try to use the Verfassungsbeschwerde in order to give himself yet another level of appeal. The Bundesverfassungsgericht has therefore enunciated a number of restrictive tests, notably by insisting that the judgment under attack must contain a 'material' or 'particularly grave' infringement of the complainant's basic rights.[5]

A frequent ground of attack on the constitutionality of civil judgments is that, in applying the rules of private law, the court did not pay

---

[4] *BVerfGE* 10, 59, 66.   [5] See *BVerfGE* 18, 85, 92f.

sufficient regard to the basic right of freedom of expression and the freedom of the press (art. 5). One well-known decision of the Bundesverfassungsgericht concerned a Mr Lüth, who had written an open letter calling upon the public to boycott a film in which Veit Harlan, a notorious director of anti-semitic films during the Nazi period, was involved. The producer of the film brought suit, and the civil court ordered Lüth to withdraw his call for a boycott. Lüth raised a Verfassungsbeschwerde, and it succeeded. According to the Bundesverfassungsgericht, the civil court's application of the rules of private law had not taken sufficient account of the special value that the Basic Law puts upon the public discussion of questions of social importance.[6]

Another case arose shortly after the construction of the Berlin Wall. Various units of the Springer publishing empire threatened to withhold supplies of Springer publications from all the news agents in Hamburg unless they stopped handling the periodicals, such as *Blinkfüer*, which gave details of radio and television transmissions from the German Democratic Republic. The publisher of *Blinkfüer* claimed damages. He obtained judgment in the lower courts, but the Bundesgerichtshof reversed.[7] The publisher then raised a Verfassungsbeschwerde, and the Bundesverfassungsgericht upheld it: the Springer publishers were certainly entitled to use their basic right of freedom of expression in order to call for a boycott of the complainant's periodical, but only if they tried to persuade people by the use of argument; if it was essentially economic pressure on which they relied to back the boycott, this would fall outside its protection and thus be wrongful.[8]

Finally, we may mention the judgment of the Bundesverfassungsgericht in the 'Lebach' affair. Four soldiers in the federal army had been killed in a surprise attack on a munitions depot in Lebach: two of the assailants were sentenced to life imprisonment, and the plaintiff was jailed for six months as an accomplice. Shortly before he was to be released, the plaintiff sought to enjoin a proposed broadcast of a documentary film which a television company had made of the incident. The civil courts rejected the suit on the ground of the basic freedom of broadcasting (art. 5), and held that the character attributed to the plaintiff in the film was generally accurate. When the plaintiff raised a Verfassungsbeschwerde the judgment of the civil court was annulled and the injunction against the television company was granted as sought. The Bundesverfassungsgericht recognized that, while it must certainly protect the freedom of broadcasting, it

[6] *BVerfGE* 7, 198 ('Lüth').   [7] BGH *NJW* 1964, 29.
[8] *BVerfGE* 25, 256 ('Blinkfüer').

must also give due weight to the plaintiff's basic right to the unhampered development of his personality (art. 2), and to the provision that requires the state to respect and protect the dignity of the human being (art. 1). What finally influenced the court was its acceptance of the expert testimony that, if the broadcast took place, the complainant would find it much harder to reintegrate himself in society.⁹

In these cases it was not the actual rules of private law that conflicted with the basic rights, but rather the way the court had interpreted and applied them. But sometimes the decision against which a Verfassungsbeschwerde is mounted, whether it be the decision of a court or of an administrative body, is based directly on a specific statutory provision. In such a case, if the complaint is upheld, not only is the decision annulled, but the *law* on which it is based is also declared to be void. The 'submission procedure' *always* involves a statutory provision. The decision of the Bundesverfassungsgericht in such cases that the statutory provision is valid or invalid has itself the force of law and is therefore published in the *Bundesgesetzblatt*.¹⁰

To hold that a statute is void may clearly have very serious political consequences. Indeed, it comes close to being a political decision in itself, especially if the constitutional standard by which the law is to be tested is broad and indeterminate. Thus in 1975 the Bundesverfassungsgericht had to decide on the constitutionality of a statute which legalized abortion under certain conditions. This was a measure which had quite recently been enacted after lively public debate by a large majority in the Bundestag, but the Bundesverfassungsgericht held that it was void as infringing the right to life and corporeal integrity (art. 2), and that in enacting it the state had not paid sufficient deference to its duty to protect human dignity (art. 1).¹¹

Naturally enough, the women's associations immediately held protest meetings, but even after due reflection it was possible to ask whether eight judges should really have the power to toss out a law passed after proper consideration by a

---

⁹*BVerfGE* 35, 202 ('Lebach').
¹⁰See §31 par. 2, Law on the Federal Constitutional Court.
¹¹For the details see *BVerfGE* 39, 1; see also the dissents by Rupp v. Brünneck and Simon.

democratically mandated legislature. Certainly there have been other cases in which Parliament, with equal deliberation, has passed laws that flagrantly infringed the rights of minorities or has enacted provisions that turned out, on closer examination, to be simply capricious; then everyone in the country was glad enough to have the judges sitting there in Karlsruhe. The Bundesverfassungsgericht itself knows full well that it must exercise judicial restraint and respect the role of the lawmaker, and that it must not simply substitute its own value-judgments for those of Parliament. It keeps emphasizing that it is not its business to decide whether the solution adopted by the legislature is the most just or appropriate one; the only question is whether it can withstand attack on constitutional grounds. Sometimes the Bundesverfassungsgericht says that a legislative provision is constitutionally acceptable for the time being, but that it may eventually have to be declared unconstitutional, unless, within a specified or reasonable period, the legislature has replaced it with a rule more consonant with the constitution.[12]

Finally, one must not suppose that, when the court is deciding on the constitutionality of statutes, it always founds exclusively on the wording of the relevant articles of the Basic Law. Those articles are often vague and sometimes virtually meaningless, but there is now an extensive case law on the most important of them; the less abstract principles and more concrete rules that are contained in these cases serve the parties and the courts as useful guides to interpretation. It is true that the judicial creativity of the Bundesverfassungsgericht is very strongly marked when it is interpreting the constitution, but this activity is not in reality very different from what other courts do when they have to apply fluid legal concepts: all judicial decisions should relate to prior holdings and to the views of writers, where appropriate, in a manner that is reasonable and comprehensible and such as to promote continuity of development.

---

[12]See *BVerfGE* 16, 130, 142; *BVerfGE* 21, 12, 39.

3

The Administration of Justice

I THE SYSTEM OF COURTS

The common lawyer who takes a quick look at the system of courts in Germany (see Figure 1) is apt to find it extremely specialized, highly decentralized, and very complicated. So it is. Apart from the Bundesverfassungsgericht or Federal Constitutional Court (see p. 20 above), there are no less than five different sets of courts, each with different jurisdictions, depending on the subject matter. As a rule, each set of courts includes courts of first instance, courts of appeal, and, in the last resort, a federal supreme court. The courts at first instance and the courts of appeal are courts of the several Länder; this means that it is for the Land to equip and maintain these courts, and to employ, appoint, pay, promote, and pension off the judges who sit in them. Only the courts of last resort in each of the five jurisdictions are federal courts, and only their judges are federal judges.[1]

Federal statutes determine the jurisdiction of each set of courts as well as the procedure to be applied in them. It is true that the Basic Law (art. 74 par. 1) provides that the constitution of the courts and court procedure fall within the realm of concurrent legislative competence, but the federal Parliament has exercised its power to legislate so fully in this area that there is virtually no scope left for the law of the Länder.

1 Ordinary Jurisdiction

Of the five jurisdictions, it is the *ordinary jurisdiction* (ordentliche Gerichtsbarkeit) with which the citizen comes into contact most frequently. This jurisdiction has by far the largest number

---

[1] There is one exception. Appeal lies from the Federal Patent Office to the Federal Patent Court (Bundespatentgericht) in Munich. Although not a court of last resort, since its decisions are appealable to the Bundesgerichtshof, this is a federal court with federal judges (see art. 96 par. 1 GG).

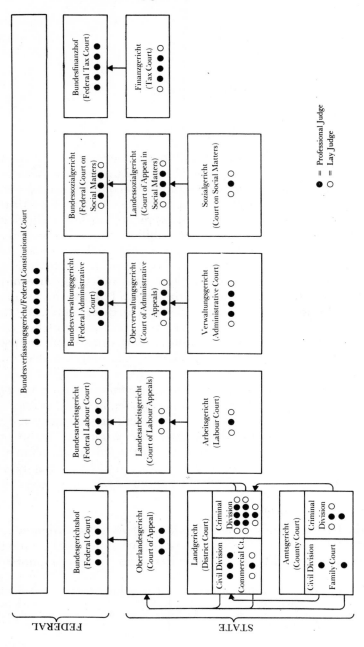

Figure 1. *Court System in the Federal Republic of Germany*

of judges, and it entertains almost all matters of civil and criminal law. The details are regulated in the Constitution of Courts Act (Gerichtsverfassungsgesetz, GVG). Here we can deal only with the general outlines (see also the survey on p. 28).

(a) Civil cases start either in the Amtsgericht, or county court, or in the Landgericht, or district court. The Amtsgericht may be said to dispense the small coin of justice. In civil matters it has jurisdiction where the value of the subject matter in dispute does not exceed DM3,000 (= £750), though certain disputes, such as disputes between landlord and tenant, matrimonial and custody cases, and claims for maintenance, fall within its competence regardless of their value. Thus divorces and the problems they generate, such as the custody of children, maintenance for the spouse and family, and property disputes between the spouses, all fall within the competence of the Amtsgericht and are dealt with there by a special division of that court called the Familiengericht, or Family Court.

In addition, the Amtsgericht is the right forum for the enforcement of judgments, and it has non-contentious business, including the maintenance of the land register and the commercial register, and the supervision of guardians, testamentary administrators, trustees in bankruptcy, and so on. On the criminal side, the Amtsgericht has jurisdiction over petty offences.

Civil cases in the Amtsgericht are decided by a single judge. So, too, are criminal cases, save that for the more serious crimes two lay judges (Schöffen) sit on the bench with the professional judge. Unlike the jurors in a common law trial, they discuss the case fully with the judge and have, at least in theory, as much say in the outcome as he does.

There are about 600 Amtsgerichte in the Bundesrepublik, which gives a rough average figure of one Amtsgericht per 100,000 inhabitants. This figure is very misleading, however; many Amtsgerichte have fewer than 20,000 citizens to cater for and so can get by with two or three divisions, whereas in a big city there are unlikely to be more than five Amtsgerichte, each of which may require dozens and dozens of divisions with a judge in each.

(b) Civil cases that fall outside the jurisidiction of the Amtsgericht start in the Landgericht, or district court. The Landgericht also hears appeals from decisions rendered by the Amtsgericht at first instance, save that if the Amtsgericht was sitting as a Family Court the appeal goes direct to the Oberlandesgericht, or court of appeal. Decisions of the Landgericht are in principle made by a division consisting of three professional judges, one of whom serves as presiding judge and as such enjoys a higher rank and a better stipend. However, unless the matter is especially difficult or raises questions of principle (§348 ZPO), a case at first instance may be delegated to one of its members to manage and decide on his own. This happens very frequently in practice, so in the Landgericht, too, litigants usually come before a single judge.

A commercial case as defined in §95 GVG is dealt with by the commercial division of the Landgericht, which consists of three judges — two experienced commercial men, who act as honorary commercial judges or Handelsrichter, and one professional judge, who presides. In large cities where the Landgericht has many divisions, it may set up special divisions to deal with particular types of dispute, such as industrial property disputes, questions of press law, or claims for damages based on breach of duty by state officials.

There are about ninety Landgerichte in the Bundesrepublik, each having, on average, about six Amtsgerichte in its circuit. In criminal matters the Landgericht hears appeals from convictions rendered by the Amtsgericht. Such appeals come before a division consisting of one professional judge and two lay judges (Kleine Strafkammer).

The Landgericht also serves as a trial court for all prosecutions unsuited to the Amtsgericht, and in such cases a division consists of three professional judges and two lay judges (Grosse Strafkammer). A few particularly serious crimes are tried before a special division of the Landgericht called the Schwurgericht, which has the same composition as the Grosse Strafkammer, namely three professional judges and two lay judges. These lay judges must not be confused with the jurors of the common law, for they join with the professional judges in considering matters of fact and law alike, and have the same voting power with regard both to guilt and extent of sentence. A majority of two-thirds of the judges is required for a conviction; and, since those who agree about guilt may differ as to sentence, no sentence can be imposed unless two-thirds of the judges approve.

(c) Since appeals from criminal convictions rendered by the Landgericht as a court of first instance normally go direct to the federal court, the bulk of the work of the Oberlandesgerichte, or courts of appeal, consists of civil matters. The Oberlandesgericht hears appeals from decisions of the Landgericht sitting as a court of first instance and from decisions of the Amtsgericht acting as a Family Court. The Oberlandesgericht sits in divisions called senates, each consisting of three judges, one of whom has a higher rank and presides. There are twenty Oberlandesgerichte in the Bundesrepublik, but they differ greatly in size, since many Oberlandesgerichte have thirty senates and other have only half a dozen. Some Länder have only one Oberlandesgericht (the Oberlandesgericht in Frankfurt is the only one in Hesse; Hamburg and the Saarland also have one each), while others have several (North Rhine-Westphalia, for example, has Oberlandesgerichte in Dusseldorf, Cologne and Hamm).

(d) The highest of the courts of ordinary jurisdiction is the Bundesgerichtshof, or Federal Supreme Court. It sits in Karlsruhe and consists of fifteen senates, ten for civil matters and five for criminal matters. Each senate consists of five federal judges, one of whom presides. The criminal senates hear appeals from the decisions of the Oberlandesgerichte.

The conditions that must be satisfied before this further appeal to the Bundesgerichtshof may be made from a decision of an Oberlandesgericht have periodically been stiffened with the aim of lightening the load of the Bundesgerichtshof. As the law now stands, an appeal lies only if the Oberlandesgericht has given its consent, on the ground either that the case is one it believes to involve a matter of principle, or that its decision deviates from a decision of the Bundesgerichtshof (§546 ZPO). It is only in cases where the appellant's interest in the appeal amounts to at least DM40,000 (= £10,000) that the Oberlandesgericht's consent is not required, but even here the relevant senate of the Bundesgerichtshof can decline to hear the appeal if it involves no issue of principle (§554b ZPO).

A litigant whose appeal the Bundesgerichtshof had refused to hear recently questioned the constitutionality of this rule in §554b ZPO by means of a Verfassungsbeschwerde (see p. 22 above). The complainant argued that the provision was in breach of the constitutional principle of the rule of law (Rechtsstaatsprinzip) in that a senate's decision whether or not to entertain an appeal may to some extent be based on its desire to lighten its workload, a criterion the application of which an appellant is wholly unable to predict and to which different senates may attribute different weight. The Bundesverfassungsgericht accepted these arguments, and held that in order to make §554b ZPO consistent with the constitution it must be interpreted to mean that a senate may refuse to hear an appeal involving no matter of principle only if at the end of the day the appeal has no chance of success.[2]

(e) The system of the courts of ordinary jurisdiction just portrayed has been the object of much criticism. Some years ago the Federal Ministry of Justice put out a proposal to amalgamate the Amtsgerichte and Landgerichte into large courts with at least 100,000 inhabitants in their circuit. The new courts would have first instance jurisdiction over all civil and criminal matters, and would generally sit in divisions consisting of a single judge. It was hoped that such a reform would make it easier for the citizen to understand the system, that increasing the size of the first instance court and the numbers of its divisions would facilitate the specialization of judges and the rationalization of court business, and that having the judge sit alone would reinforce his sense of responsibility and accelerate the procedure. The opponents of these proposals gave good reasons for doubting whether these desirable results would follow; in particular, they said that the citizen would find justice even more remote if the local Amtsgerichte were suppressed, and that it was not only desirable to have a collegiate bench, since it tended to produce better decisions, but actually inevitable, unless one were to entrust important cases to a young and inexperienced judge at the threshold of his career. The discussion continues, but quite apart from anything else these proposals for reform are very unlikely to be implemented in the near future, if only because of the enormous costs involved, for example in the reconstruction of court buildings.

---

[2] BVerfG *NJW* 1979, 151.

2 Labour Courts

A special set of *Labour Courts* deal with litigation arising out of all matters of labour law, such as the employment relationship, collective bargaining agreements and works agreements. The composition, competence, and procedure of these courts are regulated in the Labour Courts Act (Arbeitsgerichtsgesetz). Under this enactment the Arbeitsgerichte, or Labour Courts, which have first instance jurisdiction over labour law disputes, sit in divisions consisting of one professional judge, who presides, and two Arbeitsrichter, lay judges chosen from among the ranks of workers and employers respectively. Appeal from decisions of the Arbeitsgericht lies to the Landesarbeitsgericht, or Labour Appeal court, whose divisions are similarly composed. Under certain circumstances a further appeal lies from a decision of the Landesarbeitsgericht to the Bundesarbeitsgericht, or Federal Labour Court, which sits in Cassel. Although only points of law are considered at this level, the divisions of the Bundesarbeitsgericht also contain, in addition to three professional judges, two lay judges, again taken from the ranks of workers and employers respectively. Lay judges in the Länder are appointed by the top labour authority there, and in the Bundesarbeitsgericht by the Federal Minister of Labour. In each case the names are taken from lists of nominees submitted by trade unions and employers' associations respectively. Both senior employees with managerial functions and trade union officials are eligible to serve as lay judges.

3 Administrative courts

A third set of courts deals with matters of *administrative law*, except questions of tax and social insurance, which go to yet other sets of courts. Cases start before the Verwaltungsgericht, or Administrative Court, go up on appeal to the Oberverwaltungsgericht, or Administrative Appeal court, and finally, if the conditions for making a further appeal are satisfied, reach the Bundesverwaltungsgericht, or Federal Administrative Court, in Berlin. In both the Verwaltungsgericht and the Oberverwaltungsgericht a division consists of three professional judges and two lay judges; the Bundesverwaltungsgericht sits in senates of five professional judges each. In the typical case before these courts a citizen is claiming that the administration behaved unlawfully in taking a step that caused him harm or in failing to take a step that would have been of benefit to him. Such disputes may arise anywhere in the whole spectrum of public law, for example, in the law relating to building and planning, the environment, the control of trade, the civil service, the legal relations between a Land and its component communes, and much else besides. According to the Basic Law (art. 19 par. 4), 'recourse to a court must be open to anybody whose rights have been infringed by the act of a public authority'. This means, first, that the review of the decisions of authorities must be entrusted to courts whose judges have all the guarantees of judicial independence, and, second, that such judicial review applies to an extremely wide range of administrative acts, including, for example, a decision by a school that a pupil must repeat a year, or a decision by the examining board of a university that a candidate has failed an examination.

4 Social Courts

Social insurance and related matters are assigned to Sozialgerichte, or *Social Courts*, which form a fourth branch of jurisdiction. Like the Arbeitsgerichte, the Sozialgerichte contain two lay judges at each of the three levels. In the Sozialgerichte, the court of first instance, there is one professional judge; and there are three professional judges in the Landessozialgericht, or Social Appeal Court, and in the Bundessozialgericht, or Federal Social Court, in Cassel.

5 Tax Courts

Disputes of revenue law are decided by a special set of Tax Courts. Here, exceptionally, there are only two levels of court. This is because complaints against decisions of the revenue authorities first come before special committees. Since these committees form part of the authority itself, they cannot be considered as proper courts, but since they are believed to function very much like courts, it was thought that it would be sufficient to have just two levels of court, namely the Finanzgerichte, or Tax Courts, which hear appeals from the committees just mentioned, and the Bundesfinanzhof, or Federal Tax Court, in Munich.

II JUDGES, ATTORNEYS, AND NOTARIES

In discussing the people who are involved in the administration of justice, we shall concentrate on judges, attorneys, and notaries, but it should not be forgotten that at least one-third of those who practise law in the Bundesrepublik are in careers other than these. Table 1, based on estimates made in 1973, gives the number and proportion of German lawyers in the various principal careers. The foreign reader will probably be surprised by the large number of judges (see p. 40 below), but he may also think that the number of lawyers in the public administration is very high. This is because for many years lawyers have formed a fair proportion of the graduate members of the higher civil service. While this has been particularly true for the general state administration, the specialist branches, such as the ministries of health, education, and planning, have also employed a great many lawyers in addition to doctors, educationists, and architects. This has been called the 'lawyers' privilege', but if privilege it is, it has certainly weakened in recent years. Perhaps, as is often said, this is because the public administration is increasingly making plans and projections for the future and the young lawyer is not likely to have had his aptitude for this developed by the traditional training in the juridical analysis of situations that have already arisen.

1 Legal Education

A striking characteristic of legal training in Germany is that it takes the same form whatever career the adept proposes to follow. This means that all German lawyers go through the same training, whether they want to be judges, state prose-

Table 1

	Number	Percentage
Judges	13,500	18
Public prosecutors	2,800	4
Attorneys and notaries	26,000	35
Lawyers in public administration	17,000	23
Lawyers in industry, trade, and commerce	15,000	20
	74,300	100

cutors, attorneys in private practice, civil servants in the employment of the state, or lawyers in the legal or personnel department of some business. Even stranger, this uniform programme of training is designed with the judiciary in view, though less than a fifth of those who embark on it are going to be judges.

The training of German lawyers falls into two clearly separated parts, one principally theoretical and the other mainly practical. This results from the Law on the Judiciary (Richtergesetz), which provides, in §5, that in order to qualify for judicial service one must have passed two state examinations, the first taken after at least seven semesters of law study, the second after thirty months of preparatory service or internship. The law also provides that this qualification for judicial service is needed by a public prosecutor, an attorney, a notary, and a lawyer in the public service. But even if one wants to work as a lawyer in the private sector, for example in a bank or insurance company, the same legal training is still required. It is true that this requirement is not laid down by law, but in practice there is no doubt that one cannot be a 'Volljurist' without completing the training of a lawyer and having the qualification for judicial service.

In order to enrol as a student in the law faculty of a German university the candidate must have the 'certificate of maturity', which shows that his secondary education has been successfully completed. Seven semesters of law study, or three and a half years, are required before he may sit the first state examination, but in many cases no less than five years go by before the student presents himself for it. The examination is administered by the Ministries of Justice of the several Länder under local laws which regulate admission to the examination and the form of the examination itself; these laws must of course be in harmony with federal law and in fact differ only in detail. The examination generally consists of four or more invigilated papers containing a thorny problem on which the candidate must write a legal opinion; he has the text of the statutes to help him, but may not refer to textbooks or commentaries. In most Länder the candidate must also prepare a paper on his own, and be examined orally by a board chaired by a judge or senior civil servant.

Once a candidate has passed the first state examination he may be admitted to the preparatory service. This stage of training lasts thirty months. During this period the trainee is called a *Referendar*; he enjoys the status of a temporary civil servant, and he receives an allowance for the support of himself and his dependants. The essential steps of this training are laid down in the Law on the Judiciary (§5a): the Referendar works under close supervision, first as an assistant in the civil and criminal divisions of an Amtsgericht or Landgericht, then in a prosecutor's office, next in an administrative agency, and finally in the office of an attorney. He must also attend courses given by judges and civil servants, mainly devoted to the analysis of difficult practical cases.

At the end of this preparatory service comes the second state examination, which is again conducted by the Ministries of Justice of the several Länder. This is like the first state examination except that it is more orientated towards the practical, and the examiners now consist solely of judges and senior civil servants.

Towards the end of the 1960s this system of training came in for heavy criticism. The principal objection was that the training took too long: in fact, the average age of candidates for the second state examination is twenty-nine. It was also objected that too little attention was paid to connected subjects of study, and that the division between study and preparation meant that the theoretical and practical aspects of the training were insufficiently integrated. In response to this criticism the federal legislature amended the Law on the Judiciary in 1971 by introducing a new paragraph (§5b). This empowers the Länder to experiment with new methods of study that blend the theoretical and the practical, though of course the final examination must still have the same weight as the second state examination in law. At present, therefore, many universities are offering these novel study programmes in addition to the traditional form of training, which is chosen by the vast majority of students. These programmes vary quite considerably in certain ways, and it remains to be seen whether they will prove themselves and, if so, in what respects.

2 Judges

The judiciary is a career service in the Bundesrepublik, as in most other civil law countries. The successful applicant for appointment begins his career in a court of first instance, hoping that in the course of time promotion will take him to a higher court or even the presidency of such a court or of one of its divisions. Decisions on the appointment and promotion of judges in the service of a Land are for the Minister of Justice of that Land. In many Länder the Minister needs the agreement of the Committee on Judicial Appointments before he can make an appointment. This committee is usually composed of members of the parliament of the Land, a few judges chosen by their colleagues, and a few attorneys delegated by the bar association. A candidate's chances of appointment to judicial office depend mainly on his marks in the second state examination, but he will also be interviewed at length by senior officials of the Ministry, who may well have before them the opinions of the judges under whom the candidate worked during his preparatory service.

A successful candidate will be appointed to a probationary judgeship and will start off either in the Amtsgericht or as an associate judge in a division of the Landgericht. During the next five years the Minister will have to decide whether the probationer should be given a permanent appointment or should have to leave the service. In making this decision the Minister gives great weight to the reports made by senior judges on the way the young judge has acquitted his task.

Once a judge has a permanent appointment and has had some years of experience in a court of first instance, he is seconded for a few months to an Oberlandesgericht, where he sits in a senate as an 'auxiliary judge', with the same duties as an associate judge. During this period his professional capacities are evaluated by his colleagues on the Oberlandesgericht, and especially by the presiding judge of the senate to which he has been assigned. These opinions are critical for his chances of being appointed associate judge in the Oberlandesgericht or presiding judge in a division of the Landgericht, a position that carries equal weight and remuneration. Decisions on promotion are again for the Minister of Justice of the Land, but he must first take the opinion of a Committee of Judges, though he is not bound by it. This Committee is composed of the President of the Oberlandesgericht and a few other judges nominated by their colleagues on the circuit.

Decisions on promotions are based mainly on merit and

seniority; only in promotions to higher positions, especially the presidencies of Landgerichte and Oberlandesgerichte, which attract a fair amount of purely administrative work in addition to judicial functions, do a judge's political views begin to matter. This means that, when there are candidates of equal capacity for higher judicial office, the Minister of Justice will always choose the one who belongs to, or is in sympathy with, his party or an allied party. It is by no means unusual for judges to belong to a political party, but it is uncommon for them to be very active members, though even this is permitted, subject to the Law on the Judiciary, which lays down that 'the judge's conduct in and out of office, including his political activity, must be such as not to impair confidence in his independence' (§39 DRiG).

The two chambers of the Federal Parliament are much involved in the appointment of *federal judges*. On the Bundesverfassungsgericht, the Federal Constitutional Court, half the judges are chosen by the Bundestag and half by the Bundesrat (see p. 21 above), while the judges of the Bundesgerichtshof are chosen by a complex procedure in which the power of selection is shared by the Justice Ministers of the Bundesrepublik and the Länder on the one hand, and by an equal number of members of the Bundestag on the other.

Judicial independence, personal and material, is assured by the Basic Law in art. 97. Personal independence means that a judge, once appointed for life, may not be dismissed or even appointed to another post or pensioned off against his will unless a court has determined that the grounds specified in the Law on the Judiciary are satisfied in his case. Material independence means that the judge is bound only by the law, and that no instructions may be given to him regarding his decisions; this does not of course prevent the president of a court from letting a judge know when his work is dilatory or otherwise unsatisfactory in its technical or housekeeping aspects.

In reality, there is a more subtle threat to judicial independence. It lies in the fact that in a career system a junior judge's chances of life appointment or promotion are determined to some extent by his seniors' views on his qualifications and competence. A career judiciary may have many advantages, and it may even be inevitable in a country such as the Bundesrepublik, which must find many thousand of judges to staff its very comprehensive system of courts; but it certainly runs the risk that, when senior and junior judges sit together on a panel, the style of discussion may be somewhat authoritarian and that there may be generated that conformism which, in the words of Calamandrei, is 'the bastard son born of the union of fear and hope'.[3]

It goes without saying that a judicial career does not appeal to the most forceful, outspoken, and ambitious personalities, given its system of recruitment and promotion, its pervasive civil service mentality, its good but not exalted social prestige, and the fact that one rarely sits alone as a judge, at least in important cases, and never publishes a dissenting opinion, except in the Bundesverfassungsgericht or Constitutional Court. In general, those who become judges are men and women who prefer the security of life-long tenure with a modest salary to the rough-and-tumble of life and the risks of competition. While the German judiciary doubtless contains some powerful and impressive personalities, they are rather the exception in comparison with the High Court in England or the American Federal Bench. The reasons for this are obvious enough.

There were about 15,500 judges in the courts of the Bundesrepublik on 1 January 1979 (see Table 2). This must seem an incredibly large number to a lawyer from England, which manages to get by with only 200 or so judges for a roughly comparable population. Whence this difference? It is probably not that Germans are more litigious than the English, nor that courts in Germany are, or seem to be, more accessible or less expensive than those in England. In part it is that many functions which are clearly judicial in character are performed in England by persons to whom the English coyly refuse the name of judge, such as the justices of the peace who handle virtually all minor criminal cases, or the members of administrative tribunals. Again, while in Germany, as we shall see, it is the judge himself who shapes and develops a case from the very beginning, pre-trial decisions in England are made by masters and registrars. Finally, the judge in Germany probably has to work much longer on each case than his English counterpart, since German attorneys do very little of what their English colleagues would call the preparation of a case for trial, and tend to place the facts before the judge without worrying too much about the law.

---

[3]Calamandrei, *Procedure and Democracy* 44 (New York 1956).

Table 2
Number of Judges in the Bundesrepublik, 1 January 1979

	Courts of General Jurisdiction	Administrative Courts	Labour Courts	Social Courts	Revenue Courts	Total
Federal judges	269*	46	22	41	45	423
State judges	12,020	1,210	569	918	341	15,058
Total	12,289	1,256	591	959	386	15,481
including women	1,564	126	80	122	7	1,899

*Of the 269 federal judges, only approximately 100 are members of the Bundesgerichtshof, the others sitting as judges in the Federal Patent court (cf. n.1 above).

3 Attorneys

The legal profession in Germany is unified in the sense that the German attorney, or Rechtsanwalt, combines the functions performed by the barrister and the solicitor in England. Apart from those transactions which are invalid by law unless attested or certified by a Notar or notary (see p. 44 below), it is the Rechtsanwalt who advises the client on all matters of law, acts for him in transactions with third parties, and litigates on his behalf, both drawing up the necessary papers and speaking for him in court. Indeed, the Rechtsanwalt has a monopoly of such business, since a special law renders it illegal for anyone to give legal advice or represent a person in court unless he is an admitted Rechtsanwalt.[4] It is another question whether a person can dispense with the services of a Rechtsanwalt by appearing in court personally or by having someone who is not a Rechtsanwalt speak for him. In civil and commercial matters this is possible only in the Amtsgericht (§78 ZPO),[5] and even there the litigant must have a lawyer to represent him in matters that fall within the competence of the Family Court. For civil litigation in the Landgericht and the appellate courts it is invariably necessary to have an attorney.

Attorneys are admitted by the Ministers of Justice in the

[4]See Legal Advice Act (Rechtsberatungsgesetz) of 13 December 1935, as amended.
[5]The proportions of trials before the Amtsgericht in which attorneys appear are roughly as follows: for both parties: 32 per cent; for plaintiff only: 40 per cent; for defendant only: 4 per cent; for neither party: 24 per cent.

various Länder. The Minister has to consult the bar association,[6] but he is bound to admit any applicant who has passed the second state examination and therefore has the qualification for judicial office unless there is something serious against him.[7] This means that the bar associations in the Bundesrepublik neither help in the training nor control the admission of their future members. Lawyers are admitted to designated courts of ordinary jurisdiction, usually to an Amtsgericht and its associated Langericht or to one of the Oberlandesgerichte.

Can an attorney who is admitted to practise before an Oberlandesgericht be admitted at the same time to practise before another court? In some Länder such concurrent admission is permitted, as in Baden-Württemburg, Bavaria, and Hamburg. Other Länder, however, have a special bar for appeal cases, and defend it with the same arguments as the English use to justify the barristers' monopoly of litigation in the High Court. Those who are admitted to practise before the Bundesgerichtshof also constitute a special bar of their own. It follows that, while the litigant in Germany need not, as in England, retain two or even three lawyers at the same time in one and the same lawsuit, he may have to retain several attorneys one after the other, as where his case goes to the Oberlandesgericht or the Bundesgerichtshof.

Between the Rechtsanwalt and his client there is a contract of services which obliges the lawyer to use all the care that the circumstances call for in managing his client's affairs. The lawyer who falls below this standard of care incurs a liability to pay for the harm he thereby causes his client, whether his fault lay in the advice he gave, in his conduct of affairs, or in the way he represented his client in court.

[6]All attorneys admitted to the courts within the circuit of a court of appeal form a bar association (Rechtsanwaltskammer). It is a quasi-public corporation with compulsory membership and a right of autonomy to be exercised within the limits set forth in detail in the Federal Statute on Attorneys (Bundesrechtsanwaltsordnung) of 1 August 1959, as amended. Its tasks include the control of professional conduct and the nomination of attorneys to the professional disciplinary tribunals on which, at the upper levels, attorneys sit alongside judges. The power to lay down general principles of professional conduct is vested in the Federal Bar Association(Bundesrechtsanwaltskammer), i.e. the national federation of regional bar associations.

[7]This is why, in the worsening labour market for lawyers, there has recently been a dramatic increase in the number of attorneys: whereas the number of attorneys admitted in the Bundesrepublik in 1975 was 26,854, in 1980 it was 36,081, an increase of roughly 34 per cent.

As to attorneys' fees, there is a statute which lists and defines in minute detail every possible act of advice or litigation, and lays down that for every act so defined the attorney is entitled to one 'basic fee' or to a multiple or fraction thereof.[8] For instance, an attorney is normally entitled to three 'basic fees' for conducting a lawsuit, one being payable when the action starts, another when the oral argument begins, and the third when evidence is taken. The amount of the 'basic fee', under the statute, is related to the amount involved in the action or other matter handled by the attorney.[9] In principle, therefore, the attorney's fee is the same whether the case involves a lot of work or a little, whether it is complex or simple, whether the client is rich or poor, whether the attorney is well-known and experienced or not, and whether or not he is successful in the case. Where the lawyer gives advice or acts for the client in some other way outside court, there may be some scope for flexibility in putting a value on the matter, so the schedule of statutory fees may operate less rigidly; but in litigation the value of the matter will be evident from the sum claimed, or else it will be fixed by the court, as in the case of divorces and other non-pecuniary claims. An agreement to pay a fee higher than the statute prescribes has to be in writing; to charge a lower fee is considered unethical. To work for a contingent fee is not just a breach of etiquette: the Bundesgerichtshof has held such an agreement void as contrary to public policy.[10]

The advantage of this method of charging for legal services is that the client may make a reasonably accurate pre-estimate of the cost of his lawsuit, since the attorney's fee is based on the value of the claim rather than on the relatively unpredictable amount of time he may have to spend on it. On the other hand, it may be argued that the attorney will be underpaid if a case of small value involves a lot of work, while he may receive a windfall for handling a simple matter with a large price-tag. This danger is probably more apparent than real. In the former case, the attorney may stipulate for a fee above the statutory tariff, or he may do only as much work as the statutory fee justifies. As to the second situation, cases involving a lot of money and

---

[8]See Federal Statute on Attorneys' Fees (Bundesgebührenordnung für Rechtsanwälte) of 26 July 1957, as amended.
[9]An instance of the total cost of a lawsuit is given in the text at p. 49.
[10]See *BGHZ* 39, 142.

little work are fairly rare, but if one should arise, the attorney may agree to 'split the cause of action' and limit the claim to part of the sum at stake in the hope that the opponent, having lost the case, will pay the balance without more ado.

4 Notaries

As in other civil law countries, the notary or Notar is an important figure on the legal scene in the Bundesrepublik. In order to be admitted notary, one must have the qualification for judicial office; but whereas attorneys may be admitted without limit of number, notaries are admitted only in such numbers 'as are required for the proper administration of law'.[11] Notaries have an exclusive statutory right to perform certain administrative acts required by law, such as certifying a person's signature to a document (Beglaubigung) or attesting a declaration or agreement (Beurkundung). For a Beurkundung the notary draws up a document containing the names of the parties and the exact text of their declarations, which they sign, and then the notary certifies in the document that he has read these statements back to the parties and that they have approved and signed them. Notarial attestation is required by law for a number of important transactions, such as agreements to buy or sell real estate (§313 BGB), promises to make a gift (§518 BGB), and marriage contracts (§1410 BGB). As to wills, while a testator may make a perfectly valid will all by himself if he writes it out and signs it (§2247 BGB), many people prefer to have their will drawn up and attested by a notary (§2232 BGB). In making an attestation the notary is required to ascertain the true intention of the parties, to advise them of the legal consequences involved, to see that inept and inexperienced parties are not disadvantaged, and to put their statements in as clear and unambiguous a form as possible. The rule requiring notarial attestation clearly has something more than an evidentiary function: it also embodies the idea that in certain important transactions the public interest requires the presence of an impartial person who can give the parties legal advice and ensure that their intentions are properly carried out.

---

[11] See §4 of the Federal Law on Notaries (Bundesnotarordnung) of 24 February 1961, as amended, with full details about the admission and professional organization of notaries.

III ELEMENTS OF CIVIL PROCEDURE

The conclusion that legal institutions are essentially the same throughout the Western world is one that comparative lawyers often like to draw, but while it is true that certain common features will be found whenever and wherever judges try to resolve disputes in a rational manner, it is nevertheless undeniable that important differences of style and atmosphere emerge when one compares the civil procedure of Germany with that of a common law jurisdiction. Trial by jury may have practically disappeared from civil litigation in England, but it has still had a deep influence on the way civil matters are decided today. It is this tradition that helps to explain why trials tend to be so concentrated even when the judge sits alone, why the parties must be apprised of the precise issues and be fully prepared to meet and debate them when they appear in court, and why the trial is prepared so elaborately with the aid of discovery devices and pre-trial conferences.

In Germany there is no live tradition of trial by jury, and a civil suit is quite unlike the common law trial. In a civil case pending before the Landgericht there will be a number of isolated meetings and written communications between the parties, their attorneys and the judge; during the course of these meetings evidence will be introduced, testimony given, motions and rulings on procedure made, and areas of agreement and disagreement gradually marked out. The proceedings tend to be very unconcentrated.

This has a number of consequences. First, the pleadings are very general. Issues are defined as the case proceeds, and it is possible to introduce new motions, causes of action and issues with relatively few sanctions. Second, there is much less pressure on the German attorney to forearm himself with information and argument on every fact or claim that might possibly arise and prove relevant; if unsuspected facts do emerge or unforeseen allegations are made he will always be given an opportunity to search for and present additional proof. It is therefore hardly surprising that the German attorney is not equipped with the common lawyer's powers of discovery — those effective means, backed by the power of the court, for tracking down and finding out the facts that may be relevant.

Nor is the German attorney especially active in pursuing information on his own. Of course, he will have full consultations with his client and will examine his papers and files before the action starts; but he will not feel at all free to talk to prospective witnesses or take statements from them. Only in special circumstances is it thought advisable to question witnesses out of court: if a witness has previously discussed the case with counsel, a German judge might be very suspicious of his testimony.

In England the many decisions that may have to be made in preliminary proceedings before the trial really begins are made by a master or registrar. In Germany they fall to the judge himself, who will hold a series of meetings with the parties and their lawyers to raise and resolve such questions as whether the court has jurisdiction, whether certain statements of fact can be agreed, whether an expert is required, which witnesses should be called, and whether a settlement is possible and, if so, on what terms. German civil procedure lays much less stress than the common law on having everything done by word of mouth. Matters that are raised in oral conference are usually put in writing forthwith; and if one of the attorneys raises a new point orally, the other often asks the court to give him a certain period of time to put his answer in writing, a request that the court usually grants.

Nor is it by any means left to the parties to drum up as many witnesses as they choose. The German judge will do his very best to minimize the number of factual issues that can only be resolved by hearing witnesses. Significantly enough, witnesses can only appear pursuant to a special order made by the judge, stating which witnesses are to be examined and on what factual issues. Such an order is not subject to appeal and will be made only on controverted issues which the judge believes crucial to the disposition of the case.

Finally, German civil procedure has no counterpart to the highly complicated rules of the common law regarding the exclusion of various kinds of evidence. Doubtless these rules were intended to prevent the jury being misled by untrustworthy testimony, but since in German civil procedure the evidence is evaluated by professional judges, any evidence, including hearsay evidence, is admissible in principle, and it is for the judge to decide how much weight to give it.

In the last few years people in Germany have realized that the practice of having a number of isolated meetings and written communications between the parties, their attorneys and the judge, with the ensuing loss of concentration, is one of the causes of the length of time that civil suits often take. Accordingly, a new method of handling civil suits has been devised and tried out in many Landgerichte. It is known as the Stuttgart procedure (Stuttgarter Verfahren), and consists of preparing the case so thoroughly, by means of written procedure and at the most one preliminary meeting, that the matter can be resolved conclusively through a single comprehensive hearing in court. The Code of Civil Procedure was amended in 1976 so as to permit the judge, if he chooses, to adopt this method of proceeding, but of course one must wait and see how widely the Stuttgart procedure is adopted in practice.

A common lawyer who attended one of the conferences in a German civil suit would think the judge rather vocal and dominant and the attorneys relatively subdued. In part this is because the court is supposed to know and apply the law without waiting for counsel to deploy it (*'iura novit curia'*), but more importantly, the court is under a statutory duty to clarify the issues and help the parties to develop their respective positions fully (§139 ZPO).[12] Accordingly, the court asks questions and makes suggestions with the aim of inducing the parties to improve, modify, or amplify their allegations, to submit more documents, to offer further proof, to correct misunderstandings, and to throw light upon what may be obscure.

The extent to which a court will do this depends on a number of factors. Of course, the court must not commandeer the case or manage the litigation in the place of the parties, and it goes without saying that only the parties may make allegations, offer proof, and submit motions; but if a party is appearing in person or is represented by inexperienced or incompetent counsel, and the judge feels that he may be put at a disadvantage by oversight, inadvertence, or a clear misapprehension of the applicable law, the judge may make his suggestions with some vigour

---

[12]It is worth mentioning that the rules governing civil procedure are laid down by legislative enactment, i.e. by the Code of Civil Procedure (Zivilprozessordnung; ZPO) of 30 January 1877, as amended. There is no judicial rule-making in this field in Germany.

in order to reach the right result, notwithstanding any faults of advocacy. Similar considerations underlie the judge's duty to avoid 'surprise decisions' (Überraschungsentscheidungen), that is, decisions that in the circumstances the parties had no reason to expect. Thus the Code of Civil Procedure provides that, if the court is to found its judgment on a view of the law whose validity or relevance a party may well not have realized, it must first give him a chance to be heard on it (§278 par. 3 ZPO).

The way in which a witness is interrogated also indicates the somewhat inquisitorial nature of German civil procedure. The court itself asks the witness his name, age, occupation, and address, admonishes him to tell the truth, invites him to recount what he knows of the matter without undue interruption, dictates periodic summaries to the clerk, and questions the witness so as to test, clarify, and amplify what he has said. Once the court is satisfied, the attorneys are invited to ask further questions, but in general the attorney's role in examining witnesses is relatively modest, in part because the court has normally covered the ground already, in part because a lengthy examination might suggest that the court had not done its job properly.

It should be added that experts are not, technically speaking, witnesses at all. They are usually appointed by the court, act under the court's instructions, and owe the court a duty of loyalty and impartiality. Accordingly, in a German court one rarely sees head-on clashes between experts called by the parties and paid to be partial, subjected to fierce examination and cross-examination by attorneys who have just acquired a smattering of their expertise for the purpose. 'The German system puts its trust in a judge of paternalistic bent acting in co-operation with counsel of somewhat muted adversary zeal.'[13]

IV EXPENSES OF LITIGATION AND LEGAL AID

The expenses of litigation consist of attorneys' fees and court costs. Court costs, unlike those in England, are by no means

[13] Kaplan, 'Civil Procedure — Reflections on the Comparison of Systems', 9 *Buffalo L. Rev.* 409, 432 (1960) (a brilliant article).

negligible. They are fixed by statute, and consist of one or more 'basic units'.[14] The number of units payable depends on the number of stages the proceedings have run to, but not on how long the proceedings take or how complex they are or how much work they require of the court. In a fully contested case three units will be payable, the amount of the unit being fixed by the statute in relation to the amount involved in the litigation. According to the Code of Civil Procedure, the principle that 'the loser pays' applies to attorneys' fees and court costs (§91 ZPO). A party must therefore bear the litigation expenses of that part of the case as to which he turns out to be the loser. He must reimburse the winner, not only for the court costs he may have had to pay, but also for his attorney's fees, including disbursements by counsel incident to the case and other expenses. This applies, however, only to expenses that a reasonable litigant would have incurred under the circumstances, and a loser who thinks he is being overcharged in the bill presented by his opponent may ask the court to review it.

Table 3 gives an idea of the litigation expenses payable by the loser in a fully contested case in which evidence was taken.

A person who is too poor to sue or defend a suit may ask for legal aid (Prozess-Kostenhilfe) whatever the proceedings.[15] If he wishes to sue, he goes to the attorney of his choice

Table 3

	Amount in Controversy		
	DM500 (£125)	DM5,000 (£1,250)	DM50,000 (£12,500)
Court costs (3 'basic units')	DM69	DM348	DM1,386
Attorney's fees:			
3 'basic fees'	DM150	DM795	DM3,705
Expenses and taxes (approx.)	DM50	DM125	DM375
Total	DM200	DM920	DM4,080
Attorney's fees of opponent	DM200	DM920	DM4,080
Estimated cost of taking evidence (only witnesses, without experts)	DM100	DM225	DM625
Total litigation expenses	DM569	DM2,413	DM10,171
Percent of amount in controversy	114%	48%	20%

[14]See the Statute on Court Costs (Gerichtskostengesetz) of 15 December 1975, as amended.

[15]For the details see §§114–27 ZPO (as amended in 1980).

and the attorney submits a petition for legal aid along with the statement of claim. In contrast to the English system, the court itself then decides whether legal aid should be granted or not. There are two preconditions. First, there is a fairly rigorous means test; the court may require monthly contributions from a petitioner whose income exceeds the limits specified by law. Second, the court must satisfy itself at a preliminary stage of the proceedings that the petitioner's allegations disclose a prima facie case. If legal aid is refused, the petitioner may appeal. If it is granted, the court normally appoints the plaintiff's attorney as legal aid attorney and allows the plaintiff a provisional exemption from court costs.

If the assisted party succeeds in his suit or defence, his attorney recovers his full fees and expenses from the loser just as in a normal case. If the assisted party loses, his attorney recovers fees and expenses from the legal aid fund, though if the case involves more than DM5,600 (= £1,400) the fees will be lower than those normally payable. For example, in a case worth DM10,000, the legal aid attorney receives 83 per cent of the normal fees; in a case worth DM20,000, only 60 per cent, and in a case worth DM40,000, only 48 per cent. A successful opponent may claim reimbursement of his litigation expenses from the assisted party personally, but he has no claim against the legal aid fund even if the assisted party is a person of no substance at all. This is certainly unjust in cases where the assisted party was the plaintiff, for the state ought to be bound to indemnify the defendant for his expenses if its grant of legal aid enabled the plaintiff to molest him with an unfounded suit.

The present system of legal aid has met with increasing criticism in recent years. The main complaints are, first, that a poor party cannot obtain legal aid if all he wants from an attorney is legal advice rather than assistance with litigation, and, second, that legal aid is unavailable to those whose financial position is slightly above subsistence level, although they are unable to bear the full cost of litigation. A bill has been drafted by the Federal Ministry of Justice which would improve the situation substantially. (See p. 335 below).

It is worth adding that in Germany it is quite lawful and fairly common to take out insurance against the financial consequences of having to sue or be sued in certain fields, such as civil and criminal cases arising from the operation of a motor vehicle.

4

Sources, Legal Literature, and the Code

I SOURCES: THE MAIN DIVISIONS OF LAW AND THEIR CODIFICATIONS

1 The Divisions of Law

A legal system is made up of many separate areas of law (e.g., company law, criminal law, constitutional law). Often these areas are quite different in structure and style, because their historical developments have been quite different. Only once has there been an attempt to put in a single code — the General Prussian Land Law of 1794 — the entire law of a country. This attempt, a product of the Enlightenment, had neither success nor successors. Roman law had been divided into public and private law, and this division had some effect on German law, for example with regard to the jurisdiction of courts. For expository purposes, however, it is rather too coarse a division, and it gives rise to too many borderline cases.

Legal writers, and textbook writers in particular, now accept a division of the law which results from the way that legal subject matters are allocated to different courts, codes, and university courses. There is nothing mandatory about such a division, much less is it a 'system'; but while there can be variations in the order of treatment of the various topics, it has achieved a certain consensus by now, and bibliographies, library catalogues, and other publications are usually based upon it.[1]

Foreign lawyers in particular need to have some idea of the divisions and sources of German law or they may have difficulty in finding their way through it, so we shall now turn to the main divisions of German law and mention, where appropriate, the codes and principal enactments in force.

Like many other countries, Germany today is suffering from a flood of enactments. Codes are nothing like so common: they

---

[1] The same method of division is used to describe the papers in the state examinations for lawyers conducted by the Länder.

52   Sources, Legal Literature, and the Code

require much more concentrated attention from the legislator, and they take much longer to prepare — twenty years, indeed, was the gestation period for the BGB. This permits a high degree of terminological compression and coherence. Ordinary statutes today are not so consistent in their usage, and do not always maintain the same level of abstraction. Often, indeed, they are rather carelessly drafted.[2]

The outline that follows covers only those codes and statutes which are most relevant to the subject matter of this book

2 Areas of Law, Main Codes, and Enactments

The general areas of law, containing no binding rules, include:
(a) Legal history, especially German and Roman legal history
(b) Legal philosophy, general constitutional theory, legal theory; sociology of law, jurimetrics
(c) Comparative law, both general and specialized

Positive law includes:
(d) Private law and related areas
  (i) Civil Code = Bürgerliches Gesetzbuch (BGB), 18 August 1896, *RGBl* 195. Frequently amended, most recently by the Parental Care Reform Act = Gesetz zur Neuregelung der elterlichen Sorge, 18 July 1979, *BGBl* I 1061
  The Civil Code is amplified by:
  (ii) Missing Persons, Declarations of Death and Time of Death Act = Gesetz über die Verschollenheit, Todeserklärung und Feststellung der Todeszeit (VerschG), 4 July 1939, *RGBl* I 1186, reissued as amended 15 January 1951, *BGBl* I 63
  (iii) Notarial Documents Act = Beurkundungsgesetz (BeurkG), 28 August 1969, *BGBl* I 1513
  (iv) Instalment Sales Act = Gesetz betreffend die Abzahlungsgeschäfte (AbzG), 16 May 1894, *RGBl* 450
  (v) Uniform Law on the International Sale of Goods = Einheitliches Gesetz über den internationalen Kauf beweglicher Sachen (EKG), 17 July 1973, *BGBl* I 856
  (vi) Heritable Building Rights Ordinance = Verordnung über das Erbbaurecht (Erbbau VO), 15 January 1919, *RGBl* 72
  (vii) Home Ownership and Long-Term Tenancy Act = Gesetz über das Wohnungseigentum und das Dauerwohnrecht

---

[2]Kriele, *Theorie der Rechtsgewinnung* (2nd ed. Berlin 1976); Hug, 'Gesetzesflut und Rechtssetzungslehre', in Klug *et al.* (eds), *Gesetzgebungstheorie, Juristische Logik, Zivil-und Prozessrecht (Festschrift Rödig)* 1 (Berlin 1978).

Divisions of Law and their Codifications 53

        (Wohnungseigentumsgesetz, WEG), 15 March 1951, *BGBl* I 175
- (viii) Land Register Ordinance = Grundbuchordnung (GBO), 24 March 1897, *RGBl* 139; reissued 5 August 1935, *RGBl* I 1703
- (ix) Marriage and Divorce Act = Ehegesetz (EheG; Gesetz no. 16 des Kontrollrats), 20 February 1946, *ABlKR* 77
- (x) Road Traffic Act = Strassenverkehrsgesetz (StVG), 19 December 1952, *BGBl* I 837
- (xi) Law on Liability of Railway, Electricity Supply, Mining etc. Undertakings for Death, Personal Injury and Property Damage = Haftpflichtgesetz (HaftpflG), 7 June 1871, *RGBl* 207; reissued as amended, 4 January 1978, *BGBl* I 145

(e) Commercial law
- (i) Commercial Code = Handelsgesetzbuch (HGB), 10 May 1897, *RGBl* 219
- (ii) Bills of Exchange Act = Wechselgesetz (WG), 21 June 1933, *RGBl* I 399
- (iii) Cheques Act = Scheckgesetz (ScheckG), 14 August 1933, *RGBl* I 597
- (iv) Trade Regulation Ordinance = Gewerbeordnung (GewO), 21 June 1889; reissued as amended 26 July 1900, *RGBl* 871
- (v) Act against Restraints on Competition = Gesetz gegen Wettbewerbsbeschränkungen (GWB), 27 July 1957, *BGBl* I 1081; reissued as amended 3 August 1973, *BGBl* I 917
- (vi) Unfair Competition Act = Gesetz gegen den unlauteren Wettbewerb (UWG), 7 June 1909, *RGBl* 499
- (vii) Insurance Contract Act = Gesetz über den Versicherungsvertrag (VVG), 30 May 1908, *RGBl* 263

(Patent and copyright law are here omitted.)

(f) Company law
- (i) Stock Corporation Act = Aktiengesetz (AktG), 6 September 1965, *BGBl* I 1089
- (ii) Limited Liability Companies Act = Gesetz über die Gesellschaft mit beschränkter Haftung (GmbHG), 20 April 1892, *RGBl* 477; reissued as amended 20 May 1898, *RGBl* 846
- (iii) Law on Cooperatives = Gesetz betreffend die Erwerbs- und Wirtschaftsgenossenschaften (GenG), 1 May 1889, *RGBl* 55; reissued as amended 20 May 1898, *RGBl* 810

(g) Judicature and civil procedure
- (i) Constitution of Courts Act = Gerichtsverfassungsgesetz

(GVG), 27 January 1877, *RGBl* I 41, reissued as amended 9 May 1975, *BGBl* I 1077
(ii) Law on the Judiciary = Deutsches Richtergesetz (DRiG), 8 September 1961, *BGBl* I 1665; reissued as amended 19 April 1972, *BGBl* I 713
(iii) Federal Attorneys Act = Bundesrechtsanwaltsordnung (BRAO), 1 August 1959, *BGBl* I 565
(iv) Federal Notary Public Act = Bundesnotarordnung, 24 February 1961, *BGBl* I 533
(v) Code of Civil Procedure = Zivilprozessordnung (ZPO), 30 January 1877, *RGBl* 83; reissued as amended 12 September 1950, *BGBl* I 533
(vi) Foreclosure and Sequestration Act = Gesetz über die Zwangsversteigerung und die Zwangsverwaltung (ZVG), 24 March 1897, *RGBl* 97; reissued as amended 20 May 1898, *RGBl* 713

(h) Labour and social security law
(i) Collective Agreements Act = Tarifvertragsgesetz (TVG), 9 April 1949, *WiGBl* 35; reissued as amended 25 August 1969 *BGBl* I 1323
(ii) Labour-Management Relations Act = Betriebsverfassungsgesetz (BetrVG), 15 January 1972, *BGBl* I 13
(iii) Protection Against Dismissal Act = Kündigungsschutzgesetz (KSchG), 10 August 1951, *BGBl* I 499; reissued as amended 25 August 1969, *BGBl* I 1317
(iv) Labour Court Act = Arbeitsgerichtsgesetz (AGG), 3 September 1953, *BGBl* I 1267; reissued as amended 2 July 1979, *BGBl* I 853, 1036
(v) Social Code = Sozialgesetzbuch (SGB): General Part = Allgemeiner Teil, 11 December 1975, *BGBl* I 3015; General Rules on Social Security = Gemeinsame Vorschriften für die Sozialversicherung, 23 December 1976, *BGBl* I 3845
(vi) Imperial Insurance Ordinance = Reichsversicherungsordnung (RVO), 19 July 1911, *RGBl* 509
(vii) Federal Social Welfare Act = Bundessozialhilfegesetz (BSHG), 30 June 1961, *BGBl* I 815; reissued as amended 13 February 1976, *BGBl* I 289, 1150
(viii) Social Court Act = Sozialgerichtsgesetz (SGG), 3 September 1953; reissued as amended 23 September 1975, *BGBl* I 2535

(i) Criminal law, criminal procedure, criminology, and juvenile law
(i) Criminal Code = Strafgesetzbuch (StGB), 15 May 1871,

Divisions of Law and their Codifications 55

        RGBl 127; reissued as amended 2 January 1975, *BGBl* I 1
- (ii) Law on Minor Offences = Gesetz über Ordnungswidrigkeiten (OWiG), 24 May 1968, *BGBl* 1581; reissued as amended 2 January 1975, *BGBl* I 80, 520
- (iii) Juvenile Courts Act = Jugendgerichtsgesetz (JGG), 4 August 1953, *BGBl* I 751; reissued as amended 11 December 1974, *BGBl* I 3427
- (iv) Juvenile Welfare Act = Gesetz für die Jugendwohlfahrt (JWG); reissued as amended 7 January 1975, *BGBl* I 129, 650
- (v) Code of Criminal Procedure = Strafprozessordnung (StPO), 17 September 1965, *BGBl* I 1373; reissued as amended 7 January 1975, *BGBl* I 129, 650
- (vi) Law on Imprisonment = Gesetz über den Vollzug der Freiheitsstrafe und der freiheitsentziehenden Massregeln zur Besserung und Sicherung (StVollG), 16 March 1976, *BGBl* I 581, 2088

(j) Public and constitutional law, including international law and the law of supranational bodies
- (i) Basic Law of the Bundesrepublik = Grundgesetz für die Bundesrepublik Deutschland (GG), 23 May 1949, *BGBl* I
- (ii) Constitutional Court Act = Gesetz über das Bundesverfassungsgericht (BVerfGG), 12 March 1951, *BGBl* I 243; reissued as amended 3 February 1971, *BGBl* I 105
- (iii) Federal Elections Act = Bundeswahlgesetz (BWahlG), 7 May 1956, *BGBl* I 383, reissued as amended 1 September 1975, *BGBl* I 2325
- (iv) Law of Nationality = Reichs- und Staatsangehörigkeitsgesetz (RuStAG), 22 July 1913, *RGBl* 583
- (v) Constitution of Hesse = Verfassung des Landes Hessen, 1 December 1946, *GVBl* 229 (example of Land constitution)
- (vi) Treaty Establishing the European Economic Community = Vertrag zur Gründung der Europäischen Wirtschaftsgemeinschaft (EWGV), 25 March 1957, *BGBl* II 766, 1678, and *BGBl* 1958 II 64
- (viii) European Convention on Human Rights = (Europäische) Konvention zum Schutze der Menschenrechte und Grundfreiheiten (EMRK), 4 November 1950, *BGBl* 1952 II 686, 953

(k) Administrative law and procedure
- (i) Administrative Courts Ordinance = Verwaltungsgerichtsordnung (VwGO), 21 January 1960, *BGBl* I 17
- (ii) Administrative Procedure Act = Verwaltungsverfahrens-

gesetz (VwVG), 25 May 1970, *BGBl* I 1253 (parallel laws exist in the Länder)
(iii) Federal Zoning and Building Law = Bundesbaugesetz (BBauG), 23 June 1960, *BGBl* I 341; reissued as amended 18 August 1976, *BGBl* I 2257 (supplemented by building laws in the Länder)
(iv) Internal Revenue Code = Abgabeordnung (AO), 16 March 1976, *BGBl* I 613

II FINDING THE LAW

1 Approaching German Legal Literature

There are a great many law books in Germany — too many, in the view of some. One therefore needs some help in finding one's way through the writings, the decisions of the courts, and the relevant enactments. We shall give a brief indication of the special features of German legal literature which should help guide the reader's first footsteps.[3] In case of language problems, the list of publications in English on German Law given in Appendix II should prove helpful (below p. 332).

Collections of books on German law are to be found in all the faculties of law (in the libraries of institutes and seminars), in the central libraries of all universities, and in the large libraries of the Länder, for example in Berlin, Munich, and Stuttgart. The libraries of the universities and the Länder are prepared to lend their books, even to borrowers abroad. In addition, there are superb collections in the Max Planck Institutes for Comparative Law (in Hamburg for private law; in Heidelberg for public law and public international law; in Freiburg-im-Breisgau for penal law; in Munich for the law of patents, copyright and competition, as well as for social law; and in Frankfurt-am-Main for European legal history). A list of the various libraries and an indication of their holdings and how to use them will be found in the European Law Libraries Guide.[4]

In days gone by, the glory of German legal literature was the large textbook, but the *commentaries* enjoy pride of place in most

[3]For a technical treatment of German legal writing, see first Szladits, *Guide to Foreign Legal Materials: France, Germany, Switzerland* (New York 1959). Leser, *Vom Umgang mit juristischer Literatur* (Marburg 1971) may also be of use.
[4]*European Law Libraries Guide*, prepared by the International Association of Law Libraries (London 1971).

areas of law today. They follow the structure of the enactment in question, and normally give very full references to legal writings and court decisions. Many of them concentrate on the court decisions,[5] while others pay great attention to the views of scholars as well.[6] Most of these commentaries are the work of a whole team of authors, and editions usually follow each other fast enough to ensure that the information is up to date. Because of their reliability, most of these commentaries, which usually also contain a list of abbreviations, enjoy a considerable reputation.

In the last few years there has been a great increase in the number of outlines and similar booklets (nutshells) designed for the student market. The attention of the foreigner should be drawn to university writings (Hochshulschriften), such as dissertations submitted for doctorates, and Habilitationsschriften, which are required for a professorial chair. Such writings are not usually to be had in bookshops, but are reproduced in the faculties and exchanged between the university and faculty libraries. They are often of use as collating the material on special topics and in giving references to the literature. *Festschriften* form another special kind of publication, and consist of a series of articles or essays, usually on topical themes, produced in honour of a colleague, for example, on his seventieth birthday.

As to the *methods of citation*, books are cited in the normal way. Commentaries are usually cited by the name of the editor, e.g. Palandt, Staudinger, with the addition of the name of the author responsible for the paragraphs in question, e.g. Palandt/Thomas §823 N. 15 B a, or Staudinger/Coing, §164 Rdnr. 23. Periodicals are cited by abbreviated title, year of publication and page number, e.g. *NJW* 1979, 306. This means page 306 of the *Neue Juristische Wochenschrift* for 1979. The serial or annual number of the periodical is not usually given. For decoding the abbreviations used for periodicals, statutes, and other legal terms, the manual of Kirchner is of use.

All but the most familiar statutes are cited by date of publication and page number of the relevant Gesetzblatt, the official

---

[5] e.g., *Kommentar zum BGB* (edited by judges of the imperial and federal courts) (12th ed. Berlin 1974).
[6] e.g., *Münchener Kommentar zum BGB* (7 vols) (Munich 1978).

journal. Examples may be found in section I 2 above. Statutes are usually drafted in paragraphs, referred to as §; the Basic Law is exceptionally drafted in articles, referred to as 'arts.'. Sections of paragraphs are cited to 'Abs.' (for Absatz), or by Roman numerals, sentences within them by Arabic numerals. Thus the second sentence of the first section of paragraph 812 of the Civil Code is cited either as §812 Abs.I S.2 BGB, or as §812 I 2 BGB. The paragraph sign is doubled when the citation is to several paragraphs, e.g. §§812–20 BGB.

Names of parties are not used to identify court decisions as they are in Anglo-American law: indeed, one will look in vain through the published sources for the names of the parties or of the judges. Cases are cited by reference to the court that decided them and to the publication where the report is to be found, e.g. *BGHZ* 51,91,102. This means that the judgment starts on p. 91 of the fifty-first volume of the official collection of the Decisions of the Bundesgerichtshof in Civil Matters, and that the pertinent material is to be found on p. 102 of that volume. A judgment published in a periodical may be cited as, e.g., BGH *NJW* 1980, 633, 635 rechte Sp., which means that the judgment starts on p. 633 of the *Neue Juristische Wochenschrift* for 1980 and that the pertinent material will be found in the right-hand column of p. 635. Since such citations are hardly very memorable, it is now established practice to refer to leading cases by an indication of their subject matter, e.g. the 'parking-place' case (see Chapter 5, Section II below).

III LEGISLATION, CODIFICATION, AND INTERPRETATION

1 *The Supremacy of Legislation*

In his search for the solution of a legal problem, the German judge or jurist starts out from the legal rule or norm. A legal rule consists of two parts, the legal command, that is the legal outcome, and the *Tatbestand,* the facts or other circumstances in which it is to apply. To take a very simple case, §823 BGB says, in effect, that 'The person who carelessly damages the property of another is bound to compensate him for that harm.' The legal consequence here lies in the normative proposition 'is bound to compensate', so the judge only has to determine whether or not the preconditions for the application of that legal rule, the so-called Tatbestand, exist in the case. Such preconditions may well contain conclusions of law as well as matters of fact, and they are usually very much more complex than in our example, but the legal rule so formulated, typical of legislation, remains determinative of the approach adopted.

Legislation includes not only enactments which have been promulgated after going through all the proper formal stages in Parliament, but also a number of sources of differing orders of power. Thus, constitutional law takes precedence over a simple statute, while statutes take precedence over ordinances and other forms of delegated legislation. Customary law constitutes an independent source on its own. But the invariable question — and this, with certain modifications, is also true of other legal systems in continental Europe, in sharp distinction to the common law — is always: 'What is the rule of law laid down in the statute or comparable source?' even if the rule is not formulated there in so many words.

This may seem a little surprising when one considers that the Roman law, or gemeines Recht, which had such impact on continental Europe, never took statutory form and contained only a relatively small number of formulated legal rules. It was with the rise of *positivism*[7] in the nineteenth century that perspectives narrowed and the enactment was treated as the sole source of legal rules. 'Everything the legislator lays down is law, and nothing else is.' Credit for this is given to Montesquieu's theory of the separation of powers, whereby the legislature, the executive, and the judiciary are to be kept strictly separate — so as to act as reciprocal controls on each other. This was one reason, but the obvious superiority of ready-formulated rules and concepts was another; and doubtless it was the astonishing advances of the natural sciences and technology in the nineteenth century, with their causal mode of thinking, that provided the real foundation for positivism. The positivist's view of the judge as a person who simply applies the pre-formulated legal rule can be distorted into likening him to a slot-machine, where one feeds in the facts and legal rules like so many coins in an automat and gets the results delivered below.

[7] Maihofer (ed.), *Naturrecht oder Rechtspositivismus?* (Bad Homburg 1962).

Hand in hand with such one-sided positivism went an exaggerated conceptualism where ideas built on other ideas rather than on reality, the so-called *Begriffsjurisprudenz*.[8] Towards the end of the nineteenth century, and especially after the First World War, there was a strong reaction in forms such as the jurisprudence of interests and the Freirechtsschule, or school of free law, to be mentioned later. The Second World War, and the bitter experience of a criminal legislator exempt from all control, have since taught people the dangers inherent in a purely positivistic attitude to law, and there have been attempts to temper the positive law by seeing it in relation to the law of nature,[9] its theological underpinnings, and the tension between law and justice.

Thus, people today no longer envisage the legal rule as being the legislator's actual command, but rather as being dependent on the transcendental foundations of justice. By making people aware of the fact that parallel and similar systems of rules exist in other countries, comparative law has contributed much to this new understanding of legal rules and legislation. It remains true, however, that people in Germany with a legal problem to solve start out from the legal rule and look for it in legislation, even if they cannot find it in the very words of the text. Anglo-American lawyers may have the common law to turn to, but German lawyers recognize no such alternative source of law outside legislation, even though they accept that the decisions of courts amend and change the substance of the codes and statutes (see Section 3 below).

2 Codification and Interpretation

The high esteem in which legislation is held is not unconnected with the theory and practice of codification. A code, according to the notions of the Enlightenment, contains an answer for all the questions that may arise in a given sector. It must have a system, with uniform concepts founded on common principles, whose precision results from its very uniformity, and it is complete in the sense that it regulates all possible cases, including those that have never yet arisen. Such an idea was bound to

---

[8]Heck, *Begriffsbildung und Interessenjurisprudenz* (Tübingen 1932).
[9]See n. 7 above.

show cracks before long; indeed, 'the gap in the law' has been a constant theme of discussion for the past eighty years, and the problem of how to close it through interpretation runs through German legal philosophy like a golden thread.

Although the BGB is the most comprehensive of the German codes, it does not contain any rule for its own interpretation. Reliance is therefore placed on the well-known rule in the Swiss Civil Code (art.1 par.2): 'If no applicable provision is contained in the Code, the judge is to apply customary law or, in its absence, the rule which he himself would establish, if he were the legislator.'[10] It is admitted nowadays that judicial decisions are acts of will as well as of intelligence. Indeed, judges are charged with a duty to develop the law,[11] and they cannot do this without altering it. The result is that the statute or the code is stretched or changed, sometimes in the teeth of its terms. New rules often emerge from judicial decisions; but while it is unquestionably one of the principal tasks of the supreme courts to lay down such rules, it has proved very difficult to accommodate this admitted fact within the legal theory (see Section 3 below).

It is one of the most striking changes in the style of legal argument during the past century that nowadays judicial decisions play an essential, often a predominant, part in the discussion of legal problems. Professional training reflects this change, and students of law are now taught at an early stage how to analyse court decisions and to appraise their impact. The courts usually decide legal questions in terms of the relevant code or law. The first step is to interpret the law: here there is continuing tension between objective interpretation, based on what the legislator said, and subjective interpretation, based on what the legislator meant, as disclosed by the *travaux préparatoires*.[12] If interpretation is unavailing, the judge tries to fill the gap by using the general principles of the code or the accepted value judgments of the system as a source of analogy.

---

[10] The Civil Code of Louisiana (1870) also has a helpful formula in art. 21: 'In all civil matters, where there is no express law, the judge is bound to proceed and decide according to equity. To decide equitably, an appeal is to be made to natural law and reason, or received usages, where positive law is silent.'
[11] §137 GVG; *BVerfGE* 9, 338. 349.
[12] Larenz, *Methodenlehre* 298f. (4th ed., Berlin 1979).

In doing this, however, the judge is not abandoning the code, but extending or completing it, and he stays close to the law or the code even if he alters individual rules of law. He cannot resort to any other body of law, such as the common law which English-speaking judges have at their disposition. Nor could any such common law be created, because in Germany, unlike Anglo-American systems, the attitude of those who interpret legislation is basically favourable to it. Whereas English-speaking lawyers tend to treat enactments with some reserve, because the law-maker is interfering with the freedom of the judge, German lawyers believe that a rule can be charmed out of the code even when it is silent.

It is not the techniques or rules of interpretation that matter so much as the fact that, once the interpretation or completion is effected, its results are absorbed back into the code so as to form part of it thereafter. Thus the code remains the forum of legal change and experience, and this keeps the code alive and adaptable to new tasks and developments. In this form the idea of codification is still alive and fertile today.

The interpretation or completion of the code is no longer dominated by the jurisprudence of concepts, which has been succeeded, after the extreme counter-movement of the Freirechtsschule,[13] by the jurisprudence of interests. This school, started by Rudolf von Jhering[14] and now principally associated with the name of Philipp Heck,[15] has given the judge a new freedom. He no longer has to fabricate conceptualistic constructions, but may appraise the interests of the parties and effect an accommodation of them both in deciding the individual case and in formulating the general rule, though there are still, alas, many cases where the discussion is in purely formal terms.[16] Recent attempts[17] to build on these foundations include value-jurisprudence and teleological interpretation,

[13]Fuchs, *Gerechtigkeitswissenschaft* (Karlsruhe 1965); Kantorowicz, *Rechtswissenschaft und Soziologie* (Karlsruhe 1962).
[14]*Der Zweck im Recht* (2 vols.) (1st ed. 1877; 3rd ed. Leipzig 1893–8).
[15]See n. 8 above.
[16]Enneccerus/Nipperdey, I *Allgemeiner Teil des Bürgerlichen Rechts* §51 (15th ed. Tübingen 1959). For examples, see *BGHZ* 11, Anhang 85 (1953). On art. 117 par. 1 GG, contrast OLG Frankfurt am Main, *NJW* 1953, 746, and LG Giessen, *NJW* 1953, 666, with *BVerfGE* 3, 237 (same case).
[17]Larenz (n. 12 above), 128f.; Esser, *Vorverständnis und Methodenwahl in der Rechtsfindung* (Frankfurt 1972); Rüthers, *Die unbegrenzte Auslegung* (Tübingen 1968).

Legislation, Codification, and Interpretation 63

but they have not produced any generally accepted viewpoints. The development of new principles of interpretation has not, however, impaired the significance of code and law as the starting-point and main source for the judge.

3 Creation of Law by Judges?

When judicial interpretation goes to the extent of completing the law and sometimes altering it against its very words, the question is raised whether the judges are creating new law. The decision of a court has binding legal effect (Rechtskraft) only on the parties to the litigation, subject to one exception,[18] but in reality the decisions of the superior courts have a much greater impact than this. The judges' duty to develop the law is clearly acknowledged in legislation as well as in the decisions of the Bundesverfassungsgericht.[19] Indeed, no one today seriously disputes this. The new rules they lay down, often in the clearest terms, are treated just like rules of law,[20] discussed by legal scholars, taken into account by practitioners and followed by the courts. Judges and attorneys are expected to be familiar with recent decisions, as the cases on legal malpractice show.[21]

The recognition of judge-made rules as proper law is rendered difficult by Montesquieu's theory of the separation of powers which we have already mentioned. Most authors incorporate judge-made law in customary law, whose hallmarks are common adoption and long and invarying habit; to find these characteristics in judge-made law, even with modifications such as some writers adopt, is far from convincing, however, since in fact even isolated decisions of the superior courts command respect, and do so immediately rather than after an interval. Nor does the notion of a special judicial customary law,[22] which some have invoked, take us very much further.

The real question raised by the judges' power to make law is

---

[18]BVerfGG §31.
[19]See n. 11 above.
[20]e.g., the 'fowl-pest' case, *BGHZ* 51, 91, 102, on products liability (see Chapter 9 Section I below), or the 'gentleman rider' case, *BGHZ* 26, 349, 354f., on the right of personality (see Chapter 9 Section IV below).
[21]BGH *NJW* 1952, 425.
[22]Lehmann/Hübner, *Allgemeiner Teil des BGB* 21 (15th ed., Berlin 1966); Staudinger/Brändl, I *Kommentar zum BGB*, Intro. to §1. n. 42 (11th ed., Berlin 1957). *Contra*: Esser, 'Richterrecht, Gerichtsgebrauch und Gewohnheitsrecht,' in *Festschrift von Hippel* 95, 123 (Tübingen 1967).

64 Sources, Legal Literature, and the Code

not how to classify it, but how to limit their power and increase their responsibility. Admittedly, judges are bound to decide in accordance with 'statute and law' (art. 20 par. 3, Basic Law) but this is too unspecific a restraint; and attempts to use the principles of interpretation and the doctrine of legal sources as methods of limiting their room for play have been ineffective. A different consideration may perhaps prove helpful here. When the courts are developing the law, they follow the fundamental principle of the statute in question and thus seek ideally to maintain *consistency* with the law and its underlying principles. In so doing they have the support of scholars whose writings they regularly cite and discuss.[23] This lively debate and reciprocal interplay between courts and writers can inhibit untoward developments and provide the best assurance of proper progress.

IV THE CIVIL CODE AND ITS STRUCTURE

1 The Birth of the BGB

The significance and scope of the BGB give it a special position among the codes and statutes. It is not the oldest of the codes — the Criminal Code, the Code of Criminal Procedure, the Code of Civil Procedure, and the so-called Judiciary Laws[24] were embarked upon immediately on the foundation of the German Empire in 1871 — but it is the most important code for comparative lawyers and for all those who want to understand German law. Promulgated in the Reichsgesetzblatt on 18 August 1896, it came into force on 1 January 1900, and brought about the unification of private law in the German Empire which had been so ardently desired.

In contrast with public law, where individuals are subordinated to the state, the BGB regulates the relations between private persons as equals who stand on the same level. Commercial law and labour law, which form part of private law, are nevertheless separately regulated. Before the unification of private law in 1900 there was a great variety of law, more than

---

[23] See the decisions cited in n. 20 above.
[24] Criminal Code of 15 May 1871; Code of Criminal Procedure of 1 February 1877; Code of Civil Procedure of 30 January 1877; Bankruptcy Ordinance (Konkursordnung) of 10 February 1877. Further details may be found in Section I 2 above.

thirty different statutes and sources apart from the gemeines Recht or Pandects. Prussia had its General Land Law of 1794, while the French Code civil was in force in its former western provinces and also in Baden, there translated into German. Saxony had its own civil code,[25] which contributed much to the BGB, and there was partial codification in Bavaria. Unification of law was urgently required for practical purposes, though it needed an amendment to the Constitution of 1871. The valuable documents[26] of the two Commissions that sat consecutively after 1880 and produced Drafts I and II for the code are still referred to today.

2 The Characteristics of the BGB

The purpose of the BGB was to unify and clarify the existing private law without undertaking any basic reforms. The main influence was the gemeines Recht in the form of the Pandects, but the other codes made their contributions, such as the French Code civil of 1804, the Austrian Civil Code (ABGB) of 1811, and the Saxon Civil Code of 1863, to mention only the most important. On the other hand, the decisions of the Reichsgericht, set up in 1879, and of its predecessor, the Reichsoberhandelsgericht (1871–80), had surprisingly little influence. It was not that there was any shortage of published decisions, but it was not in tune with the times to benefit from the teachings of experience. The consequent gaps in the code became apparent shortly after it came into force: for example, there are no rules for positive breach of contract (see Chapter 6, Section II 1 below), and the treatment of contract as a unit was quite insufficient (see Chapter 6, Section VI 2).

The BGB has been severely criticized for containing little or nothing in the way of socially protective rules. One of the earliest protective laws in the private area, the Instalment Contracts Act of 1894, dealing with hire-purchase and instalment sales, was left outside the code altogether, significantly enough, and while the BGB does regulate the contracts of

---

[25]Civil Code of Saxony (1863), in force from 1865.
[26]Vierhaus, *Die Entstehungsgeschichte des Entwurfes des BGB* (Leipzig 1888). The five volumes of Mugdan, *Die gesamten Materialien zum BGB* (Berlin 1899) contain the conclusions of the First Commission, the discussions of the Second Commission, the text of Drafts I and II, and the introductory monograph.

employment and lease, it does so with hardly any safeguards for employees or tenants. The existing rules of property and family law were simply consolidated; the reforms not carried out then had to be undertaken later, either by way of complementary statutes,[27] or by partial revisions of the code itself.[28]

The consistently austere conceptualism of the BGB and its abstract language bear the stamp of Pandectism, and to some extent of Begriffsjurisprudenz as well. The code is the work of theorists, and it took at least twenty years of exposition before the courts had it at their fingertips. Its strength lies in the formalism of its rules, which gives them elasticity, in the balance of the structure it gives to its institutions, always taking the interests of both sides into account, and in the general clauses which it incorporates, such as §§138, 157, and 242 BGB. These general clauses have proved invaluable in adapting the code to changed circumstances. The language of the BGB, unlike that of the Swiss Civil Code of 1907, for example,[29] is technical and juridical, but despite its abstractness it is powerful, clear, and precise. Terms are used with the same sense throughout, and the system of cross-referencing is well developed, to say the least. For all these reasons the BGB has maintained its position as the centrepiece of all legal education.

3 Structure

Apart from the Introductory Law (Einführungsgesetz, EGBGB), which contains a handful of positive rules for private international law, the BGB consists of five books, with a total of 2,385 paragraphs.

Book One is the General Part (Allgemeiner Teil) of the BGB. This is a characteristic product of the Pandectist view that abstraction should be carried as far as possible: all rules which are not specific to a particular institution, say, the contract of sale or the contract of services, are 'factored out' and put at the

---

[27]e.g., in property law, the Erbbaurechtsverordnung of 15 January 1919 and the Wohnungseigentumsgesetz of 15 March 1951 were respectively designed to facilitate the construction of dwellings and the acquisition of homes, aims in which they have been successful. For further details see Chapter 10, below.
[28]Most recently in family law: see p. 67 below.
[29]On this see Rabel, 'Bürgerliches Gesetzbuch und schweizerisches Zivilgesetzbuch' (1910) in Rabel (ed. Leser), I *Gesammelte Aufsätze* 141 (Tübingen 1965).

beginning so as to render them of general application. The principle has been carried so far as to render the BGB less easy to work with, for matters that go closely together from the point of view of substance are often widely separated. Such extreme abstraction no longer seems necessary to people today, because they are readier to generalize rules from a single institution.

The General Part begins with a treatment of the subjects of law, natural, and legal persons. Its general provisions about capacity to act (age of majority), legal transactions, especially contracts, defects of intention (avoidance for mistake or duress), agency, and prescription apply not only through the five books of the BGB but outside it as well.

The rest of private law is to be found in the following four books, divided in a manner familiar from the classical codes, the French Code civil and the ABGB of Austria.

Book Two of the BGB deals with the law of obligations (Schuldrecht), the obligation being treated as a means of exchange. Here, too, one speaks of a General Part and a Special Part, the latter containing the different institutions and special types of contract (Chapter Seven, Book Two, §§433–853). Obligational contracts in their various forms receive the most attention, but unjustified enrichment and delict, as other sources of obligation, are also to be found here.

Book Three contains the law of property, possession and ownership of land and moveables, and also security rights therein. In contrast with the rather dynamic law of obligations, the law of property is more static and durable.

Book Four contains family law. This part of the BGB has been amended more than any other, with reforms clearly reflecting social change. Important developments since the Second World War include the equiparation of man and woman, the introduction of the principle of breakdown of marriage as the ground of divorce, the reconstruction of the law of matrimonial property, and, most recently, a reform of the relationship of parent and child.

Book Five contains the law of succession, testate and intestate.

In addition to a number of transitional provisions, most of which are now obsolete, the Introductory Law to the BGB contains the few patchy rules of private international law (arts.

7–31 EGBGB). Unlike the BGB, which is divided into paragraphs, the Introductory Law is composed of articles.

4 Fundamental Concepts

The fundamental concepts, which are so characteristic a feature of the BGB, took their essential shape at the height of Pandectism, but further additions to this treasure-house of ideas were made even after the code was enacted. Limitations of space make it impossible to give any proper impression of them here, but as an example we may take the *individual right* (subjektives Recht), a concept that includes all the different forms of entitlement.

A fundamental distinction is drawn between relative and absolute rights. What arises out of a contract, a tort, unjustified enrichment, or the like is described as an obligational relationship (Schuldverhältnis), which includes both rights and correlative duties. Usually only two parties (subjects of law) are involved in such an obligational relationship, and the relationship exists between them and them alone. This obligational relationship is therefore described as 'relative'; there is said to be a bond of obligation between the parties. Contrasted with this are absolute rights, commonly rights of dominion over a thing; ownership is the prime example, but rights of security are also included, as are patents (intellectual property), and also personal freedom. In these cases the right is vested in an individual person, and that person is protected in his title and the enjoyment of his right against any other person who questions it. Thus an absolute right does not depend on a relationship between two persons, but is attached to a single person alone. Absolute rights are found mainly in the law of property. The importance of the distinction between the two types of right can be seen in the law of tort; under §823 par. 1 BGB absolute rights are protected against invasion, but relative rights arising out of obligations are not.

Suppose, for example, that an actor's performance has to be cancelled because he has been injured in a collision owing to the carelessness of a taxi-driver. The actor can claim compensation for the damage to his health and property (his clothes, for example), for these are absolute rights which attract such protection. On the other hand, the theatre, which had to abandon the production because their contract with the actor has been interfered with (but without rendering him liable to them), has no claim against the taxi-driver, for what is involved here is only a relative right, a contract between the theatre and the actor, which has effect only *inter partes*.

Under the heading of 'Freedom of Contract' in Chapter 5, Section IV below, one will see how people are free to invent new forms of contract and contractual rights, these being relative rights. But whereas this power of creation exists in obligational relationships, a so-called *numerus clausus* exists with regard to the absolute rights of property law. This means that absolute rights are at the disposition of individuals only in a limited number of given forms: new forms of property rights cannot be created by the parties. This restriction is connected with the fact that absolute rights are protected against everyone. All these rights must be well-known, so that each member of society can take account of them and avoid invading them, but this is not necessary in the case of relative rights, for their effects are limited to their parties.

As our second example, we may instance the very thorough way in which contracts and legal acts are analysed into their component elements. The diagram overleaf illustrates this in the everyday case of the purchase and sale of a packet of cigarettes.

The principle of abstraction, which has been known to cause problems even to German students, lays down that, while contract and transfer are connected in the sense that the latter is a performance of the former and that the former constitutes a justification or *causa* for the latter, they are nevertheless to be treated as separate or abstract, even though they often constitute a single unit in legal reality. The effect of this principle is indicated in the sketch below by the lines that separate the creation of the duties from their performance.

The principle of abstraction has practical consequences when it comes to the avoidance of the contract of sale for mistake. The contract of sale is avoided retrospectively (§119 in conjunction with §142 BGB), but the conveyance remains valid and effective. It is not included in the avoidance. Matters are put to rights by means of the law of enrichment: the recipient whose continued possession no longer has any legal justification must return what he received (§812 BGB).[30]

Stage One

Offer (declaration of will) — Acceptance (declaration of will)

Contract of Sale (§433 BGB)

> This is an obligational transaction which gives rise to duties: the vendor has the duty to deliver the goods, and the purchaser the duty to pay the price.

(Separation by the Principle of Abstraction)

Stage Two

The performance of these duties by means of real contracts (either immediately or after an interval).

> Transfer of the cigarettes (agreement and delivery — §929 BGB) as performance of the vendor's duty.
>
> Transfer of the money (agreement and delivery — §929 BGB) as performance of the purchaser's duty.

---

[30]For more details on avoidance of contracts, see Chapter 5 Section III 3 below.

5

Contract: Capacity, Formation, and Freedom of Contract

I PERSONS

1 Persons with Rights

The opening provisions of the BGB relate to persons, natural and legal. These persons are the actors in private law, the subjects of rights and duties, the ultimate beneficiaries of all private legal interests. But before they can acquire property, or become liable for the torts they commit, or enter into contracts with all the ensuing consequences, the subjects of law need legal capacity (Rechtsfähighkeit), to which may have to be added capacity to act (Handlungsfähigkeit).

In the case of natural persons there is no problem in determining legal capacity: legal capacity is an attribute of every human being without exception or qualification. Legal persons, however, being mere constructs, need first to be recognized as such by the legal system; indeed, for them legal capacity is the touchstone of separate legal personality. Only two types of legal person with separate legal personality are covered in the BGB, the Verein (corporation by membership) and the Stiftung (foundation) (§§21–88 BGB). There are much more important types of legal person in commerce, such as the Aktiengesellschaft (stock corporation) and the Gesellschaft mit beschränkter Haftung (limited liability company); since these creatures are features of company law, they will be dealt with in Chapter 14 below, along with the elements of corporations and foundations.

2 The Legal Capacity of Human Beings

'The legal capacity of a human being begins at the time of his birth' (§1 BGB). This short and pithy rule determines the moment when the legal capacity of natural persons begins, but it neither defines nor describes legal capacity itself: its nature is simply assumed. Even a code, as this instance demonstrates, cannot and does not spell out every concept: it has to depend to a large extent on the ideas worked out by legal scholars. It is a disputed question whether the legal capacity of a human being is accorded to him by the legal system or is rather a pre-existing attribute, a kind of natural right,[1] but the question is hardly of practical importance since all people have legal capacity. Legal capacity is generally held to be inalienable, although individual rights may be waived.

Even before birth, the child in the womb, the so-called nasciturus, is treated for certain purposes as if it had legal capacity already: for example, it may acquire a right of inheritance (§1893 par. 2 BGB) and a right to damages for the wrongful death of the person bound to provide maintenance (§844 par. 2 BGB). These rights mature only at birth, when the child attains full legal capacity. The statutory rules which confer these rights are rather exceptional, but the child in the womb has received further protection from the courts.

In a case decided by the Bundesgerichtshof in 1952, a woman who had gone to a clinic for a blood transfusion was infected with syphilis which the clinic had culpably failed to diagnose in the donor of the blood (*BGHZ* 8, 243). Three years later she gave birth to a child with grave handicaps which were attributable to the infection. The clinic was held liable in damages to the child as well as to the mother, notwithstanding the fact that the child was not born at the time of the tort. The clinic tried to argue that the child's health could not be said to have been affected since the child had never had any health to affect, having been ill from the very outset, but the Bundesgerichtshof rejected this argument, on reasoning rather reminiscent of natural law. For example, the court said that even in the womb the human child has a 'justified and protected expectation of healthy growth'. This holding was later confirmed when the Bundesgerichtshof held that a spastic child could sue the motorist responsible for the collision that had seriously injured his mother when she was six months pregnant with him (*BGHZ* 58, 48).

On this point compare the Congenital Disabilities (Civil Liability) Act 1976 in England and the comparable law in Canada and the United States.

---

[1]Stoll, 'Zur Deliktshaftung für vorgeburtliche Gesundheitsschäden', I *Festschrift Nipperdey* 739 (Munich, Berlin 1965); Wolf-Naujoks, *Anfang und Ende der Rechtsfähigkeit des Menschen* (Frankfurt 1955).

No provision like that of §1 BGB fixes the end of human life. Such a text might have been placed among the rules of succession law, but when the code was being drafted the question was hardly a real one. Only recently, in connection with the removal of organs for transplantation, has the medical determination of death begun to be a problem. The recipients of organs benefit if death is fixed at the earliest moment, whereas donors would prefer it to be fixed as late as possible, just to make sure. Donors and doctors alike need the protection of a statutory rule that determines the moment of death, and proposals for some such enactment are afoot.

3 Capacity to Act

Although legal capacity and the capacity to act (Handlungsfähigkeit) are very closely linked, the two notions can be distinguished. The capacity to act is the power through one's own acts and conduct to acquire entitlements, to assume duties, and to incur liabilities. For a natural person this depends on his physical and intellectual powers; for a legal person, on its structure and organs. In order to make contracts, a special form of capacity to act is required, namely Geschäftsfähigkeit, or the capacity to do business: such capacity is lacking in children under the age of seven, in persons mentally disturbed, and in persons under interdict (Entmündigte) (§104 BGB). Liability in tort depends on what is called delictual capacity, which the rules of tort law treat as a form of ability to be at fault (§§827–9 BGB), and there are special rules for criminal liability whereby the accused must have been able to appreciate the wrongfulness of his conduct.

Capacity to do business is acquired on reaching the age of majority, that is on the completion of one's eighteenth year (§2 BGB). A legal act done by a minor between the ages of seven and eighteen is not void *ab initio*: its validity is indeterminate until his parents have either accorded or refused their consent. Indeterminate validity (§§106–11 BGB) is a technical and sophisticated concept which is used elsewhere in the BGB as well, for example in the case of the false agent (§§177–9 BGB). While the protection of the minor is its basic and predominant principle, the doctrine of indeterminate validity makes a careful accommodation of the interests of all the parties.

When the only result of a minor's legal act is to give him a legal advantage, he needs no protection; such a legal act is therefore valid *ab initio* (§107 BGB). The definition of 'advantage' in this connection is, however, formal rather than economic: it is counted as a legal disadvantage if one gives up any legal right or incurs any legal liability. Here one differentiates between the obligation that is incurred under the contract (for example, in sale — §433 BGB) and the subsequent disposition by conveyance that the contract calls for (for example, the transfer of property in the goods and the transfer of the money price — §929 BGB). This is consistent with the principle of abstraction (see Chapter 4, Section IV 4 above), but the results are a little surprising. Even if the contract of sale has been disapproved, the minor still becomes owner of the goods that are delivered to him, because the efficacy of the transfer is independent of the validity of the contract; but if he pays over the price, the payment is ineffective because it constitutes a disadvantage for him. The law of unjustified enrichment has to be invoked in order to resolve these groundless transfers: §812 BGB permits the adult vendor to demand the return of the thing delivered to the minor without legal cause.

II DECLARATION OF WILL AND FORMATION OF CONTRACT

1 Declaration of Will, Juristic Act, and Interpretation

In dealing with the formation of contracts, the BGB uses the concept of Willenserklärung, the declaration of will or intention. This is one of the general abstract ideas elaborated by the Pandectists, and it embraces not only offers and acceptances by persons negotiating a contract, but also unilateral declarations such as giving notice or effecting a cancellation, and even voting for a resolution. Indeed, almost every utterance of legal significance is a declaration of will.

The cardinal feature of a declaration of will is that it is directed towards a specific legal result. In this it resembles another general concept, the Rechtsgeschäft or juristic act, which is one of the basic units of private law. A contract is an example of a juristic act, normally consisting of two declarations of will. Other elements may be required for the validity of a juristic act: for example, the transfer of a motor vehicle requires not only agreement on the transfer of ownership, but also the actual delivery of the vehicle (§929 BGB), and the transfer of land calls for an entry in the register or Grundbuch (§873 BGB) as well as the formal agreement or Auflassung (§925 BGB). Only when all these requirements are satisfied does a juristic act have its legal effects, although it is true that there are some juristic acts, such as exercising a right of termination or putting a debtor in default, which consist of a declaration of will and nothing else. The notions of juristic act and declaration of will are sometimes juxtaposed in the BGB, and sometimes used interchangeably.

A declaration of will has two aspects. One can distinguish between the internal aspect, sometimes called the subjective aspect, which is the intention itself, and the external or objective aspect, which is its perceptible manifestation. In cases where there is a mismatch between the subjective aspect (what was intended) and the objective aspect (what was expressed), one speaks of a defect of intention or Willensmangel. Under certain circumstances this may justify the rescission of the contract (see Section III below).

The external aspect of a declaration of will consists of words, signs, or meaningful conduct. For instance, a sale in a self-service store can take place in complete silence: taking the goods to the cashier is conduct that amounts to a declaration of will; it wordlessly expresses (external aspect) the internal will to achieve the legal consequence, namely the acquisition of the goods by purchase.

To ascertain the content of a declaration calls for the process of construction or interpretation (Auslegung). Here §133 BGB provides that 'In interpreting a declaration of will one must seek out what was really intended and not adhere to the literal meaning of the words used.' This suggests, as was certainly the aim of the legislator, that in the process of interpretation the subjective intention should trump the objective expression (will theory). The modern interpreter, however, looks to the objective expression as the recipient would have understood it; the objective meaning is now determinative of the content of the declaration, in order to give stronger protection to the justified reliance of the addressee and of commerce generally (declaration theory). The content of the declaration is what the recipient would normally have understood by it.² In order to support this generally accepted view, §157 BGB is invoked. This paragraph, which concerns the interpretation of contracts rather than of individual declarations of will,³ applies the standard of fair dealing and normal practice, which relates to the external aspect of the declaration. A divergence between the inner will of the person expressing himself and the objective meaning of his expression as ascertained through interpretation constitutes a defect of intention which may lead to the rescission of the contract.

2 Formation of contract — Meeting of the Minds

In order to produce a contract, offer and acceptance must correspond. There is no express provision to this effect in the BGB; once again we see how codes rely on basic doctrine, for the formation of contract is assumed by the BGB.

The distinction between offer and acceptance is only one of timing, the offer being the prior declaration of intention and the acceptance being the subsequent one. In principle the offer should be so phrased that it can be accepted by a simple 'Yes', but this is hardly ever the case in practice, at least in transactions of any importance. If, as usually happens, the contract is the last of a large number of separate stages, the distinction between offer and acceptance loses its signifiicance.

In Germany, unlike England, the offeror is bound by his offer for a reasonable period of time (§145 BGB). German law has nothing like the Anglo-American doctrine of consideration to prevent it from allowing a person to bind himself in this way. In practice the offeror frequently excludes the binding effect that the law normally attributes to an offer by adding express words such as 'freibleibend' (subject to change) or 'widerruflich' (revocable). In such cases, indeed, there is very often no real offer at all, but only what is called an invitation to make offers; and when a declaration is directed to a large number of recipients, as in the case of a catalogue, price list, newspaper advertisement, or display in a shop window, it is more realistic to hold that it constitutes an invitation to make offers rather than a

---

²*BGHZ* 36, 30, 33; *RGZ* 131, 343, 350–1.
³Lüderitz, *Auslegung von Rechtsgeschäften* (Karlsruhe 1966); Kramer, *Grundfragen der vertraglichen Einigung* (Munich 1972).

binding offer in itself. Unless offer and acceptance are given orally or by telephone, they must *arrive* (zugehen) at the recipient's address (§130 par. 1 BGB). The same rule applies to both types of declaration. In German law the moment of formation of contract is not accelerated as it is under the 'mailbox' theory of Anglo-American law.

A few declarations of will do not need to arrive and are effective on dispatch, but they are exceptional (for example, §377 par. 4 HGB). For a declaration to arrive, as is normally required, it must have reached the recipient's sphere of control. This occurs, for example, when the letter containing the acceptance is placed in the addressee's letter-box or is delivered to his house. It is not necessary that the addressee should have read the letter: it is enough that he should have had the means of doing so.

This principle was applied by the Reichsgericht as early as 1902 (*RGZ* 50, 191). A lottery firm sent a letter offering to sell a numbered lottery ticket to a workman who had often purchased tickets from them before. The letter with the ticket arrived at the workman's lodgings one morning after he had left for work. Towards midday the lottery firm learnt that the ticket they had offered had won a prize, and before the workman returned from work they persuaded his landlady by some pretext or other to hand the letter back to them. The Reichsgericht held that the offer had 'arrived' since it had reached the workman's zone of control though he did not know of it, and that the seller was therefore bound by his offer; a revocation would only have been effective if it had arrived simultaneously or sooner (§130 par. 1 BGB).

Absent further requirements, such as some special form, a contract is brought into being by timely acceptance. A belated acceptance is treated as a fresh offer (§150 par. 1 BGB), as is an acceptance that contains any variation or addition (§150 par. 2 BGB). Thus if the recipient of an offer accepts it on his own general conditions of business, he is treated as making a fresh offer. This gives the original offeror the last word, and he can decide whether to accept the new offer or not.

3 The Legal Effect of Silence in Negotiations

It is not always necessary for the declaration of acceptance to reach the offeror. This provision, in §151 BGB, makes it easier for contracts to be formed. Sometimes it is enough if the intention to accept is expressed in some way other than by direct communication with the offeror, and such expression can often be inferred. Suppose, for example, that a mail-order house, having received an order for goods from its catalogue, does not write to acknowledge or accept the order, but simply dispatches the goods: the dispatch of the goods itself constitutes acceptance of the offer. If unsolicited goods are sent to a person, he can be treated as accepting them when he starts to use them. In these cases it is only the arrival of the declaration of acceptance, not the expression of the intention to accept, that is dispensed with.

Silence is different: in the absence of a declaration of intention, or of conduct that can be construed as such, there are normally no legal consequences at all. But important breaches in this principle have been allowed to occur in German law. A provision in the Commercial Code (§362 HGB) requires a merchant in certain circumstances to make it clear that he is not going to fill an order he has received: if he says nothing the contract is treated as having come into force. Here silence is deemed to constitute acceptance of the contract.

The courts have applied this rule very widely to *commercial letters of confirmation*. Quite often a business contract, not necessarily a contract between merchants in the technical sense, is formed orally by telephone and is then forthwith confirmed in writing. If the letter of confirmation coincides with the terms of the prior agreement, it only serves as evidence of it, but if it deviates from the prior agreement and is accepted without demur, the agreement is treated as modified or amplified in accordance with the letter of confirmation. In the latter case the letter of confirmation is said to have 'constitutive effect': the contract is valid on the terms of the confirmation.[4] If the letter of confirmation proceeds on the assumption that a contract has been formed, a binding contract may result even if no contract was actually reached in the negotiations.

A case before the Bundesgerichtshof in 1953 involved negotiations between the agent of a metal-processing firm and a scrap-dealer for the purchase of large quantities of scrap metal on the basis of a list which stated the amounts and prices of the items for sale, as well as

[4]*BGHZ* 54, 236, 240.

the date and method of delivery. Two days after the negotiations the metal-processing firm wrote to confirm that it had purchased the quantities offered at the stated prices. The scrap-dealer did not reply. Nor did he deliver the goods. His excuse was that no contract had ever come into existence. The Bundesgerichtshof, however, held that there was a contract, because the scrap-dealer's silence counted as acceptance of the letter of confirmation (*BGHZ* 11, 1).

The recipient of a letter of confirmation must refute it forthwith if he wishes to escape this result. It is true, of course, that the letter of confirmation must fall within the area of the negotiations and that it must not deliberately contain anything false or unexpected.[5] The striking results which the courts have reached by deeming silence to be a declaration of will have not escaped criticism from legal writers.[6]

4 The Factual Contract

The assumption that the terms of contracts result from free negotiations between the parties is characteristic of the individualistic attitude of the BGB but it is quite out of harmony with the realities of modern business methods. When a person uses the underground or takes a bus, when he subscribes for electricity or gas, or when he uses a paying parking-place, no special individuated bargain is struck, because the tariffs are always fixed, and essentially there is nothing more than conduct on his part that evinces a desire for the service. This is as far as the consumer's will usually goes. It has therefore been suggested that in such cases the will, hitherto the formative element in a contract, could be replaced by what one might call 'typical social conduct' (Larenz).[7] This view can be traced back to the theory of the factual contract (faktischer Vertrag),[8] according to which the voluntary elements in the formation of contract can often be wholly replaced by factual conduct, with the result that one could dispense with the declaration. The courts have struck out on this path.

---

[5] *BGHZ* 40, 42, 45.
[6] See BGH *NJW* 1974, 991–2, with reference to criticisms; Bydlinski, *Privatautonomie und objektive Grundlagen des verpflichtenden Rechtsgeschäftes* (Vienna 1967).
[7] Larenz, *Allgemeiner Teil des deutschen Bürgerlichen Rechts* §28 II (4th ed. Munich 1977).
[8] Haupt, *Über faktische Vertragsverhältnisse* (Leipzig 1941).

In 1956 the Bundesgerichtshof decided the first of the parking-place cases (*BGHZ* 21, 319). A person drove his automobile into a paying parking-place which was clearly marked as such, but refused to pay the attendant the stipulated fee on the ground that he believed himself entitled by custom and usage to park there free of charge and had no intention of making any contract. The Bundesgerichtshof found that a factual contractual relationship had arisen on the parking of the car in the parking-place, and that the individual's contrary intention, forcefully expressed though it was, did not stand in the way of this conclusion.[9]

The courts have not, however, persevered in this line of development.[10] The question whether conduct can take the place of the declaration of intention in such cases goes to the very roots of private law, and the debate is by no means concluded. The modern trend is to deal with the novel problems of mass business by adapting the notion of contract and differentiating its legal consequences rather than by dropping the requirement of a declaration of will and consequently abandoning the realm of private autonomy altogether.

III DEFECT OF INTENTION AND AVOIDANCE OF CONTRACT

1 Deceit and Duress

German lawyers speak of a Willensmangel, or defect of intention, when the objective meaning of a declaration of intention and the subjective intention behind it do not coincide. Apart from a few exceptions (§§117–18 BGB), the declaration of intention remains valid in the sense inferable from its objective expression until the person who made it successfully impugns it on the ground that the expression did not represent his true intention. In such a case the declaration is avoided, and avoided with retroactive effect.

This is clearest in the case where the declarer's freedom of decision has been subjected to improper influence such as deceit or unlawful duress (§123 BGB). Here the law affords protection to the declarer whose freedom of choice has been affected; the party deceived or threatened is still required, in the

---

[9] Accord, *BGHZ* 23, 177ff. (electricity supply contract case I).
[10] BGH *NJW* 1965, 387, 388 (parking-place case II); BGH *MDR* 1968, 406 (electricity supply contract case II).

interests of clarity, to impugn his declaration, but if he does so he can return to the *status quo ante*.

Duress is constituted by the threat of some evil over which the person making the threat has some control, including a threat directed against third parties, such as hostages. The threat must be unlawful, either in respect of its purpose, as where the threatener has no right to the declaration he seeks to exact, or in respects of the means employed, when the evil threatened is contrary to law. The unlawful quality may also result from a combination of these two elements. For example, it is improper for a creditor who wants immediate payment of a sum due to threaten to lay an information against the debtor for drunken driving, and the threat is unlawful although the information would be accurate and the act of laying it intrinsically justifiable:[11] but if the creditor simply threatens to bring a civil suit for payment, this is a lawful means of pressure and there is no unlawful duress.

Deceit arises when a contractor creates a false impression or mistake in the mind of the other party and thereby causes him to make the declaration, or at least to make it in the form he did. The deceitful party must be conscious of what he is doing, but he need not intend to cause any harm. If the deceit is practised by someone other than the recipient, a declaration of will can be impugned only if he knew or should have known of the deceit or if the third party was representing him in some way.

Deceit may result either from a positive act or from an omission, that is, from the presentation of false facts or the suppression of true ones. Omission constitutes deceit only if there is a duty to inform. Such a duty may arise from the contractual negotiations, espcially if the contractor asks questions, but there is no general duty to volunteer information about the characteristics of merchandise being sold. The courts have, however, held that such a duty exists in the case of sales of used cars, for example, to the extent that the purchaser cannot inform himself of these characteristics on the spot. In particular, this is true of the accident history of a well maintained car and of the accuracy of the mileage indicated on the odometer. A car salesman who knows of an accident or of an

[11]*BGHZ* 25, 217, 220–1.

alteration of the odometer must inform the purchaser, even if he does not inquire about it, or he will be held liable for deceit by omission and the purchaser may rescind.[12]

2 Mistake

Rescission for mistake is permitted if the objective meaning of a declaration fails to reflect the subjective intention behind it (§119 BGB), so far as that subjective intention relates to the content and effect of the declaration. If rescission is allowed, the other party whose expectations have been disappointed has a claim for compensation (§122 BGB).

The legislator has tried to limit the scope of rescission for mistake by drawing a distinction between mistake as to the content and mistake as to the expression. This distinction was based on a view of psychology current in the late nineteenth century but no longer accepted.[13] Today rescission is allowed if the person making the declaration was mistaken as to the content or the reach of his declaration as he made it, but not if he made the declaration because of ill-founded expectations or preconceptions. There is a mistake as to the expression, for example, if a person who means to buy two machines puts '12' instead of '2' in the order-form by mistake, or if the vendor hands over the wrong goods, or if a person offers to pay £1,000, but means Egyptian pounds rather than pounds sterling, the objective meaning of the symbol. But if the twelve machines are ordered because the purchaser expects a favourable export contract which does not eventuate, or because he erroneously thinks they are a special bargain, the purchaser will not be able to rescind, for here the mistake is not so much in the expression as in his motive, as it is called. Error of motive is a ground of

---

[12]On the duty to inform, see *BGHZ* 63, 382, 387. The employees of a second-hand car firm cannot simply rely on what the previous owner tells them about a prior accident. They are under a comprehensive duty to check and inform. Even if the car has been admirably repaired, the fact of its involvement in a serious accident is a defect to which the guarantee of freedom from defects applies (see Chapter 7 below). The buyer can therefore elect whether to rescind the contract on the ground of deceit, or rely on the guarantee of freedom from defects. The decisions of *BGHZ* 53, 144ff. and *BGHZ* 57, 137ff. are important, especially with regard to the consequences of breach, and are still disputed. For further references see Leser, *Der Rücktritt vom Vertrag, Abwicklungsverhältnisse und Gestaltungsbefugnisse bei Leistungsstörungen* (Tübingen 1975).

[13]See Rothoeft, *System der Irrtumslehre als Methodenfrage der Rechtsvergleichung* 92ff. (Tübingen 1968).

rescission only in two exceptional cases in the code (§119 par. 2 BGB): when the mistake relates to the characteristics of the person or of the thing involved in the contract. These two classes of error of motive are important ones, and it is not always easy to delimit them. The rescission of contracts of sale, however, is severely restricted in practice by the rule that mistakes concerning the thing sold and its qualities must be dealt with exclusively by the special texts relating to defects in the goods (see Chapter 7 below).

3 Rescission

When all the elements required for rescission are present — German lawyers speak of an Anfechtungsgrund, or ground for avoidance — the party deceived or mistaken may rescind or not, entirely as he chooses. If he fails to rescind, or waives his right to do so by confirming the contract, his declaration of intention remains wholly valid, but if he does rescind, his declaration of intention and with it the contract disappear retroactively as if they had never been (§142 BGB). Rescission is effected by means of a unilateral declaration addressed to the contractor or other person to whom the original declaration was addressed (§143 BGB), this being one of those constitutive declarations (Gestaltungserklärungen), like giving notice or cancellation, that can alter a legal relationship unilaterally. Such a declaration is irrevocable, and must be unconditional.

Where the rescission is for mistake, the other party, who expected that the declaration would continue in force, has a claim for damages to compensate him for his disappointment (§122 BGB). The existence of this duty to make compensation, more or less as the price of rescission, explains why rescission on the ground of mistake is so freely permitted. Negligence has no effect on the right to rescind, for rescission is possible even for a mistake that was the party's own fault.

In the case of a contract that involves a conveyance or other transfer of real rights (such as a contract of sale, which is forthwith performed by the delivery of the goods), rescission for error usually applies only to the contract and not to the conveyance. The conveyance, here the delivery of the goods, is certainly a performance of the contract, but is regarded as independent of it in accordance with the principle of abstraction.[14] Thus the rescission does not usually apply to the transfer of the goods, though conveyances and real contracts are just as susceptible of rescission as obligational contracts. The rights of the parties in such cases are adjusted by the law of enrichment (§§812ff. BGB). In cases of deceit the principle of abstraction is not applied so sternly. Here there is an increasing tendency to allow rescission of the conveyance as well, even if the declaration induced by the deceit was in the obligational contract. If both the obligational and the real transactions are rescinded, both of them, contract and conveyance, fall away. The transferor can then claim the property back just like any other owner (§985 BGB).

IV FREEDOM OF CONTRACT AND GENERAL CONDITIONS OF BUSINESS

1 Freedom of Contract[15]

Hitherto we have been dealing with the more technical aspects of contracts, such as their formation. Now we shall turn to the function of the contract institution itself. No feature of private law is more important for the autonomy of the individual or his power of self-development: freedom of contract is a basic right protected by the constitution as part of the general freedom of action (art. 2 par. 1 Basic Law). A flexible tool which is constantly adapting itself to new ends, contract is also an indispensable feature of a free economy: it makes private enterprise possible and encourages the responsible construction of economic relationships. Freedom of contract is thus of central significance for the whole of private law.

A distinction is drawn between the freedom to form contracts and the freedom to give content to them (Abschlussfreiheit: Inhaltsfreiheit). The freedom to enter a contract or not is limited whenever there is a monopoly, legal or factual. For example, a supplier of electricity in an area has both a factual and a legal monopoly for the supply of current to consumers

---

[14]See Chapter 4 Section IV above; the obligational contract of sale which gives rise to the duty to convey is distinguished from the conveyance transaction which constitutes its performance: they are treated as 'abstract'.

[15]Raiser, 'Vertragsfunktion und Vertragsfreiheit', I *Festschrift Hundert Jahre Deutscher Juristentage* 101 (Karlsruhe 1960); Reinhardt, 'Die Vereinigung subjektiver und objektiver Gestaltungskräfte im Vertrage', *Festschrift Schmidt-Rimpler* 115 (Karlsruhe 1957).

there, so it is therefore bound to enter contracts. The same is true by statute for the federal railways and for other means of public transport. The duty to contract may exist even where there is no legal provision, but this is much less common; liability for breach of this duty is based on §826 BGB.[16] More important is the freedom to decide what one's contracts are to contain. It is this freedom that allows the fullest scope to human inventiveness, reinforced by the fact that most of the rules of the General Part and the Law of Obligations of the BGB are dispositive rather than imperative in nature. This freedom of types means that new species and kinds of contract may be created *ad libitum*. In the law of property, family law, and the law of succession, on the other hand, the number of possible forms of legal institutions is limited in the interests of legal security (on this see Chapter 4 Section III above).

But there must be some limits even to the freedom to fix the content of one's contracts. Freedom of contract can exist only within the limits of constitutionality and legality. Morality provides another limit, as we shall see in relation to §138 BGB. A late product of Enlightenment and Liberalism, freedom of contract has never existed in unlimited form. The important question in practice has always been where the limits are to be drawn.

2 Limits to Contractual Freedom

Statutory prohibitions constitute the first of the limits to freedom of contract (§134 BGB). As examples of transactions discountenanced by the legal system, one can mention criminal conspiracies, dealings in prohibited drugs, and so on. Obviously, the courts will not allow their apparatus to be used for the enforcement of such agreements. But there are a great many legal prohibitions, and it is not always easy to determine when they apply so as to strike at legal transactions as well as acts. Clearly, this will be so in the case of dealings in prohibited drugs or of agreements to divide the spoils of a bank robbery, but not in the case of a sale of goods outside permitted opening hours or in the sale of medicaments without the required prescription. Between these extremes lie many debatable cases where the

---

[16] *RGZ* 48, 114, 127; Nipperdey, *Kontrahierungszwang und diktierter Vertrag* (Jena 1920).

decision must turn on the gravity, the danger, or the turpitude of the transaction.[17]

The prohibition of usurious contracts constitutes a further limit to the freedom of contract (§138 par. 2 BGB). To extract a promise of an an unfair economic advantage by exploiting the inexperience or need of another is as ineffective in Germany as it is in most other legal systems.[18]

An important limit to the freedom of contract is set by §138 par. 1 BGB: 'A legal transaction is void if it is contrary to good morals.' Here the legislator uses the general canons of ethical conduct in order to fix the limits within which contractual freedom is vouchsafed. This is a general clause which needs to be fleshed out by the judges. The standard of good moral behaviour has often been described as the 'feeling of propriety entertained by all right-thinking people',[19] but this does not take us very much further. It is average sensibility that is in issue here, not the unduly lax or demanding ethics of any particular group or judge. One may try to invoke the values underlying the Basic Law, but in practice it is the judges who must render specific the content of general clauses like this by distinguishing the various types of case that may arise. These types of case may be found in the current commentaries, and only with reference to them can a particular case be decided. As examples we may mention contracts that oppressively restrict personal and economic freedom of movement,[20] contracts to pay people for changing their religion,[21] and contracts that benefit one creditor to the undue detriment of others.[22]

Another extremely important limit to the freedom of contract is cognate with the prohibition of immoral transactions. This is the requirement of good faith and fair dealing (Treu und Glauben) in §242 BGB, which applies to all obligations and contracts. This general clause, like that of §138 BGB, has given rise to many different types of case in practically all areas of the

---

[17]BGH *NJW* 1968, 2286f. (medicaments without prescription); *BGHZ* 14, 25, 30–1 (transaction to evade tax void only if tax evasion principal purpose).
[18]*RGZ* 86, 296 (exploitation through disproportionate counterprestations); *RGZ* 57, 95 (invalidity of obligational and conveyance contracts).
[19]*BGHZ* 10, 228, 232; *BGHZ* 69, 295, 297.
[20]*BGHZ* 22, 347, 355.
[21]RG *Seuff Arch* 69, no. 48.
[22]*BGHZ* 55, 34, 35 (1970); *BGHZ* 30, 149, 153; *RGZ* 143, 48, 51.

law. It limits freedom of contract by rendering invalid any part of a legal transaction that is inconsistent with good faith and fair dealing as specified by the courts. For a description of the way in which §242 BGB has been developed and of the various types of case to which it has been applied, reference should be made to Chapter 8 below.

Some of the rules of the BGB are imperative and incapable of alteration by contract. These rules constitute a further general limit to freedom of contract. Examples may be found in §276 par. 2 BGB (invalidity of exclusion of liability for intentional fault) and §248 par. 1 BGB (invalidity of prior agreement to pay compound interest). There are also limits to the permissible kinds of arrangements in other branches of law, such as property law (closed list of property interests, such as ownership and mortgage), family law (matrimonial property arrangements), or the law of succession (life tenancies and remainders). Here only the forms laid down by the law are available: they cannot be modified at will or replaced by others.

Finally, one must mention the provisions as to formalities, which are of practical importance, such as the requirements of §§126–9 BGB, and §313 BGB, which is especially important as relating to sales of land; these provisions are normally imperative, and failure to respect them will render the transaction invalid (§125 BGB).

In the last analysis it becomes clear that freedom of contract is hedged about with numerous limitations and prohibitions. Does this leave sufficient room for flexibility in private law? Perhaps one could liken freedom of contract to a game reserve, where pains are taken to minimize external dangers so that the denizens within its bounds may move freely and fend for themselves. It will be a question for the future whether the limits are so drawn as to permit the inhabitants to develop themselves to the maximum within them.

3 General Conditions of Business[23]

The analogy we have just employed may make it easier to understand recent developments designed to control what are

---

[23]Raiser, *Das Recht der Allgemeinen Geschäftsbedingungen* (Bad Homburg 1935; reprinted 1961); Löwe-Graft v. Westphalen-Trinkner, *Kommentar zum Gesetz zur Regelung des Rechts der Allgemeinen Geschäftsbedingungen* (Heidelberg 1977).

called general conditions of business (allgemeine Geschäftsbedingungen). Ever since the First World War insurance companies, banks, large firms, and associations have tended to rationalize their business by abandoning the practice of tailoring contracts to the individual customer and by adopting a standard uniform and preformulated pattern of general conditions of business. The operation of many branches of industry, trade, and commerce would be inconceivable today without such uniform terms; they make mass transactions possible and facilitate the use of computers. While they made legal transactions uniform with terms appropriate to the specific problems of the different kinds of transaction, this was not their only effect. General conditions of business were increasingly used in order to deviate from the rules implied by law and to produce a form of contract that cast all the risks and disadvantages on the other party. The other party was usually quite unable to resist this one-sided shifting of the risk, for his contractor was hardly ready to renegotiate his general conditions of business individually. The customer of a bank or a subscriber for electricity was generally powerless to insist on modifications, and if the party using the general conditions of business had acceded, he would have lost the advantages of uniformity. Only a party with equal or greater economic strength could insist on a special contract. If one of the contractors could use his economic power to dictate unfair and one-sided terms to the other, especially terms relating to breach of contract, the freedom of contract on which the general conditions of business themselves rested needed some supplementary protection. In order to redress the balance the courts intervened by using the principle that general conditions of business must be construed in favour of the other party. This was far from being sufficient, however, so the courts invoked §138 BGB, with its prohibition of immoral transactions, and §242 BGB, with its requirement of fair dealing and good faith in the execution of all contracts, in order to strike down unfair conditions as invalid.

There were many instances of such revisions of the terms of contracts in favour of contractual justice, and they had clear practical effects.

A decision of the Bundesgerichtshof in 1956 may serve as an example (*BGHZ* 2, 90, the facts being simplified by the omission of the role of a finance house in the hire-purchase contract). A young married couple bought a suite of bedroom furniture in a discount store. Before long, the drawers stuck, the veneer peeled off and other defects became manifest. The couple refused to pay the balance of the price until the furniture was repaired, but the discount store's attempts to effect the repairs were unavailing. The general conditions of business excluded all rights arising out of the contract of sale except the right to have the goods repaired. The Bundesgerichtshof decided that the purchaser of new furniture could not be deprived of all his rights in the event it proved defective, and if the only unexcluded right — the right to repair — was useless, he must be able to resort to other rights arising from the contract of sale. The relevant right here was the right to have the price reduced or, on returning the goods, repaid (Wandlung); and the exclusion of these rights by the general conditions of business was held to be invalid. The Bundesgerichtshof emphasized that the terms of this contract diverged too far from the terms implied by law, but it limited the scope of its decision to the purchase of new objects, leaving it open to decide otherwise in the case of sales of used goods.

General conditions of business have now been regulated by a separate statute which came into force on 1 April 1977.[24] It is designed to protect the weaker party and thereby to ensure contractual justice in the context of freedom of contract. Apart from consolidating in statutory form the principles and standards developed by the courts,[25] it also seeks to regulate other aspects of general conditions of business. It invalidates a large number of clauses in general conditions of business that deviate from the balanced model of the dispositive law; when such clauses are invalidated, the terms implied by law take effect. The statute also confers on bodies like consumers' associations the standing to complain of individual clauses. This law has certainly solved some problems, but it has also thrown up many new questions, and the full impact of the limits it puts on contractual freedom in the interest of protecting the weaker party remains to be seen.

---

[24]General Conditions of Business Act (Gesetz zur Regelung des Rechts der Allgemeinen Geschäftsbedingungen) of 9 December 1976 (AGBG) (*BGBl* I, 3317).

[25]Thus in §11 no. 10b it adopts the constraints on the exclusion of remedies for defects in goods that were developed by the Bundesgerichtshof in *BGHZ* 20, 99ff., as explained above.

6

Breach of Contract

I THE CONTRACT AS A BLUEPRINT FOR PERFORMANCE

1 The Contract as a Blueprint

German legal theorists have rightly stressed that the conventional or obligational contract gives rise to many different duties. Thus the principal duties generated by the contract of sale, for example, are on the one hand a duty to deliver the thing sold, and on the other hand a duty to pay the purchase price (§433 BGB), but they are attended by numerous complementary or ancillary duties regarding the carriage of the goods, their packaging, their clearance through customs, or, on the purchaser's side, the provision of confirmed credit, payment in advance, and so on. These duties, which all have their roots in the contract and can very often be determined only by the process of construction, are best seen as being designed to achieve a goal, namely *performance*. Performance is the end, in every sense, of the obligational relationship. The transaction has a task to perform, such as the satisfaction of a need or the acquisition of means, and the contract and the duties it engenders are designed to achieve it. The duties are a device or technique for describing and directing the various steps which are required on the way to optimal performance and for allocating to parties the consequences of behaviour that conflicts with any of them.

It may help to consider the contract as a blueprint[1] or plan, such as a route plan. The contract lays down, as in a plan, the stages and fixed rules by which the parties are to arrive at the future result on which they are agreed. The contractors intend to keep to the programme and expect that it will be kept to, implemented, as a matter of legal technique, by the duties in the

---

[1] Leser, 'Die Vertragsaufhebung im Einheitlichen Kaufgesetz', in *Kolloquium zum 65. Geburtstag von E. v. Caemmerer* 1f. (Karlsruhe 1973).

contract and the rules regarding the allocation of risk in any goods involved.

If the programme proceeds without a hitch and ends in normal performance (§362 BGB), the numerous duties and ancilliary duties will have done their job and can be dispensed with: they expire, unless there are some after-effects to be regulated (see Section I 3 below).

Straightforward performance raises few problems, but the rules become more difficult and more important when there are deviations from the programme, hitches in its planned development — anything that German lawyers call 'Leistungsstörungen', or irregularities in performance. Such problems are a central part of any theory of the law of obligations, and put the efficacy of any system of private law to the test. In Germany the statutory regulation of what in England is called breach of contract is unduly complex without being comprehensive, but the courts have solved the problem quite satisfactorily.

2 The Completion of the Programme: Time and Place of Performance

Before we turn to irregularities in the contractual performance we take a look at the suppletive rules that amplify the terms of the contract and assist its performance. The debtor — and in bilateral contracts both parties are debtors — will need to make preparations before he can perform. If the obligation is a *generic one* (Gattungsschuld), relating to things of a kind and description available on the market, the debtor has to have done everything necessary for performance before the obligation 'concentrates' on particular articles (Konkretisierung: §243 BGB). There is then no difference between such a generic obligation, certainly the commonest form of contract today, and an obligation relating to specific goods (Stückschuld), where the very article that constitutes the object of the contract has been identified from the beginning; contracts of this latter type are essentially limited today to dealings in real property, in the art and antique trade, and in second-hand goods, especially motor cars.

So far as the *time of performance* is concerned, §271 BGB lays down the rule that 'in case of doubt performance is due immediately'. This is simply a dispositive rule, as its terms show, and yields to any other contractual arrangements, even if they are inexplicit and have to be inferred. The rules regarding *place of performance* are more important, for they have an impact on the allocation and distribution of risk in the thing to be delivered. In the absence of any contractual provision, §269 BGB states that the place of performance is the domicile of the debtor, so that in case of doubt the creditor must go and fetch what is due to him. Here, too, any other provision that emerges from the construction of the contact takes precedence.

Under the rules relating to risk, the purchaser who has to fetch the goods from the seller once they are ascertained must bear the risk of accidental destruction from the time they are handed over (see §§446, 447 BGB on sale); should the purchaser be late in fetching the goods (Gläubigerverzug, or delay by creditor), and they have been made ready for delivery, he bears the risk even if they are still in the hands of the seller when they are destroyed (§243, or §300 par. 2 BGB).

An example may make this clearer. P orders one thousand cuckoo clocks from V, which P is to collect from V's factory on 1 April. V crates the clocks and puts them aside for P to collect (concretization under §243 BGB). P does not come until 3 April, only to find that the clocks have been destroyed the previous night in a fire in V's factory caused by reasons for which V is not responsible. According to §§300 par. 2 and 324 par. 2 BGB, P must pay for the clocks since he was late in coming to collect them.

For money obligations there are different rules, since people can be expected to take the simple step of sending money, whereas going to fetch it could involve disproportionate expenditure. Thus for money debts §270 BGB lays down that the debtor must, at his own risk and expense, send the money to the creditor's domicile.

These ancillary duties that arise from special texts constitute only a small proportion of the duties that fill out the contractual blueprint. Much the larger part result from the *construction* of the contract under §157 BGB and sometimes §242 BGB as well (for details see Chapter 8 below). As a method for implying such ancillary duties and directing and specifying the performance of the contract, it has proved generally successful.

3 Performance

If performance is duly rendered, the network of duties becomes otiose: the duties expire with performance, though some ancillary duties may remain, such as to give a receipt or to return a document of indebtedness (§370 BGB). The obligational relationship also comes to an end if the debtor tenders performance of something other than what was originally required and the creditor accepts it as constituting performance (§364 BGB). Even after performance the obligational relationship may have legal *after-effects*: only when the goods have been delivered can the guarantee of their quality and the claim for their improvement play their roles. There are also ancillary duties that expand the scope of performance, as by requiring a party to give information or to abstain from conflicting conduct.[2]

II IRREGULARITIES IN PERFORMANCE

1 Basic Principles

(a) If there is a hitch in the programme and the contract is not duly performed, rules are needed to deal with the deviation, and especially to allocate between the parties the harm, loss, and disadvantage caused thereby. Damages for non-performance can provide compensation for harm actually suffered as well as for expectations such as profits forgone, but one also needs to be clear what happens to the unperformed contractual duties, and how benefits already rendered are to be returned, should one party withdraw from the contract.

For this difficult task the BGB has not just one remedy, but a number of institutions taken over from the Pandectists of the late nineteenth century and from Roman law itself. The ensuing system of regulation is rather complicated and difficult to grasp: even today it gives many a German student a headache.

The BGB has no comprehensive notion like 'breach of contract' in Anglo-American law, and it does not deal with irregularities in performance in a unitary manner. Only in the 1920s did legal scholars adopt the comprehensive notion of Leistungsstörungen, or irregularities of performance, but the specific

---

[2]*BGHZ* 16, 4, 10 (using the same pattern for other customers); *BGHZ* 71, 144, 148 (architect's duty to investigate and advise after the building is completed).

independent rules remain in force, so that whenever there is a hitch in the programme one must first ask what kind of irregularity it is, then check whether the preconditions for the appropriate remedy are satisfied, and finally see whether damages or rescission is available. Each kind of irregularity has different legal consequences, and many difficult marginal cases arise.

(b) German lawyers speak of Verzug, or *delay*, if performance is not rendered at the proper time and the debtor is responsible for the delay: an example would be failure to deliver on the due date. If performance cannot now be rendered at all, whether because the thing has been destroyed or for some other reason, including legal obstacles, they speak of *impossibility*, or Unmöglichkeit, which may render the debtor liable in damages or release him from his obligation, according as he is or is not responsible for the impossibility. These rules apply to contracts of all kinds, but where the contract is for sale or services and the goods delivered thereunder are defective or not of the proper quality, there are special rules that provide for the return or reduction of the purchase price, regardless of whether the vendor was aware of the defect or was at fault in permitting it to arise. Again, there are other special rules if the goods were guaranteed to have a certain characteristic and do not have it.

But even this did not cover all the possibilities: a famous gap became evident immediately after the BGB came into force.

A person stabled a horse he had bought with thirty other horses he owned. The newly purchased horse had an infectious disease which caused five of the other horses to sicken and die. The new acquisition recovered completely.

Under the BGB the purchaser in such a case had no contractual claim, but the courts used the device of positive Vertragsverletzung, or positive breach of contract, to fill the gap: it is now applied in all cases of irregular performance which cannot be classified as impossibility or delay, and therefore fall outside those specially regulated areas. In the next few sections we shall consider these different remedies in more detail.

2 The Contribution of the Pandectists and of Roman law

(a) But first we must take a look at Roman law, more particularly Roman law as refined by the scholarship of the late nineteenth-century Pandectists, for it was this that formed the ground plan for the BGB. The general rules almost all developed from the contract of sale, and the contract of sale focused on delivery of a thing. Even if there were irregularity in performance, the claim for delivery of the thing itself could remain unimpaired. If the specific object could no longer be delivered, perhaps because it had been destroyed, the buyer might, as a substitute or *surrogate*, claim damages instead of the thing: monetary compensation was seen as replacing the thing rather than as sanctioning the non-performance of the promise as it does in Anglo-American law.[3] Thus the kind of irregularity depended on what had happened to the thing rather than on whether the promise had been broken. The promise was, as it were, unaffected by the disappearance of the thing: the thing was simply replaced by a sum of money.

The fact that the claim for performance continued to exist meant that the solution of *rescinding* the contract did not develop until late, and even then not in Roman law: it was unknown even in the later stages of Pandectist legal science and came into the BGB indirectly through commercial law.[4] The novelty of the provision explains why the lawyers had some difficulty in applying it.

(b) Furthermore, the Roman lawyers thought in terms of the unilateral obligation rather than the bilateral contract; that is, for example, of the vendor's duty to deliver the goods as an entity in itself, rather than in conjunction with the concomitant duty of the purchaser to pay the price. The duties were treated as separate obligations, held together by only a few rules of interdependence, the *synallagma* of Greek origin. The BGB retained the idea of the contract as a collocation of two obligations, a conception that soon gave rise to practical problems, for example in quantifying damages.[5]

3 Impossibility and Delay as the Only Categories of Irregularity

In 1853 Mommsen propounded the view that all forms of irregularity of performance could be attributed to impossibility

---

[3] See Rheinstein, *Die Struktur des vertraglichen Schuldverhältnisses im anglo-amerikanischen Recht* (Berlin 1932).
[4] See Leser, *Der Rücktritt vom Vertrag* 10f. (Tübingen 1975).
[5] See Section VI 2 below.

or delay.[6] Before long this view had attracted a great many adherents. Impossibility was understood primarily in a physical sense, as turning on the existence and availability of the thing rather than on the breach of contract or non-performance of the promise. This greatly narrowed the realm of impossibility as a category. The only other form or irregularity was to be delay, or non-delivery on time. This exclusive duality of treatment dominated the discussions that led to the BGB, and it was only in 1902 that another, less specific, way of dealing with contractual situations which had gone awry, namely positive breach of contract, was adopted by the Reichsgericht.

III IMPOSSIBILITY OF PERFORMANCE

1 The Forms of Irregularity

(a) The first matter to discover, as we have seen, is what has happened to the specific thing. A thing that has been physically destroyed can no longer be used to fulfil the contract. If the picture has been destroyed by fire, it can no longer be delivered, either by the artist who painted it or by anyone else. A contract that calls for such delivery is a contract to do the impossible and is therefore void (§306 BGB). Also included are cases of so-called legal impossibility, as where the authorities have banned trading in goods of the kind in question or have confiscated them. If performance is still notionally possible, but only by efforts out of all proportion to the result, people speak of factual or practical impossibility: for instance, if a ring sold by a jeweller falls into a river while in transit, performance is still notionally possible because the ring still exists, but diverting the river or conducting an underwater search would be out of all proportion to the end to be achieved. Such cases may therefore be classed as impossibility, though the test of disproportionality is extremely severe.

In some cases, where one cannot speak of such a disproportion between expenditure and result, the debtor will nevertheless be involved in unforeseeably high expenditure if he is made to perform his contractual duty. These cases are very

---

[6]Mommsen, 'Die Unmöglichkeit der Leistung in ihrem Einfluss auf obligatorische Verhältnisse', I *Beiträge zum Obligationenrecht* (Brunswick 1853).

difficult, for in principle the debtor must devote all his energy and resources to performing his obligation, even far in excess of the promised counterpart. It is controverted whether one should speak of economic impossibility here, for many observers think that such cases should be treated in relation to the duty of good faith (Treu und Glauben, §242 BGB)[7] as instances of disappearance of the basis of the transaction, and not as cases of impossibility at all.

The different types and degrees of impossibility make it a complex matter, as even this brief treatment shows.

(b) To determine the legal consequences of impossibility as an irregularity in performance, one must qualify it further. First, one distinguishes according to whether the impossibility is *objective*, that is, whether all debtors would be prevented in the same way, or, if it is only *subjective*, where this particular debtor cannot perform, though someone else could. Suppose that V is unable to deliver a picture he has sold: if this is because the picture has been destroyed, then delivery is objectively impossible, but if it is because the picture belongs to O, the impossibility is subjective, for although V cannot transfer ownership in the painting, O could do so. This distinction is of importance with regard to V's liability.

(c) A further distinction drawn by the legislator depends on whether the impossibility is *initial* or *subsequent*, whether it existed at the moment the contract was formed or arose only later (§275 BGB). Here the connection between the fate of the thing and the duties engendered by the contract is clear. If the impossibility arises after the formation of the contract, there is an already existing duty to be affected, whereas different considerations must be invoked where the impossibility existed at the outset.

(d) The third dichotomy relates to responsibility for the impossibility, that is, to the *principle of fault* (treated in detail Section VII 1 below). A debtor is liable to pay damages only when he is responsible for the irregularity or deviation from the programme. §276 BGB lays down that the debtor is always responsible for deliberate or careless misbehaviour. In German law, unlike Anglo-American law, this principle is of quite general application.

[7]See Chapter 8 below.

The classic case of responsibility is where the irregularity in performance is due to blameworthy and faulty conduct on the part of the debtor himself. The debtor is responsible for such behaviour and must be liable for the outcome: as a free individual he must shoulder the consequences of his actions. The principle of fault dominates the law of damages and in a certain measure protects the debtor.

But in several instances the legislator has departed from this principle, and the courts have considerably extended its scope by the way they have applied it. Under what is called Erfüllungsgehilfenhaftung, the debtor is liable not only for his own shortcomings, but also, without possibility of exculpation, for those of the people whose services he uses in the execution of his obligations (§278 BGB; see Section VII below).

Liability is further extended by another departure from the principle of fault. If the debtor is unable to perform an obligation that concerns not a specified individual object, such as a work of art, a plot of land, or a used vehicle, but rather goods of a described class available in the market, he is always responsible (§279 BGB). These so-called generic obligations are by far the commonest of transactions, and the debtor bears the risk of his inability to satisfy them on the market. This also involves that a debtor is never excused for a shortage of funds: inability to pay never justifies non-performance. The situation changes only when goods of the contractual type have been set aside for the particular performance (Konkretisierung: §243 BGB; see Section I 2 above). Once individualized by concretization, a generic thing is treated as a thing specified: liability is limited to the actual thing so individualized, and risk passes to their acquirer on delivery (§§446–7 BGB, on Sale).

In addition, a contractor may extend his liability beyond personal fault by undertaking a duty to procure, or by an assumption of responsibility (see Section VII below).

2 Effect on Unilateral Obligations

In applying these three pairs of notions to any case where the irregularity of performance takes the form of impossibility, one must remember the legislator's assumption that a normal bilateral exchange contract is just a combination of unilateral obligations. Before we deal with the effect that irregularity in the performance of one duty may have on the duty of the other contractor (Section III 3 below), we must first consider the duty of the first party in isolation, here the duty to deliver.

Figure 2 illustrates the legal effect of the different situations on the unilateral obligation. It will be noted that, for subjective impossibility *before* the contract is formed, the debtor is liable without regard to his responsibility for it, whereas he is only liable for events *after* the formation of the contract if he is responsible for them. If the impossibility at the outset is objective, the contract is void (§306 BGB). This rule, *impossibilium nulla obligatio*, was taken over from Roman law and has proved very difficult to harmonize with the other rules of liability, so it is very restrictively construed.[8]

3 Effect on Bilateral Contracts

We now turn from the complex rules on the unilateral obligation to the effect that irregularity on one side may have on the duties of the other side. This matter is dealt with separately in the BGB (§§320–7 BGB) as a reaction to the breach of duty. The link between the obligations in the exchange is called the *synallagma*. If one party is backward in his performance, the other party can withhold his own performance until it is forthcoming (§320 BGB), and can demand a judgment to this effect if he is sued (§322 BGB). If a party is not responsible for non-performance, and is therefore not liable, his partner is released from his duty to provide the counterpart (§323 BGB); anything already rendered may be reclaimed in accordance with the law of unjustified enrichment.

If one party is responsible for the irregularity in his performance, the other party has a claim for damages under §325 BGB on the ground of non-performance of the contract and is normally released from performance of his own duties. Instead, he may wish to rescind the contract and thereby free himself directly from his own obligations. The remedies of damages and rescission are dealt with in detail in Section VI below.

When it is the buyer rather than the seller who is responsible for the impossibility of delivery, which is admittedly exceptional, the seller retains his claim for the counterpart, that is,

[8] A view already expressed by Rabel, 'Unmöglichkeit der Leistung', in *Festschrift Bekker* (Weimar 1907), I *Gesammelte Aufsätze* 1f.

100 *Breach of Contract*

Initial impossibility, i.e. before contract formed	Objective impossibility: obligation void — §§306, 307	Subjective impossibility: strict liability for performance, so invariable liability (non-statutory)
Subsequent Impossibility, i.e. after contract formed	Distinction turns not on subjective or objective impossibility, but only on whether or not debtor is responsible (Vertretenmüssen): *Responsible* for §276: deliberate or negligent breach §278: due to contractual helpers §279: performance of generic obligations — Otherwise, *not responsible*	
Legal consequence	Liability for nonperformance under §280	Debtor released under §275 par. 1

Figure 2.

payment of the price (§324 BGB). Thus, if a builder agrees to erect a house for a site-owner, and it transpires that the building plot is wholly unsuitable, the site-owner is responsible for the impossibility of building, so that the builder, notwithstanding that his performance has become impossible, may claim the contractual sum, less any expenses of performance saved (§324 BGB).

4 Summing Up

The rules regarding impossibility of performance are complex, but one can perhaps sum them up as follows. A debtor will be liable for all irregularities of performance for which he is to blame or is responsible; even if he is not to blame, he may be liable because the people who helped him perform were at fault, or by reason of the strict guarantee which exists in contracts to procure and deliver generic goods, or because he has assumed the risk by contract or, exceptionally, because his obligation was subjectively impossible when the contract was formed. The end result is not so different as might at first sight appear from the solution reached in Anglo-American law with its strict liability for breach of contract. It is mainly the differentiation of the forms of irregularity of performance that causes the difficulty.

Of the legal consequences of liability, essentially the claim for damages and the claim to rescind the contract (on which see Section VI below), it is the claim for damages that has proved the more flexible throughout.

Cases of partial non-performance raise the question whether the creditor still has any interest in the balance of what was originally promised: if so, he may claim damages as to part, if not, damages for non-performance of the whole contract.

IV DELAY

1 Time and Responsibility

(a) The second basic type of irregularity in performance, namely delay, is very much simpler. The difficulties in the case of impossibility arose from the fact that the type of irregularity (initial, subsequent, etc.) depended on what had happened to the thing: here, our one concern is failure to perform on time, that is, non-performance of the contract: what has happened to the thing is relevant only indirectly, to the question of responsibility for such delay. Delay has therefore proved the more important remedy in practice, and rightly so; many more cases of irregularity in performance are dealt with under this heading than under the heading of impossibility. Here, too, as in the case of impossibility, the legislator starts out from the unilateral obligation. It survives the delay and attracts a claim for damages for the loss attributable to it (§286 BGB). Thereafter the draftsman turns to the reciprocal contract, and gives the creditor the extremely important power of fixing an additional time for performance once the delay has occurred (Nachfrist), and then freeing himself from the contract entirely (§326 BGB).

(b) Before delay in the legal sense arises, performance must have been *due* and the creditor must have put the debtor *on notice*, that is, must have reminded the debtor of his duty to perform. If, but only if, the contract fixes a date for delivery which can be precisely identified by the calendar can one dispense with this notice. Other clauses, even if to much the same effect, are not sufficient: thus a promise by a car salesman to deliver 'as soon as possible' or 'with all speed' is less effective than a promise to deliver 'on 1 March' or 'three days before Easter'.

(c) A further requirement is that the debtor be *responsible* for the delay. Here, too, as in the case of impossibility, the classic instance of responsibility is where the debtor is himself to blame for the delay, and it has been extended in the same manner (see Sections III 1 above and VII below). A debtor is responsible if he deliberately delays performance in order to gratify another customer, or if he omitted to procure the necessary raw materials when they were available on the market, or if he failed to employ enough workmen to ensure delivery on time. The debtor who is late in delivering because of shortage of funds is always responsible, just as in any case of generic obligations (§279 BGB). He is also fully responsible for any fault on the part of those he gets to help him perform (§278 BGB; see Section VI below). But if the delay is caused by *force majeure*, such as blockades or unavailability of transport, or by mere mischance, such as fire or theft, the debtor is excused.

2 Consequences of Delay

(a) The debtor's delay gives the creditor a claim for compensation for the harm attributable to the delay (§286 BGB). This claim exists *in addition to* a claim for performance, if it can still be rendered, albeit late. Damages are to compensate for all disadvantages caused by the delay. If the market has collapsed and the buyer has to resell at a lower price, he must be awarded the difference between the market price now and what it was when delivery was promised. If the delay has rendered him liable to one of his customers, this again is harm attributable to the delay, as are the costs of giving notice, legal action, and so on. Interest for delay in the payment of money is specified by law (§288 BGB), even if the creditor is not paying any interest himself: more important nowadays is the compensation the creditor may claim for any interest he has actually paid to his bank during the period of delay.[9]

(b) In addition, the debtor who is in delay becomes more strictly liable: although the accidental destruction of the object required for performance would not normally render him liable under the rules relating to impossibility, the debtor will be liable if he is in delay when it is destroyed.

3 Fixing Time and Freeing Oneself from the Contract

(a) §326 BGB, for *bilateral contracts*, is in practice the most important of the provisions regarding delay. It allows the innocent party to free himself from the contract and claim damages for non-performance, or to rescind the contract. This possibility, which was unknown to Roman law and came into the BGB from the Allgemeines Deutsches Handelsgesetzbuch of 1867, is actually used to resolve most cases of irregular performance. It lets the innocent party make a positive decision to free himself from the obligations of the contract and to redeploy his assets. Whereas in a case of delay the contract and its duties of performance normally remain in existence and the creditor must accept performance however late it is tendered, here he can rescind the contract. Both parties are then freed from their original obligations and the contract is 'stopped short'. Instead

---

[9] This is the general practice. See BGH *MDR* 1978, 818.

of being performed by order of the court, the 'sick' contract is now dismantled (Abwicklung).

(b) First there must be a Nachfristsetzung, or setting of an additional time for performance. Once the debtor is in delay, the creditor can give him a reasonable period of time within which to perform, declaring at the same time that he will no longer accept performance after the expiry of this period (§326 BGB).[10] This gives the debtor a last chance to do his duty. If he fails to do it, the creditor can no longer claim performance. The contract can now only be wound up, either by a claim for damages for non-performance, or by a claim for rescission (see Section VI below). The innocent party must opt, by means of a unilateral declaration, for one or other of these remedies, for the claim to performance is now excluded. This way of solving the difficulties of a contract gone awry has proved extremely satisfactory.

4 Anticipatory Refusal to Perform

In contracts where performance is to take place only after a period of time, there may be an anticipatory refusal to perform. This is the place to deal with it. Of course if the time for performance has not yet arrived there can be no question of performance being overdue, but there ought still to be some possibility of dismantling the contract now, if the debtor makes it unequivocally clear that he is not going to do his duty when the time comes.

A comparable class of case is resolved by §326 par. 2 BGB, which states that, if delay has robbed further performance of all interest for the innocent party, he need not fix a time for performance: if Xmas trees are delivered to a retailer on Boxing Day, or if Easter eggs arrive after Easter, or if the taxi comes to fetch a traveller when his plane has already taken off, the lapse of time has so fouled up the contract that its execution is now pointless. In such cases the debtor too must know that performance is futile, so the setting of a period for performance is dispensed with.

In cases of *anticipatory refusal* to perform, the irregularity

---

[10]The threat to refuse to accept can be associated with the notice to perform, *RGZ* 93, 181. In other respects the requirements of clarity for such declarations are very strong; see, for example, BGH *NJW* 1968, 103.

consists in the debtor's own announced decision. Here it makes no more sense to set a time for performance than it would do to wait for it. The debtor having abandoned his duty, the victim can substantially lessen the prospective harm by seeking cover elsewhere. To allow the innocent party to abandon the contract right away and make other arrangmeents is to protect rights and prevent waste. The Uniform Law on International Sales adopted this remedy in art. 76, and was right to do so.

In a case of serious refusal to perform, German municipal law allows the innocent party to demand damages for non-performance immediately, or to rescind the contract.[11] In other words, there are the same legal consequences as are provided by §§325, 326 BGB for cases of irregular performance. This is not laid down in the BGB, but it is generally recognized in the decisions of the courts.[12] Some jurists see in this remedy an instance of positive breach of contract imperilling the creditor's claim (Section V below). This is not so. It is a classic instance of breach, by the debtor's express repudiation, of a duty to perform, to which the general principles of irregularities of performance directly apply.[13] The creditor has an option: he can adhere to the contract or withdraw from it, but if he withdraws, he is bound by his election and can no longer return to the contract.

V POSITIVE BREACH OF CONTRACT AND CULPA IN CONTRAHENDO

1 Impossibility and Delay Cease to be the Only Categories

(a) We have seen that under the leadership of Mommsen,[14] the Pandectists held that any irregularity in performance must be a case either of impossibility or of delay. This narrow position was adopted by the BGB, but it was unable to do justice to the wide variety of possible breaches of contract, as was soon seen in a case decided by the Reichsgericht in 1902.[15]

[11]The repudiation normally takes the form of imposing extra conditions, raising the price, and so on. See *BGHZ* 65, 375.
[12]*BGHZ* 11, 80, 84; *BGHZ* 49, 56, 59.
[13]See Leser, 'Die Erfüllungsverweigerung', in II *Festschrift Rheinstein* 649f. (Tübingen 1969):
[14]Above n. 6.
[15]*RGZ* 54, 98. Staub, *Die positiven Vertragsverletzungen* (Berlin 1904; 2nd ed. Berlin 1913).

A gravel merchant contracted to sell and deliver to a building contractor the gravel he needed for the construction of a large bridge, payment to be in terms of the size of the completed bridge. The building contractor used the gravel for constructing the approach roads as well as building the bridge, and this entirely falsified the basis for computing the price. The Reichsgericht held that the building contractor had committed a 'positive breach of contract', and that the gravel merchant could rescind.

Thus people realized quite soon that it is not only by impossibility and delay that a contract may be broken or impaired.

We have already noted the case which soon became the textbook example: a purchaser buys a horse which he has no means of knowing has an infectious disease, and puts it in his stables, with the result that five of his other horses are infected and die, whereas the new acquisition recovers. If the seller knew, and did not warn the purchaser, that the horse he sold was infectious, he would be contractually liable for positive breach of contract, despite the absence of any such provision in the code. This is another instance of irregularity of performance which is neither delay nor impossibility. The rules regarding defects in goods cannot be invoked here, for the defective horse, having fatally infected the others, himself gets well. The purchaser's claim for damages arises from the vendor's breach of a protective duty that he owed to the purchaser.

(b) The scope of positive breach of contract has been much expanded by the implication into contracts of a *network of duties* designed to protect the interests of the parties. These duties of care and protection arise through the process of construction, and stand alongside the main obligations, such as the duty to deliver the goods and pay the price in a contract of sale. Remedies have been devised for breach of these duties of care and protection which are in line with the remedies that the Code itself provides for breach of the main obligations: the rules for non-performance have been extended to these ancillary duties. But there is still a difference between principal and ancillary duties as regards the legal consequence of breach: only breach of a principal duty leads to damages for non-performance of the contract as a whole or to rescission of the contract; in other cases damages are awarded to compensate for the harm actually suffered, which can admittedly exceed the value of performance.

(c) Liability for breach of any of these duties of care and protection depends on culpable behaviour, responsibility being extended in the manner indicated earlier. Though not contained in the BGB, positive breach of contract is constantly being applied in practice and is generally recognized as an example of judicial customary law.

2 Positive Breach of Contract as a Residual Category

(a) While it is generally accepted today that positive breach of contract is truly a general category of irregularity in performance, the exigencies of the system mean that it is available only if, and to the extent that, none of the specific statutory remedies apply. As a ground of liability, positive breach of contract is *subsidiary*. All other categories of irregularity must therefore be checked, to see whether they apply or not: impossibility, delay, defects in goods (see Chapter 7 below), disappearance of the basis of the transaction (see Chapter 8 below). Concurrent and independent claims may, however, arise if conduct that constitutes a positive breach of contract is also tortious and gives rise to a remedy in tort.

No complete list of the forms of positive breach of contract can be given, since they are extremely varied and manifold. At most one can give examples of the different types, which must be kept fluid precisely because its function is subsidiary and residual.

One group consists of cases of misperformance that cause the recipient material harm going beyond mere non-performance. The case of the infectious horse belongs here, as does the supply of poisonous fodder that kills the purchaser's horses,[16] or the delivery of fuel that damages the engines in the purchaser's vehicles.[17] Performance in these cases is defective, but as the statutory remedies for defective performance do not cover the damage that occurs, one needs to invoke positive breach of contract as a basis of compensation for this extra harm.

In another group of cases the performance itself is not defective, but damage is caused because inadequate information is given or because the instructions are defective, as where a machine supplied without the instructions needed to operate it

---

[16]*RGZ* 66, 289.   [17]BGH *NJW* 1968, 2238.

caused great damage when it was set in motion.[18] Further examples are the gravel case of 1902 (above p. 106), the case where a vendor delivering heating fuel carelessly pumps it into a pipe leading to the purchaser's water tank, or where a customer who has just purchased a carpet and is having it packed up is badly injured by another roll of carpet which falls on her.

The borderlines are often difficult and not always logical.

3 Culpa in Contrahendo

There is a close connection between the recognition of positive breach of contract and the extension of contractual liability to *precontractual contacts*.

The celebrated linoleum case is the starting point.[19] Here the plaintiff was not, as in the case just mentioned, a customer who had already bought a carpet, but one who was in the shop looking for a roll of linoleum to buy when she was injured: the purchase never took place because of the accident. The Reichsgericht granted her claim in terms of 'precontractual contact': contractual standards of liability could be applied when one was on one's way towards a contract. The customer, according to the argument of the court, had already put herself in the protection of the shop, and the employee whose mishandling of the linoleum caused the injury had thereby broken a duty of care that would have been generated by a contract had one been concluded. Since the customer had gone to the shop in order to make a purchase, and relied on it for her safety, the contractual duty of care should be applied to the contractual overtures. The plaintiff thus obtained damages on the basis of *culpa in contrahendo*.

This is a kind of proleptic contractual effect: the damage that occurred as the result of a breach of a protective duty arising in a precontractual situation was remedied on contractual principles. In particular, responsibility attaches under §278 BGB for Erfüllungsgehilfen, or those whom one uses to perform one's obligations. This bypasses the possibility of the defendant's disculpating himself under §831 BGB from tort liability. Indeed, this was one of the reasons for which the court extended contractual liability in this way.

[18]*BGHZ* 47, 312.   [19]*RGZ* 78, 239.

Culpa in contrahendo is now generally recognized as a remedy. Though it may be seen as a prelude to contractual liability, only the plaintiff's negative interest is covered: he is put back into the situation as it was before the contractual negotiations began.

VI REMEDIES FOR BREACH OF CONTRACT

1 Specific Performance and Rescission

Owing to the influence of Roman law, the claim to performance (Erfüllungsanspruch) remains even where there is irregularity in performance. Except where performance is actually impossible, the creditor may choose to stand by his claim to performance and need not invoke the remedies that exist for dismantling the contract. This is seen most clearly in the case of delay, for a claim for damages for delay lies *in addition* to a claim for performance.

In practice, however, this combination is rather unusual. What usually happens when there is an irregularity in performance and the contract does not proceed normally is that performance no longer has any interest for the innocent party, so he does not wish to claim it. What he really wants is to regain his freedom to dispose of his resources and to satisfy his requirements elsewhere. Unfortunately, the draftsman of the BGB did not recognize the *liberative* effect of the irregularity in performance as an independent element, though it figures, more or less tangentially, in the different remedies for dismantling the contact. If performance has become impossible by reason of the physical destruction of the thing (such as a work of art), then the parties are necessarily liberated from their duties to deliver and accept delivery. For cases of delay, the legislator has devised §326 BGB, whereby a party can free himself from his duty to perform, with binding effect for both parties. And if rescission is adopted as the method of dismantling the contract, the parties will be liberated from any duties still unperformed and will have to return any benefits already rendered. Even in claims for damages, the most important of the methods for resolving a contract, the idea that one is free from one's duty to perform has, as we shall see, had its effect, though in rather a disguised manner. It is therefore possible to say that in cases of irregularity in performance there is usually some way, though

2 Damages for Non-performance

(a) In cases of impossibility and delay, the creditor may, if he chooses, claim damages for non-performance (§§325, 326 BGB); this is also true in cases of positive breach of contract. The formula is that the innocent party should be put in the position he would have been in if the contract had been duly performed. This is a unilateral claim for money, but the practical problem of calculating the amount of money remains.

The courts were very early faced with the following simple case.[20] A customer to whom a vendor had made regular deliveries of petrol failed to pay for them and was becoming insolvent, so the vendor proceeded under §326 BGB, i.e., he gave the purchaser a time for payment and indicated that he would not accept it thereafter. Was the vendor thereby liberated from his duty to deliver, or was he bound to continue delivering, as he could, at the risk of losing through the purchaser's insolvency?

The difficulty here came from the Roman habit of seeing obligations *in isolation*, treating a contract of sale as only a conjunction of two otherwise independent obligations (see Section II 2 above). The legislator had taken no position on this question, and it provoked a famous dispute.[21] Supporters of the Austauschtheorie, or exchange theory, accepted the division into two obligations and therefore insisted that, before the vendor could proceed against the purchaser, he must have made the exchange, here the delivery of the petrol. Adherents of the Differenztheorie, or balance theory, on the other hand, appraised the ruined contract as a whole as at the time of calculation, embracing all the claims arising out of the contract, and taking into account all the advantages and disadvantages that the contract envisaged for the innocent party, and setting them off against each other. This produces the difference or balance between the existing situation and the situation that would have existed had the contract been duly performed. This difference or balance, which is also called the 'net value of the

---

[20] *RGZ* 50, 255.
[21] Leser, *Der Rücktritt vom Vertrag* 122f. (Tübingen 1975).

contract', is the object of a unilateral claim by the innocent party: it tells us how much he can claim as damages for non-performance. To use this method of calculation entails accepting that the innocent party is *liberated* from any duty still outstanding.

This is the theory originally adopted by the courts[22] and they have kept to it, with the support of most commentators, subject to one slight modification. In some cases the innocent party has an interest in rendering his performance, notwithstanding the other party's delay, perhaps because he wishes to dispose of it. In these cases the balance theory is modified so as to permit the creditor to perform his own side of the bargain and then count it into the calculation of the balance, i.e. to include the price of the thing in his claim for damages (modified balance theory). If the innocent party has already performed, the damages claim does not permit him to reclaim his performance as such; for this, he must rescind the contract.

(b) In computing the balance, that is, the amount of damages to be awarded, one must take account of all deviations from the situation that would have existed had the contract been properly performed. Causal relationship is the only limitative criterion here. Different bases of computation may be adopted — concrete, i.e. actual, or abstract, i.e. notional. If the computation is done concretely, one takes the difference between the contract price and the actual cost of cover or the actual price on resale; as profit forgone, one may include an especially favourable subsale to a customer; further damage, such as compensation paid to sub-purchasers, warehousing costs for the goods, and so on, is taken into account. But the creditor may alternatively have his harm computed on a notional basis, and claim the difference between the contract price and the market price; here, too §252 BGB allows him to claim profit forgone if in the natural course of events he would probably have made such a profit. The courts have presumptions that facilitate proof and computation of loss.

3 Rescinding the Contract

Instead of bringing a claim for damages, the creditor can nor-

---

[22] *RGZ* 50, 255; *BGHZ* 20, 343. For a general view see Larenz, *Schuldrecht Allgemeiner Teil* §22 (12th ed. Munich 1979).

mally rescind the contract (§§325, 326, 327 BGB). As compared with a damages claim, rescission is rather a limited remedy, appropriate where the contract has not yet been fully executed, for it brings unperformed duties to an end and requires restitution of benefits already rendered.

It used to be thought that the effect of rescission was to avoid the contract, but nowadays the dominant view is that rescission produces a liquidation relationship within the original contractual framework. The purpose of rescission is to bring to an end contractual duties still unperformed and to procure the restoration of performances already rendered or exchanged; for this there is a subtly worked-out system of mutual rights and duties stemming from the model of rescission laid down in §§346ff. BGB. The special problems of rescission arise where the benefit rendered no longer exists as it was, owing to damage or destruction. Rules are needed to allocate the risk in performances already rendered. The aim of rescission is to achieve the position as it existed *before* the contract was formed, rather than the position that would have arisen if the contract had been executed, so the remedy it offers is rather limited as compared with damages. Some of the details are still quite unclear, since rescission was a novel legal institution when it was adopted by the BGB.

VII THE FAULT PRINCIPLE AND LIABILITY FOR CONTRACTUAL ASSISTANTS

1 The Fault Principle[23]

(a) As we noticed in connection with impossibility, liability for irregularity in performance very generally depends on the debtor's responsibility for it. The irregularity must have been due to his behaviour, and his behaviour must have been culpable. Though §276 BGB makes him liable for intentional and careless misbehaviour, that is, for fault in any form, the debtor is not responsible for irregularities that are purely accidental. As in tort law, fault is determined in relation to the Haftungstatbestand, or element of liability — here the breach of the specific contractual duty, not the harm or the way it arises. So far as

---

[23]von Caemmerer, 'Das Verschuldensprinzip in rechtsvergleichender Sicht', 42 *RabelsZ* 1f. (1978).

causal connection is concerned, the criterion of adequate cause is employed. Rather like proximate cause in Anglo-American law, this criterion is very extensible and hardly leads to any genuine limitation of liability. Nor does German law have a rule like that of *Hadley* v. *Baxendale*, which limits liability in damages to the risks that were foreseeable to the debtor at the time the contract was formed. Contractual liability therefore goes very far. Omission is, on general principles, equated with an act when there is a legal duty to act.

(b) The principle of *fault* has been so far extended in the law of contract that in large areas one can speak of objective responsibility. For example, the expression Vertretenmüssen, or responsibility, as used in §§325, 326, BGB, covers very much more than personal fault. Many circumstances other than fault can induce liability. This is true of the fault of those whose services one uses for the performance of the contract; Erfüllungsgehilfen, as they are called (§278 BGB; see Section VIII 2 below) render the debtor liable automatically, according to a principle recognized by most legal systems. It is true, also, of the procurement of the commodities and resources required for performance, as has already been mentioned in relation to impossibility (Section III 1 above). If the contract calls for the delivery of fungible goods, available on the market, the debtor bears the entire *risk of procurement* (§279 BGB), and the same is true for the availability and disposition of the necessary funds: the debtor cannot do like the Roman emperors and defend himself with the famous *exceptio caesarea* that the Treasury is empty. He is responsible for his ability to perform,[24] and must keep himself ready to perform, make the necessary arrangements, and have the necessary means to hand.

(c) Another factor that tends to sharpen responsibility is that the requisite care is specified by *objective standards*, not in relation to the characteristics of the individual debtor. The question is whether the average accountant, doctor, or gardener would have recognized the probability of the harm that occurred and would have averted it, rather than whether the actual debtor, who may be a remarkably inept or inefficient person, would

---

[24]But the limit is reached when goods of the contractual description cannot be procured on the market even with the expenditure of more money: *RGZ* 88, 172; *RGZ* 107, 156.

have done so. It does not help the debtor to be of less than average competence.

Nor must the debtor undertake an action for which he is unqualified: a goldsmith who undertakes a heavy engineering job will be expected to have the necessary expertise, as will a physician who undertakes an operation. In the absence of the necessary average qualification it is culpable to undertake the job at all, so that one need not look for fault in its execution.

The tendency of all these considerations is to heighten or extend responsibility; or, to put it the other way round, the gradual weakening of the principle of fault has brought German law much closer to other legal systems, which do not require fault but allow a possibility of exculpation.

2 Liability for Others

(a) As in practically all other legal systems, a person is liable for those he employs to act for him. Here German law draws a very clear line between liability for employees in tort (§831 BGB; see Chapter 9 below) and contractual liability for assistants (§278 BGB). In tort, the employer's liability depends on a presumption that he himself was at fault, a presumption he can rebut by proof of exculpation, whereas in contract, liability for assistants is a strict liability for the fault of others without any possibility of exculpation.

Liability for one's assistants rests on the consideration that, since the principal is using assistants in order to expand his range of business activity, he should bear the risk of the division of labour he creates for his own benefit.[25] To impose strict liability in this area had proved generally satisfactory.

(b) A precondition of liability is that the debtor's assistant should have acted under the contract, that is, in order to help with its execution. The contractual assistant must have the principal's consent to his co-operation in the contractual performance, and his conduct must be orientated towards performance. The assistant may be any of the principal's workforce or family, or even an independent third party brought in by the debtor to perform his duties. Sometimes it is difficult to draw the line, for example where the debtor uses a bank or a carrier.

[25] *BGHZ* 62, 124.

If a seller instructs the manufacturer or a third party to make delivery direct to the purchaser, it is possible to see the manufacturer or third party as the vendor's contractual assistant, but this viewpoint is disputed and not supported by many decisions.[26]

The contractual assistant must act in purported execution of the debtor's duties — not necessarily the main duty to perform, but also collateral duties, especially duties of care.

The 'protective ambit' of the contract may include third parties who stand in a close relationship to the creditor.[27] If the contract is broken, they may acquire their own contractual claim against the contractor, although they are not really parties to the contract but are only brought within its protective ambit.

The contractual assistant must have acted culpably in the sense described above. If so, the principal's liability under §278 BGB is strict. The contractual assistant himself will be personally liable only if he is guilty of a delict or, in special cases, *culpa in contrahendo*. There is the same liability for statutory representatives as for contractual assistants (§278 BGB).

---

[26] BGH *NJW* 1968, 2239.
[27] Especially members of the family. If a gas boiler is being repaired and a child of the customer is injured by escaping gas, the child has a personal contractual claim for damages against the repairman. This is the contract with protective effect for third parties (Vertrag mit Schutzwirkung für Dritte). See *BGHZ* 61, 233; *BGHZ* 66, 51.

7

Sale

I FUNDAMENTALS

1 Historical

Of all exchange transactions, sale is the most important. In primitive economies property is bartered for other property, but at a later stage it is sold for money, and this forms the very basis of a functioning market and money economy. Sale has lost none of its overriding importance, and it seems to be indispensable even in governmentally controlled economies. In a working market the basic institution of sale has many variants, found throughout the whole system of distribution from the producer through the wholesaler and retailer to the ultimate consumer.

The German law of sale contains elements of both Roman law and Germanic law, along with a few components from the law merchant, and the blend is not entirely homogeneous, so that today there are still certain features that do not fit very harmoniously.

It is from the law of sale that most of the rules of contract law have come, a process already well under way at the time of the Pandectists in the nineteenth century. In the BGB, characteristically, the rules have been generalized so far as possible, and 'factored out'.[1] Thus in the general law of obligations, contained in Chapters 1–7 of Book Two of the BGB, many of the rules have been extrapolated from sales, and the same is true of the General Part of the BGB. When it comes to applying the rules, one needs to know how far one is to proceed from the specific to the general; this can be quite tricky, and it certainly demands a knowledge and grasp of the whole range of rules.

2 Sale as a Special Type of Contract

(a) The rules on sale are placed, as befits the most important of

[1] See Chapter 4 Section IV 3 above.

the obligational contracts, at the beginning of what is called the Special Part of the Law of Obligations, Chapter 7 of Book Two of the BGB. This chapter contains many other types of contract which the law makes available to the citizen. If a thing is made to order, say a building or a ship, or even a portrait, this will be a Werkvertrag, or contract for services (§§631ff. BGB). Then there is Miete, or lease, which is of great practical importance. It applies to immoveables and to chattels such as a car. Where the lease is of a dwelling, there are laws to protect the tenant, as the socially weaker party, against eviction and excessive rent.[2] The contract of employment (Dienstvertrag, §§611ff. BGB) has outgrown private law and now forms the core of labour law. Other types of contract include loan (Darlehen), deposit (Verwahrung), the brokerage contract (Maklervertrag), and the modern contract of tourism (Reisevertrag), to name only a few. Freedom of contract applies to them all; they are flexible and adaptable in that the parties may determine the content of their arrangements as they choose. But there is no room here to go into details.

(b) The rules special to sale in §§433ff. BGB are supplemented, as we have seen, by the rules in the General Part of the Law of Obligations and in the General Part of the BGB. Provisions in the Commercial Code will also apply if the sale is a commercial one (§§373ff. HGB). Thus, if both parties are merchants, the solution of contractual difficulties is speeded up by requiring the buyer to give prompt notice of any shortcomings in delivery, and by enabling the buyer to effect cover and the seller to resell. In addition, the Law on General Conditions of Business (AGBG)[3] and the Law on Instalment Contracts (AbzG)[4] are often relevant to sales, not to mention the all-pervasive principle of good faith (§242 BGB; see Chapter 8 below).

---

[2] Protection against Eviction from Dwellings Act (Wohnraumkündigungsschutzgesetz) 18 December 1974, *BGBl* I, 3603. Furthermore, §§535ff. BGB have often been amended in order to strengthen the protection; the provisions on eviction are §§556a–c BGB.

[3] Law of 9 December 1976, *BGBl* I, 3317. On this enactment see Chapter 5 Section IV 3 above. This enactment is itself an example of the way the law of sales acts as a source of general rules, since general conditions of business apply predominantly to sales contracts.

[4] Law of 16 May 1894, *RGBl* 450. The Act is dicussed in Section V below.

118   *Sale*

(c) Special mention should be made of the Uniform Law of International Sales, which provides a model for the sales contract regardless of its country of origin. To reconcile the fundamental differences between continental and Anglo-American laws of sale took long and tedious proceedings at the international level, but the agreement was ratified at The Hague conference of 1964 and was brought into force in the Bundesrepublik, the United Kingdom, and elsewhere by municipal enactments.[5] It applies only to sales contracts that cross national borders, and has not yet had a great impact on practice, but recent international conferences[6] suggest that it may prove possible to adapt, and consequently extend, this unitary form of sale. In any case the German law of obligations has already benefited greatly from the new model and the preparatory work it required.[7]

3 *The Duties in a Contract of Sale*

(a) As one would expect from the way German law is constructed, the only duties that a sales contract imposes on the parties are obligational: the vendor must deliver the goods and transfer ownership in them, while the buyer must pay the price and take delivery of the goods (§433 BGB). The performance of these duties is regarded as a wholly distinct matter: the act whereby performance takes place may be a real transaction, such as the handing over of the goods and the handing over of the price, or an obligational transaction, such as the assignment of the right that has been sold. Once again, the principle of abstraction[8] makes a sharp division between the contractual duties on the one hand and the transactions that fulfil them on the other.

In the everyday cash sale these different legal transactions

---

[5] Uniform Law on International Sales of Goods of 17 July 1973, *BGHl* I, 856, and the Uniform Law on the Formation of International Sales Contracts of the same date, *BGBl* I, 868. On the preparation, documentation, and text see Dölle (ed.), *Kommentar zum Einheitlichen Kaufrecht* (Munich 1976). Great Britain ratified the treaty, with reservation, on 31 August 1967; for further details see Graveson and Cohn, *The Uniform Laws on International Sales Act 1967* (London 1968).

[6] Most recently, the United Nations Conference on Contracts for the International Sale of Goods in Vienna in March and April 1980.

[7] See the commentary of Dölle, n. 5 above, and the numerous articles there mentioned.

[8] See Chapter 4 Section IV above and Chapter 10 below.

occur together and the observer can hardly distinguish the obligational duties from the real transactions by means of which they are executed. The distinction must nevertheless be clearly maintained, for it is reflected in the fact, mentioned in relation to the principle of abstraction, that rescission of a contract of sale for mistake affects only the obligational contract and not the title acquired by the purchaser. In German law, therefore, unlike French law,[9] a contract never operates as a conveyance.

(b) This split between duty and performance involves the danger that property may be *sold twice*. Nothing prevents the vendor from undertaking several conflicting duties to deliver the same thing. Such duties do not confer any right to the thing itself: German law knows no *ius ad rem*. If the vendor transfers the thing to the second purchaser, all the first purchaser can do is sue the vendor for damages. Only in exceptional cases can the purchaser claim the thing itself from the second purchaser by means of a tort claim.[10] Nor does the purchaser have any right to the thing if the vendor goes *bankrupt* before performance: all he has is a claim in bankruptcy, which is unlikely to be very satisfactory. Only in land law is there a security device that assures the purchaser's personal right to conveyance; such a Vormerkung,[11] as it is called, must be entered in the Grundbuch.

(c) In addition to the principal duties, the sale contract gives rise to numerous *collateral duties* (Nebenpflichten); some of them derive from the Code, some from the contract, and others from the principle of good faith (see Chapter 8 below). In this way one can ascertain precisely how performance is to be rendered, without forfeiting the necessary flexibility. For example, the seller must draw the buyer's attention to any special dangers that lurk in the property sold,[12] and refrain from competing with the purchaser of his business,[13] and procure the tax certificate which the buyer requires.[14] The buyer, too, may have numerous other duties, but one cannot give a tally of them,

[9]Art. 1583, Code civil, whereby the property passes to the purchaser on the formation of the contract of sale. See also arts. 711 and 1138, Code civil.
[10]In cases of inducing breach of contract, *BGHZ* 12, 308, 318; see also *RGZ* 108, 58.
[11]§§883ff. BGB.
[12]BGH *NJW* 1975, 824.
[13]*RGZ* 163, 111. See also Chapter 8 Section II 1 below.
[14]OLG Hamm, *MDR* 1975, 401.

since they keep changing in response to the demands of the particular situation.

II THE DUTIES OF THE SELLER: GENERAL REMEDIES OF THE BUYER

1 What can be sold

According to the Code, a sale may have as its object a thing or a right. Things include the whole range of moveables and land. Whichever it is, the seller must make the buyer owner and possessor, but if it is a sale of land it has to be notarized (§313 BGB). Rights include debts and other claims arising from obligations, mortgages, patents, and shares in partnerships and companies. Here the seller's duty is to vest the right in the buyer. How this is to be done depends on the nature of the right in question, for a claim may be assigned by mere agreement (§398 BGB), whereas for the transfer of a mortgage entry in the Grundbuch is required (§873 BGB). But the requirements of the sale remain the same.

Actually it is not just things and rights that may be sold, but virtually anything of economic interest or value, including electricity, heat,[15] know-how, and the chance of winning a lottery. A combination or complex of things, such as an inheritance or a business, may also be sold by a single contract, although the appropriate mode of conveyance must be employed for the different articles that comprise it. For a long time ethical considerations made it doubtful whether a doctor or lawyer could sell his practice along with patients or clients, but it is now admitted that he can.[16] Sale is often a component of the so-called *mixed contracts*, that is, transactions that contain elements of different types of contract. Examples would be a contract to book a room in a hotel (lease, sale, and employment), or to produce goods for a cannery (sale and services).

2 Performance and Transfer of Risk

(a) The function, or programme, of a contract of sale, like other obligational contracts, is performance or execution. Once performance is rendered, the principal duties expire; collateral duties may occasionally have subsequent effect. Whenever

---

[15]The Wärmeversorgungsvertrag: see *BGHZ* 64, 288.
[16]*BGHZ* 43, 46.

there is a deviation from the programme, a Leistungsstörung or hitch in performance, much turns on *when* performance takes place. Preparation for the requisite act, for example delivery of a thing, may be necessary; such preparation is for the vendor alone, for the thing is in his control. Should the thing be destroyed when it is in his control, he, as the owner, must bear the risk. But suppose that the goods are stolen from the seller after they have been prepared for collection and the buyer should have come to fetch them? At a certain point in time the risk in the object of sale must pass from seller to buyer. At the latest this will be on full performance, for from that moment the buyer owns the thing and the risk of its loss will be on him.

The Code contains general rules on the allocation of risk in cases of what is called delay by the creditor (Gläubigerverzug; §§293–304 BGB), though this is not a true case of Verzug, or delay, since it does not depend on the creditor's responsibility for it. If performance is tendered and the buyer does not accept, he is in delay in taking delivery of the goods (§293 BGB). From this moment the risk of destruction of the goods is on the buyer, whether it is accidental or due to the slight negligence of the seller or his employees (§300 BGB). In the example given above, the buyer would have to bear the loss due to the theft of the goods which he failed to collect in time. Here the risk passes to the buyer, at least in part, before performance is completed; the justification is that the seller has done everything required of him.

(b) Two further important exceptions to the general principle that risk passes at the time of performance are specific to the law of sales. According to §446 BGB, the risk of accidental destruction of a thing passes to the buyer at the *time of delivery*, meaning the delivery of possession, whether or not title is also transferred. For the risk of physical destruction, the critical element is possession of the thing, that is, the exercise of factual control over it. When the buyer acquires control of the thing delivered to him, any loss is his affair. This provision is vital in cases of reservation of title (§455 BGB), where the vendor delivers the thing but retains ownership until payment of the whole purchase price.[17] Possession is delivered, ownership to

---

[17]On the use of reservation of title as a security device, see Chapter 10 Section V 1 below.

vest later. If the risk did not pass on delivery of possession to the buyer, there would be very little point in reserving title in this manner.

(c) When the goods sold are to be sent (Versendungskauf), the moment when risk passes is determined by §447 BGB. Who is to bear the risk in goods in transit where the seller dispatches them from his place of business to a purchaser in another town? The BGB puts this risk on the buyer if the goods were dispatched on his orders: when the seller has dispatched the goods, he has done everything that is required on his side, so the risk of carriage lies on the buyer. There are, however, many exceptions to this, for example, where the seller has clearly contracted to deliver the goods at the buyer's domicile. This is easily inferred when it is impossible for the buyer to effect the carriage of the goods: the seller of heating oil, for example, must deliver the goods to the buyer's premises, since the buyer does not have the necessary means of transport; but it would be different in the sale of a picture, for under §269 BGB it is for the buyer to fetch it from the seller, and so if he instructs the seller to send it, the buyer must bear the risk.

(d) The transfer of the risk means that the buyer must sometimes pay for a thing he has not received. This happens when he as creditor is in delay, or when goods are delivered under reservation of title. In a sale by dispatch, (§447 BGB), the transfer of risk occurs even earlier: from the moment the seller has dispatched the goods, any loss falls on the buyer, and he must pay for the goods even if they are stolen *en route* and never reach him. This applies not only when the goods are lost but also when they are damaged.

3 Remedies for Non-performance by the Seller

(a) If the sale contract is not properly performed as agreed by the parties in their programme, the provisions regarding irregularities in performance are applicable. They are to be found in the General Part of the Law of Obligations (see Chapter 6 above), but there are additional rules in the law of sales, especially as regards defects in the goods. This means that any irregularity in performance must first be categorized.

In the next few paragraphs we shall survey the different categories of irregularity or disturbance of performance that

The Duties of the Seller

may occur in an individual contract of sale, and indicate the legal consequences. We shall be more specific than in Chapter 6, but some duplication of what was said there is inevitable.

(b) The duty to perform a contract lasts from the time the contract is formed until the time it is performed. In the case of the sale of a thing — ignoring collateral duties for the moment — this *claim to performance* (Erfüllungsanspruch) calls for transfer of title and delivery of possession. This claim is often capable of surviving a disturbance of the programme, and it is not automatically converted into a damages claim. But one must analyse the situation carefully since the type of disturbance may be such as to exclude the claim to performance, as happens where the debtor is in delay and the creditor has fixed a period for performance under §326 BGB (Nachfrist). Normally, however, the claim to performance remains available as an alternative to the remedies for non-performance. It may be exceptional in Anglo-American law to claim specific performance, but in Germany it is a standard remedy.

(c) If the seller fails to perform, supposing that the risk has not passed, the rules on *impossibility* and *delay* apply (§§323–6 BGB). Details may be found in Chapter 6 above.

The legal consequences of impossibility essentially depend on whether the seller is responsible for it or not. If he is so responsible, the buyer can claim damages for non-performance of the sales contract (§325 BGB), including damages for lost profits. In the exceptional case where it is the buyer who is responsible for the impossibility of performance, as where he has damaged the goods while inspecting them prior to delivery, the seller retains his claim for the price and need not perform (§324 BGB). If neither party is responsible for the impossibility, they are both released from their obligations, so that the buyer to whom no delivery has been made need not pay the price but may not claim damages.

The rules for delay are akin to those for impossibility (§326 BGB), though here the buyer may fix a final date for performance and thus give the seller the option of performing the contract or letting the final date pass and having the contract wound up by damages or rescission. The options of the innocent buyer are investigated in detail in Chapter 6 Section VI above, as is the computation of damages in cases of non-performance;

as we saw there, they may be estimated on an abstract basis or proved on a concrete basis, with damages for lost profits available in either case.

(d) Delivery of only part of the goods constitutes partial non-performance, and the rules for disturbance apply analogously to the undelivered balance. Only if the buyer has no interest in the partial delivery, as where the roof tiles that have been delivered will cover only half the roof, is the whole contract treated as disturbed; then the buyer can return the part delivered and claim damages for non-performance of the entire contract.

4 Legal Defects

It is a case of *legal defect* (Rechtsmangel) if the thing is delivered to the buyer but he does not become owner, or where the thing is subject to encumbrances of which, according to the contract, it should be free. Such a defect is treated as a partial non-performance of the contract on the ground that possession has been transferred but not ownership or full ownership.

The buyer's first remedy is to claim performance, to insist that ownership be transferred to him or that the encumbrance be removed. Other remedies stem from the general rules on disturbance of performance, to which there is a cross-reference (§§440 BGB). Legal defect is a partial non-performance of a principal duty and gives rise to a claim for damages for non-performance or to rescission (§§323–6 BGB). The seller is normally responsible for a legal defect, since it is usually due to an initial subjective impossibility, existing at the time the contract was formed. This kind of disturbance of performance is not explicitly covered by the Code, but the courts have long since filled the gap.[18] Thus no form of fault or responsibility is normally required in the case of a legal defect; the seller is liable on the ground of the guarantee to be found in the contract. The consequences of §325 BGB therefore ensue. In the assessment of damages only the harm that the buyer has actually suffered through the legal defect is taken into account; this is the limit of the indemnity he can claim (§440 par. 2–4 BGB).

The buyer of a thing that had a legal defect in the hands of the

---

[18] Guarantee liability has been accepted since *RGZ* 69, 355.

seller may nevertheless acquire full unencumbered ownership through acquisition in good faith.[19] If so, he has no claim for damages: good faith acquisition leads to complete performance of the contract.

III THE DUTIES OF THE SELLER: SPECIFIC REMEDIES OF THE BUYER

1 Defects

(a) If goods delivered by the vendor are not of the quality called for by the contract, the programme has not been correctly fulfilled. The buyer may have acquired both ownership and possession, but he cannot use the property as planned; indeed, it may be altogether useless. There is a defect in the thing purchased (§459 BGB): the watch does not go, the canned food is inedible, the house has dry rot. In all these cases there has been delivery of possession and transfer of ownership, but the buyer's expectations have not been answered, and the balance between price and value has been upset. The general rules on disturbance of performance are not applied here; instead, there are special rules for *liability for defects*, which go back to the practice of the market courts in ancient Rome.[20] The basic idea is that, whatever may be prescribed by the rules on non-performance, the buyer of defective goods can return them against repayment of the price (Wandlung) or else have the purchase price diminished (Minderung). There is no need to show that the vendor was in any way responsible for the defect, but if less need be proved for this remedy than under the general provisions for disturbance of performance, the rights arising out of defects are correspondingly less far-reaching: the transaction may be undone or the price reduced, but no damages are allowable on this count.

These rules have a wide and important area of application, but they are not easy to reconcile with the rules on disturbance of performance, and there are still some debated points where the two sets of rules intersect.

(b) This special liability depends on there being a *defect in the thing* (§459 BGB). The defect must be present in the thing, at

---

[19] For details see Chapter 10 below.
[20] In Roman law these were called the *actio redhibitoria* and the *actio quanti minoris*. The remedies provided by the BGB reflect these models even in the details.

least in embryo, at the time that the risk passes, and it must adversely affect the value or usefulness of the thing. According to the prevailing view, the test of whether a thing is defective or not is a subjective one,[21] and turns on the use to which both parties to the contract of sale contemplated that the thing was to be put. For example, if a buyer orders crates of specified dimensions rather than crates for a specified purpose, crates which are adequate for normal purposes will not be defective even if they are not sturdy enough to carry books. But if the buyer orders book crates, the purpose of the thing has been determined, and if the crates which are delivered prove to be too weak for the purpose, the buyer can give them back and reclaim the price he has paid. Any deviation in quality that affects the value or use of the thing constitutes a defect: a picture which is sold as the work of an old master but turns out to be a forgery,[22] a plot of land which cannot be built on because the subsoil is unsuitable[23] or because a public highway is to go through it,[24] or gas which can only be used with great difficulty because its pressure is not constant.[25] A thing may, of course, be in very poor condition and still be fit for the contractual purpose, such as a house sold for demolition.

2 Wandlung and Minderung

(a) Where the object has a defect, the buyer has a choice between the remedies of Wandlung and Minderung, that is, rescission and abatement of the price (§462 BGB). But he cannot demand delivery of proper goods on tendering back the ones he received unless they are fungible and were sold by description (§480 BGB). In commercial sales, where buyer and seller are both merchants, a *duty to notify* of defects is imposed on the buyer. He must inspect the goods immediately on arrival and notify the seller of any defect (§373 HGB), on pain of losing any claim based on the defect in the goods if he fails to complain

---

[21]von Caemmerer, 'Falschlieferung', in *Festschrift Wolff* 3f. (Tübingen 1952). It is the purpose which emerges from the contract that is critical.
[22]*RGZ* 135, 340f.
[23]BGH *NJW* 1965, 532.
[24]*BGHZ* 67, 136 and frequently elsewhere. On the other hand, a restrictive building covenant in private law is treated as a defect of law to which the rules for non-delivery are applied.
[25]*RGZ* 117, 317.

in good time. The provisions on defects in goods are applied analogously when the goods delivered are of the wrong type or in the wrong quantity (§378 HGB). There is no statutory right to have the defect removed and the goods put right, though this has often been proposed; such a right is nevertheless of practical importance, since it is often incorporated in contracts or imported through General Conditions of Business. Contracts often limit the rights of the buyer to the replacement of any defective component, and a text introduced into the BGB in 1977 provides a foundation for this (§476a BGB).

(b) In *Wandlung* the exchange of goods for money is put into reverse. The Code makes the rules on rescission of contracts applicable here (§§462, 476 BGB). The contract is undone by replacing the exchange relationship by a liquidation relationship, whereunder each of the parties must return what he has received, and the unperformed duties expire. Liability and risk as regards performances already rendered are regulated, with a few gaps, by §§346–60 BGB.[26]

(c) The buyer may, if he prefers, bring a claim for *Minderung*, which lies under similar conditions. He can even switch from one remedy to the other (the *ius variandi*), unless the seller is counting on the remedy first selected. This claim is for reduction of the purchase price (§472 BGB), so as to redress the balance between the agreed price and the value of the purchased property, which the defect has falsified. In order to maintain the contractual relation between price and value the reduction is made proportionally. For the rest, the sale contract with all its rights and duties remains unaffected by a claim for Minderung.

3 Guaranteed Attributes

(a) The vendor's liability for defects is limited to returning the price or suffering some reduction in it: no liability in damages arises under this head. But §463 BGB makes an important exception, unknown to Roman law, for two cases: where the vendor guarantees that the object sold has a specific attribute, and where he fraudulently, that is intentionally, keeps quiet about some defect of which he is aware. The buyer in such cases

---

[26]See Leser, *Der Rücktritt vom Vertrag* (Tübingen 1975).

may still opt to rescind or seek abatement of the price, but he may also claim damages. Where the vendor guarantees an attribute of the object sold, he is liable *on the guarantee,* independently of fault or responsibility; the general remedy of damages here attaches itself to the rules on defects. Liability is imposed for deceitfully suppressing a defect because this constitutes a particularly serious instance of fault in the formation of contract.[27]

Though the purpose to which the thing is to be put is very relevant to both guaranteed attributes and simple defects, they must be kept distinct. The guarantee of the attribute must have become part of the contract: unilateral declarations by seller or buyer are insufficient. The courts are insistent that the guarantee be quite specific, and it must be clear from the contract that the vendor is ready to accept responsibility for the absence of the guaranteed characteristic and its consequences.[28] While it is admitted that such a guarantee need not be explicit, but may be inferred by construing the contract, implied guarantees are hardly ever established. Neither trade descriptions nor advertisements suffice in themselves:[29] there must have been specific representations as to the qualities of the property or its aptness for particular purposes.[30] Special factors, such as the expertise of the vendor who gives advice, may make a difference. Such guarantees are found fairly readily in the sale of used cars.[31]

(b) Damages are supposed to put the buyer in the position he would have enjoyed had the contract been duly performed. There are two ways of achieving this. Either the purchaser keeps the thing and computes the harm he suffers from its having the defect and being less valuable: this amount can on occasion exceed the purchase price. Or he can return the thing and compute the damages on the basis of non-performance of the contract, just as if the thing had not been delivered at all.[32]

[27]Most recently *BGHZ* 60, 321. For more on *culpa in contrahendo* see Chapter 6 Section V 3 above.
[28]*BGHZ* 59, 160.
[29]*BGHZ* 51, 100; *BGHZ* 70, 356, 361f.
[30]*BGHZ* 50, 200; *BGHZ* 59, 161.
[31]The seller of a used car must explain the reading on the mileometer; breach of this duty is treated as a deceitful failure to disclose a defect. See the instructive decisions in *BGHZ* 53, 144; *BGHZ* 57, 137.
[32]BGH *NJW* 1979, 812. The repayment of the purchase price already paid is treated as the lower limit of damages: *BGHZ* 71, 238.

The Duties of the Seller 129

Difficulties can arise in cases where an indemnity is sought not just to make up for the defect in the goods but also for damage which the defective goods have caused to other legal interests of the buyer.

A firm guaranteed that the adhesive it sold and delivered was suitable for affixing ceiling tiles, but shortly after the plaintiff buyer used the adhesive as instructed, the tiles fell off. He consequently had to do expensive repair work for his customers, costing more than DM5,000.[33] The Bundesgerichtshof not only allowed the plaintiff to claim back what he had paid for the glue, but awarded him damages of DM5,000, many times more than the purchase price. This was in respect of the Mangelfolgeschäden, or consequential harm caused by the defect. The guarantee of the attribute must have been intended to protect the buyer against the risks that eventuated,[34] but that requirement was satisfied here because the adhesive had been used as instructed. Should the harm that occurs not be covered by the guarantee, there may still be a case of positive breach of contract (see immediately below).

4 Remedies for Defects and General Provisions

(a) As soon as the risk in the goods has passed, the rules relating to defects take precedence over the general rules relating to irregularities in performance. This is because of the special nature of the rules on defects. Should the buyer discover before they are tendered that the goods have a defect, he may refuse to take delivery of them; indeed, he can probably demand delivery of undefective goods, though there is some disagreement about this. As soon as the risk passes, however, his rights are limited to those arising out of defects.[35]

(b) It often happens that, where goods are defective, the buyer is *mistaken* about one of their essential characteristics. In principle such a mistake could be a ground for rescinding the contract under §119 par. 2 BGB, but no one doubts that rescission is excluded if the more specific remedies for defects can be applied.

The Code contains no solution for the case where both parties are mistaken about a quality of the thing. Such a *common mistake*

[33] The contact adhesive case, *BGHZ* 50, 200.
[34] This was subsequently emphasized in *BGHZ* 57, 292, 298.
[35] *BGHZ* 34, 37.

is treated as an instance of collapse of the foundation of the contract, to which §242 BGB applies. Many details are still controverted.[36]

(c) It is important, but by no means easy, to ascertain the interrelationship of liability for defects and liability for positive breach of contract, which is one of the forms of disturbance of performance.[37] Given that the preconditions of a positive breach of contract are satisfied, including proof of a culpable breach of contractual duty, the appropriate remedy coexists with the remedy for defects, except in respect of the defect itself. In this way the buyer can often obtain compensation for the harm that the breach of contract has caused to his other interests, though not for the loss consisting of the thing's being defective. Compensation by way of damages for such a Mangelschaden, or harm caused by defect, can be had only under §463 BGB, that is, where the attribute has been guaranteed; failing such a guarantee, there are only the limited remedies of Wandlung and Minderung. As may be imagined, the concurrence of these remedies gives rise to serious borderline disputes.

In one case, where the plaintiff nursery had ordered a replacement part for the boiler which heated its greenhouses, the ventilator supplied by the defendant seller was defective. Many flowers were consequently damaged by frost.[38] The Bundesgerichtshof held that because the seller had been negligent in installing the defective ventilator, he must indemnify the buyer for the loss of his flowers. Other instances of positive breach of contract were given in Chapter 6: the horse that infected, and the fodder that poisoned, the purchaser's animals.[39] In such cases the buyer must bring two claims, though they may be consolidated. Basing himself on the defect as such, he can only claim the return of what he paid, but he can get compensation on the ground of positive breach of contract for the harm caused to his other protected interests, provided that he can show that the seller was at fault. Jurists try to keep the two claims distinct, but they do not always succeed.

The rules relating to defects have been found to be too

[36] See Chapter 8, Section II 3 below.
[37] See Chapter 6 Section V above.
[38] *BGHZ* 60, 9, 12.
[39] *RGZ* 66, 289; see also Chapter 6 Section V above.

restrictive, so the courts have quite recently[40] sought to supplement them by invoking the rules of tort, notably by applying to the direct relationship of purchaser and vendor the principles that have been developed in product liability cases.[41]

5 Prescription

The period of prescription for claims arising out of defects is extremely short: six months from the time of delivery for moveables, and one year for immoveables (§477 BGB). This contrasts strikingly with the thirty years which is the general period of prescription (§195 BGB). The buyer who has not yet paid the price can rely on a defect as a defence to the seller's action, whenever it is brought, provided that he notified the seller of the defect before the period of prescription had run. Claims for damages based on the absence of a guaranteed attribute also fall under this short period of prescription.

The short period of prescription also applies to positive breaches of contract if the damage in issue is due to the defect in the purchased thing itself;[42] but the regular period of thirty years would apply if the breach was of a collateral duty arising out of the contract of sale and unconnected with any defect in the goods.[43] A claim in tort retains its own period of prescription, namely three years (§852 BGB), even when it concurs with a claim arising out of a defect in the goods.

IV THE DUTIES OF THE BUYER: REMEDIES OF THE SELLER

1 The Duty to Pay the Price

According to §433 par. 2 BGB, the buyer is bound to pay the purchase price and take delivery of the goods. The *money debt* so created is to be paid at the domicile of the seller, or by sending the money to him there (§270 BGB). But payment in cash is not normal nowadays, and a seller is taken to agree to the buyer's paying in forms other than cash, for example by crediting his account, if the seller indicates his account number on the in-

---
[40]*BGHZ* 67, 359.
[41]See Chapter 9 below.
[42]*BGHZ* 66, 317, 321.
[43]*BGHZ* 66, 208, 214 (badly packed battery causes fire); see also *BGHZ* 47, 319.

voice or elsewhere.[44] In such a case the crediting of the seller's account constitutes performance of the buyer's promise. Since there are no special rules regarding the duty to pay, the general rules on disturbance of performance apply. Impossibility can hardly ever arise as an obstacle to performance in the case of a money debt, since the debtor must keep himself solvent,[45] so it is usually a matter of delay and the setting of a period for payment under §326 BGB. It is, however, laid down there that a claim to performance is excluded once the period allowed has elapsed, so the seller can no longer claim performance of the promise to pay the price of the goods he has delivered, but only damages. This claim for damages will admittedly amount to the purchase price at the very least.[46] The possibility that at this stage the seller can insist on delivering the goods, even if the buyer does not want them, has already been mentioned;[47] the seller can then claim full damages, including the purchase price.

2 Accepting Delivery

Whereas payment is a principal duty, the duty to accept the property, that is, to take physical possession of it, is usually just a collateral one. Its purpose is to 'unburden' the seller. If the buyer does not honour this duty, the seller can bring a standard claim for performance and demand that the property be accepted. In a case of delay, he can also claim compensation for any harm he has suffered thereby, and if it is a commercial sale, the seller can help himself by reselling the goods (§373 HGB). He cannot, however, bring the contract to an end by setting a period for performance under §326 BGB, except in the unusual case where the duty to accept the property is a principal duty.

Acceptance of the property can be made into a principal duty if the seller has some special interest in getting rid of it, for example if the goods are bulky,[48] such as coal or corn, or if he is clearing a warehouse.

If the buyer is in delay in such a case, the seller may exercise his rights under §326 BGB: he can fix a period for performance,

---

[44] *BGHZ* 6, 122.
[45] See Chapter 6 Section III 1.
[46] Established case law, *BGHZ* 20, 343.
[47] See Chapter 6 Section VI 2.
[48] *RGZ* 57, 112.

and if performance is not forthcoming before that period expires he may free himself from his obligations and either claim damages for non-performance or rescind the contract.

V VARIANT FORMS OF SALE

1 Instalment Contracts

The economic importance of sale is reflected in the wealth of its variants. Here we can deal only with instalment contracts (Abzahlungsgeschäfte)[49] and the relevant legislation.

Sales of moveables which are to be paid for by instalments are covered by a law originally enacted in the nineteenth century, not incorporated in the BGB, and frequently amended and extended. As its purpose is to protect the socially weaker party, it does not apply to merchants.

The law on instalment contracts has three ways of protecting the purchaser: by attending to the *formation* of the contract, by controlling the *content* of the terms of credit, and by supervising the *liquidation* of the contract if it goes wrong. As to formation, such a contract must be in writing and must clearly state the effective yearly interest. For seven days after the contract has been formed, the buyer has a *right of cancellation,* which he can exercise without giving any reason. The buyer must be told of this right and of the period within which he may exercise it, or else the seven days start running only on payment of the final instalment. As to the content of the contract — and this was one of the first matters covered by the law — clauses whereby sums paid may be forfeited are rendered invalid. If the seller repossesses the goods, which are usually delivered under reservation of title, this is tantamount to rescission: both parties must restore what they received, so the seller has to pay back any instalments he has collected. The seller's claim for expenses is limited, and any contractual penalty may be reduced by the court at the request of the buyer.

2 The Financed Instalment Sale

The enactment of this law caused a sharp drop in the number of

---

[49]Law of 16 May 1894. *RGBl* 450. On what follows see Marschall von Bieberstein, *Der finanzierte Abzahlungskauf* (Karlsruhe 1980).

instalment sales that were made, so it very early became clear that its protection must be extended to analogous devices in order to forestall evasion. In the financed instalment transaction, for example, the unitary exchange transaction is broken down into a contract of sale and a financing agreement, typically with a bank; the credit is now extended by the bank rather than by the seller, and apparently separate contracts are formed with the seller on the one hand and the financier on the other. In order to safeguard the instalment purchaser when the contracts are severed in this manner, courts and scholars have worked out effective solutions whereby the bank's claim may be met with defences based on defects in the goods or disturbances in the contract of sale.[50] Legislative reform of the instalment sales law is at present under way.

---

[50]The indirect defence (Einwendungsdurchgriff), *BHGZ* 47, 241 and 253.

8

The Principle of Good Faith: §242 BGB

I FUNDAMENTALS*

1 Origins

Unimpressive though it looks, §242 BGB is one of the most astonishing phenomena in the Code. Out of a general clause concerned with how to perform contracts has grown a 'super control norm' for the whole BGB, and indeed for large parts of German law outside it. The wording of the text hardly suggests why: 'The debtor is obliged to perform in such a manner as good faith requires, regard being paid to general practice.'

The aim of the legislator was 'to make people conscious of the true content of the contractual obligation', and the scope of the provision was to be 'confined to regulating the manner and method of the duty to perform'.[1] §242 BGB is still used to spell out what performance entails, for example, to show that one need not accept delivery at an inconvenient time, but that one must pack goods properly and provide instructions for use with them. This function is, however, overshadowed by its recognition as a statutory enactment of a general requirement of good faith,[2] a 'principle of legal ethics',[3] which dominates the entire legal system.

---

*Apart from textbooks and commentaries, the following may usefully be consulted: Diederichsen and Gursky, 'Principles of Equity in German Civil Law', *Festschrift René Cassin* 277ff. (Brussels 1973); Schmitthoff, *Die englische Equity, Festschrift Ernst von Caemmerer* 1049ff. (Tübingen 1978); Wieacker, *Zur rechtstheoretischen Präzisierung des §242 BGB* (Tübingen 1956).
[1] See I *Protokolle zum BGB 303.*
[2] *RGZ* 85, 108, 117 (1914); for the formula used now, see *BGHZ* 58, 146, 147.
[3] Larenz, *Methodenlehre* (4th ed. Heidelberg 1979). The dominant view today is that its content is reliance, which acts as an integrating element in an organized legal culture, especially reciprocal reliance, which takes account of the interests of others that deserve protection. See on this Fikentscher, I *Methoden des Rechts* 109f., 179f. (Tübingen, 1975).

2 Development

Other norms with ethical reference are found elsewhere in the BGB. One is the rule of §157 BGB that contracts are to be construed 'as good faith requires, regard being paid to general practice'; and another is the general clause in §138 BGB, which renders a legal transaction void if it conflicts with good morals.[4] Foreign codes, such as the French Code civil,[5] the Louisiana Civil Code,[6] and notably the Swiss Civil Code,[7] to mention only a few, also contain provisions relating to ethical principles, but they have never generated a system of control such as has happened in German law.

In order to understand this development, we must consider the relationship between decision-making and the new codal text. During the period of positivism, which can be seen as extending to the First World War, the judge was very strictly tied to the text of the Code.[8] If he was to fit the new complex and comprehensive Code to the social scene, he needed express authorization and also a flexible tool. During the period when §242 BGB was being increasingly applied, the discussion about how to fill gaps in the law was yielding to the recognition that judges create law in construing enactments. The motive forces can be seen in the development from the jurisprudence of interests to the jurisprudence of values of today.[9] §242 BGB has served as a vehicle for judicial development of private law and as a way of keeping the law receptive to additional elements of order.

---

[4] See Chapter 5 Section IV 2 above on §138 BGB as a constraint on freedom of contract, and Chapter 5 Section II 1 on the construction of contracts.

[5] 'Contracts 'must be executed in good faith', Code civil (1804), art. 1134 par. 3. For the view that this provision has very little real meaning, see Lyon-Caen, 44 *Rev.tri.dr.civ.* 45f. (1964).

[6] 'In all civil matters, where there is no express law, the judge is bound to proceed and decide according to equity. To decide equitably, an appeal is to be made to natural law and reason, or received usages, where positive law is silent'; Civil Code of Louisiana (1825, 1870), art. 21.

[7] 'In the exercise of his rights and in the performance of his duties everyone must act in good faith'; Swiss Civil Code (1907) art. 2 par. 1.

[8] See Chapter 4 Section III 1 above.

[9] See Chapter 4 Section III 2 above.

3 Function

The importance of the provision clearly lies less in its actual normative content, which is rather difficult to discern, than in its function in giving legal force to broad ethical values. The phrase, 'regard being paid to general practice', suggests that its content might be rooted in empirical fact, but this has not happened: its function is to justify the value-judgments of the judge.

It is through §242 BGB that the old *exceptio doli* of Roman law has entered the system, although the draftsmen of the BGB thought that the new Code could do without it.[10] §242 BGB has also been described as the gateway for natural law to enter the BGB.[11] It certainly provided a very convenient framework in which to supplement the Code in the balanced manner advocated by the jurisprudence of interests. At present many of the value-judgments that are being transformed into principles of private law come from the Basic Law; the result is to give protection not only against the state but also against private parties (so-called Drittwirkung).[12]

But the courts do not just decide as the fancy takes them. From the very beginning they have been especially careful to make their decisions coherent with results and decisions already arrived at, and in this way to develop general principles from the individual cases. Indeed, the extremely numerous decisions rendered under §242 BGB are more comparable with the case law in the Anglo-American systems than anything else in German law. Of course, §242 is not the only source of legal developments: the general right of personality, for example, and the rules of products liability have evolved without reference to it.

---

[10]Staudinger/Weber, *Kommentar zum BGB* §242 N. d 5 (11th ed., Berlin 1961).

[11]Wieacker, *Zur rechtstheoretischen Präzisierung des §242 BGB* 8f. (Tübingen 1956); Wieacker, *Privatrechtsgeschichte der Neuzeit* 476 (Göttingen 1967).

[12]This Drittwirkung of basic rights, that is, their transformation and adoption into the BGB through the general clauses, is a slow process. See Leisner, *Grundrechte im Privaterecht* (Munich 1960); Nipperdey, 'Die Würde des Menschen', in Neumann, Nipperdey and Scheuner (eds), II *Die Grundrechte* 1f. (2d ed. Berlin 1968); Nipperdey, 'Freie Entfaltung der Persoñnlichkeit', in Bettenmann and Nipperdey (eds), IV 2 *Die Grundrechte* 741f. (2d ed. Berlin 1972). It is principally the basic rights, such as the principle of equality (on which see Raiser, 111 *ZHR* 75), freedom of conscience, dignity of man and so on, that are involved.

138   *The Principle of Good Faith: §242 BGB*

4   Does it Provide a System of New Remedies?

The various attempts to schematize the applications and implications of §242 BGB[13] are bound, in view of the function we have ascribed to it, to be merely provisional, illuminating certain aspects only; many decisions contain the warning that only the individual case can justify the application of the principle.

All we can do here is to indicate how it has been used to spell out the duties of contractors, even *after* performance, a matter close to the legislator's original purpose. Then we can look at how it is used to adapt and supplement contracts, notably in the revalorization decisions and generally under the doctrine of the collapse of the foundation of the transaction. Finally, there are cases where §242 BGB has been used to curb or control the exercise of admittedly existing rights; here a flexible limit is set to the law's intervention to protect individual interests. The main examples are the control of general conditions of business, the modification of leonine provisions in form contracts, and the doctrine of loss of rights by estoppel (Verwirkung). As yet there is no system: court practice is crucial, since the provision acquires meaning and significance only through its application in the cases.

II SPECIFICATION OF CONTRACTUAL RIGHTS AND DUTIES

1   Construing and Completing Contracts

The first matter to which §242 BGB was applied was the mass of specific duties which, as we have seen, are favoured by the structure of German contract law. When the contract contained too little for §157 to bite on as a matter of interpretation, §242 BGB was used to construe the contract creatively and produce the new specific duties that were called for.

This may be seen in a case decided in 1931.[14] The plaintiff had for ten years been operating a jewelry business in one of two adjoining houses owned by the defendant, his landlord, when the defendant let a shop in the neighbouring house to another jeweller. The plaintiff sued to make the defendant renounce this contract, and the

---

[13]Wieacker (above n. 11); Roth, 'Das Problem der Rechtsprognose', in *Festschrift Bosch* 573f. (Bielefeld 1976).
[14]*RGZ* 131, 274.

Specification of Contractual Rights and Duties 139

Reichsgericht held that, while rental contracts did not generally contain any implied promise by the owner of several properties not to let any of them to a tenant's competitors, such a collateral duty could nevertheless arise 'out of the speical circumstances of the case'. Reference was made to §242 BGB as well as to §§133 and 157 BGB.

If this decision still smacks of contractual construction, the following case from 1923 clearly oversteps the borderline.[15] The defendant had infringed the plaintiff's trademark, but the plaintiff needed certain information from the defendant before he could quantify his claim for damages. The court held that the plaintiff was entitled to this information: there was a principle of law, stemming from §242 BGB, to the effect that a person who is excusably ignorant about the existence or extent of his rights may demand the necessary information if the other party can easily supply it. §242 BGB has often been deployed in the area of competition: a commercial agent may not act for any of his principal's competitors during the period of his contract.[16] When duties are imposed after a contract has been performed, it is even clearer that the contract is being added to: when a doctor moves out of tenanted premises at the end of the lease, his ex-landlord must for a reasonable time allow him to display a notice with the new address of his surgery.[17] Good faith also requires a contractor to refrain from conduct which would either destroy or sensibly diminish the advantages that the contract has conferred on his partner. Thus a man who has charged a very high price for a plot of land because of its remarkably fine view must not build on an adjacent plot of his if that would destroy the view;[18] and if a proprietor sells a business with its goodwill on the terms that he is to participate in the profits for ten years thereafter, he must not compete with the business for at least that period.[19]

The numerous duties that have been created in this way include duties of care, duties to supervise the manner and form of the principal performance, duties to assure performance,

---

[15] *RGZ* 108, 1, 7.
[16] *BGHZ* 42, 59, 61.
[17] *RGZ* 161, 330, 338 (*obiter dictum*).
[18] *RGZ* 161, 330, 338. This is the famous Venusberg case, which approaches the doctrine of the collapse of the foundation of the transaction.
[19] *RGZ* 117, 176, 179.

duties of co-operation, and duties of information and explanation.

2 Judicial Reconstruction of Contract: The Revalorization of 1923

The following examples will show how the courts, quite apart from construing contracts creatively, were prepared, where necessary, to reconstruct them entirely.

Between 1919 and 1923 unparalleled hyperinflation in Germany reduced the value of the mark to one-billionth of its 1914 value. This led to numerous problems in long-term exchange contracts such as leases and instalment sales (see Section II 3 below, on the collapse of the foundation of the transaction).

During this period of headlong devaluation, long-standing debts were often paid off in currency that was almost entirely valueless. This was very unjust, especially where a vendor of land had a mortgage for the unpaid part of the purchase price. If the buyer could free himself from this debt with valueless money, he was virtually speculating at the vendor's expense, contrary to the whole purpose of security in land, which is designed to secure a part of the capital value. As the legislature of the day was incapable of providing the necessary solution, the Reichsgericht leapt into the breach with a famous decision in 1923:[20] mortgages were revalued in terms of the value of the money at the time of their creation. Even if the debtor had repaid the debt, he was now bound again for part of it, and bound in terms of the new 1923 mark.

Adherence to the basic principle that one mark is as good as another had always been strict theretofore, but so acute was the conflict between nominalism and good faith that the forced currency had to yield to the exigencies of §242 BGB. The action of the courts evoked general approval, but it did show the judges acting openly as legislators, albeit in a situation of necessity.

The question of devaluation is still actively discussed. In principle one adheres to the rule that one mark is as good as another. Index clauses which may serve as a safeguard against

[20] *RGZ* 107, 78, 86.

Specification of Contractual Rights and Duties 141

devaluation need the approval of the Bundesbank or the central banks of the Länder. Such approval is usually given only to secure claims for pensions and long-term leases, not for contracts for the sale of goods or purchases of capital equipment. The cost-of-living index is normally used; gold clauses were once fashionable, but have now largely disappeared.

3 The Collapse of the Foundation of the Transaction

But the revalorization of mortgages paid off with 'valueless' money was only a small part of the very large problem posed by hyperinflation. In all contracts of any duration it falsified the equivalence of price and performance, and thus put strain on the principle *pacta sunt servanda*. The unforeseen collapse of the currency and the restriction of commerce in the First World War, quite beyond the control of the parties, affected the 'underlying basis of the transaction'. Like the old *clausula rebus sic stantibus* of Roman law, this means the circumstances vital to the contract which the parties impliedly treated as its preconditions. At root it is a matter of disappointed expectations and irregularity in the performance of the contract.

The courts were originally very reticent in dealing with such disturbances. Even in 1915 the Reichsgericht refused to allow a circus proprietor to terminate his long-term lease although he could not hold any performances because of the war.[21]

The breakthrough came in a judgment of 1920 on revaluation. A long-term rental contract dating from 1912, not itself in issue, provided for the supply of steam for heating purposes at a fixed price. The cost of steam had, however, risen by a large factor since then. The Reichsgericht raised the price for the heating steam above what the contract provided, 'since otherwise the situation would be simply intolerable and a mockery of the principles of good faith and every mandate of justice and equity'.[22]

The old doctrine of *clausula rebus sic stantibus* was expressly endorsed in the important Vigogne–Spinnerei decision of 1922.[23] Here the fall of the currency induced the court to revalue the real property belonging to a partnership which a partner

---
[21] *RGZ* 86, 397; *RGZ* 99, 258 is similar.
[22] *RGZ* 100, 129, 132.
[23] *RGZ* 103, 328, 331, 333. See also Oertmann, *Die Geschäftsgrundlage* (Leipzig 1921).

was leaving. This was to recognize as the foundation of the transaction those facts on which, in the view of both parties, the contractual intention was bottomed (P. Oertmann). Today people accept a distinction made by Larenz[24] between the subjective and objective foundations of the contract. The former is akin to common error, and is resolved by rescinding the contract or modifying the contractual obligation.[25] In cases of objective collapse of the foundation of the contract, one distinguishes between factors that disturb the parity of exchange, where the courts are still loth to intervene in view of the power of the statutory principle that one mark is as good as another, and events that destroy the purpose of the transaction. The latter come close to irregularities in performance, and help to enlarge the unduly narrow category of impossibility (see Chapter 6, Section II 1).

Such cases provide a parallel to the coronation cases in English law. Further distinctions have to be drawn. If A's engagement to B is broken off, A cannot compel the vendor of the ring to take it back; nor can the lessee of a shop cancel his lease because his turnover is not what he hoped.[26] In every case it must be established that *both* the parties regarded the use of the property as the foundation of the transaction. If a hall is booked for a recital by a contralto and she falls ill, this is a case of collapse of the basis of the transaction,[27] and similarly, a person who rents a boat-house can abandon the contract if sailing on the lake is banned shortly thereafter.[28]

There is certainly much dispute over the details. What is clear is that the courts will not depart from the principle that contracts must be honoured and risks lie where the parties have agreed except when such a departure is necessary to avoid intolerable results which are irreconcilable with law and justice. In other words, it must be grossly unfair to expect the debtor to perform the original contract.[29] If so, the contract is adapted, to

---

[24] Larenz, *Geschäftsgrundlage und Vertragserfüllung* (3rd ed. Munich 1963).
[25] See BGH *NJW* 1976, 565, 566 (transfer of a football contract player who was actually barred: contract held void).
[26] BGH *NJW* 1970, 1313.
[27] OLG Bremen, *NJW* 1953, 1393, noted *NJW* 1953, 1751.
[28] BGH *WPM* 1971, 1303.
[29] *BGHZ* 2, 188.

Controls on the Exercise of Rights 143

the extent that the obligations are susceptible of adaptation,[30] or particular provisions may be treated as void; only in extreme cases is the whole contract rescinded.

III CONTROLS ON THE EXERCISE OF RIGHTS

1 Control of General Conditions of Business

As we saw when discussing freedom of contract (Chapter 5 above, Section IV 3), several other provisions in the BGB besides §242 are used to monitor the exercise of rights. The system of controls that the courts have built up in connection with general conditions of business is most impressive. As an instance we mentioned the decision that the right to complain of defects in new furniture could not be wholly excluded by the contract of sale,[31] general conditions of business being construed against the party who uses them.[32]

In a decision of 1957[33] the Bundesgerichtshof had to consider a contract for the sale of a plot of land and the construction of a house upon it. The vendor's general conditions of business excluded his guarantee of the quality of the house but assigned to the purchasers any claims he might have against the actual builders and other third parties involved in its construction. The Bundesgerichtshof gave this clause a restrictive interpretation. To the extent that the assigned claims did not *in fact* provide the purchasers with an indemnity, it was inconsistent with good faith for the vendor to avoid his personal liability: decency and justice required that the vendor bear the risk that the indemnity might not be effective.

Many such limitations on general conditions of business, imposed by judicial decision, have now been incorporated in the law of 1976 (Gesetz zur Regelung des Rechts der Allgemeinen Geschäftsbedingungen — AGBG).

2 Formal Requirements and §242 BGB

§242 BGB can even modify the direct effect of a statute. This is particularly clear in cases where a transaction has been effected without the prescribed formalities, such as notarial attestation

---

[30] *BGHZ* 47, 48, 52.
[31] *BGHZ* 22, 90. See Chapter 5 Section IV 3 above.
[33] BGH *DB* 1975, 682.

of a contract for the sale of land (§313 BGB). The consequence of non-observance is enunciated in §125 BGB: such a legal transaction is void. But §242 BGB nevertheless comes into play if the lack of form is attributable to the fault of one party, and the other runs the risk of being improperly deprived of his rights.

The famous Edelmann decision of 1927 is an example.[34] An employee of company lived as such in a company house which he had a right to buy for RM120,000. Later he was told that the house was to be his as a Christmas bonus, since cash was not available. On the employee's inquiring whether this was a serious promise, he was assured by the company that 'the notarial act was a mere formality, and that the word of honour of the company representative was as good as a contract'. When the plaintiff claimed the actual conveyance, the company declined on the ground that the notarial attestation required by §313 BGB was missing. The Reichsgericht held that this defence was contrary to the principles of good faith and awarded damages, but it stopped short of granting the plaintiff's claim to performance. Performance has, however, been ordered in subsequent cases,[35] and today such a contract, though void for want of form, may practically be treated as valid.

3 Estoppel, or Sleeping on One's Rights

The passage of time may disable one from exercising his rights. In German law prescription or time-bar is treated as a matter of substance: the general rules are to be found in §§194ff. BGB, in the General Part. §195 BGB makes thirty years the normal, as well as the longest, period of prescription, but situations can arise well before thirty years have elapsed such that it would be unfair and inconsistent with good faith to make use of one's right. Here the principle of good faith anticipates the repose prescribed by law. If the creditor's conduct over a period of time leads the debtor to believe that no claim will be made, the debtor's reliance will be protected.[36] The idea that in legal intercourse one must not blow hot and cold, which was known to the gemeines Recht as *venire contra factum proprium*, underlies many other examples of the application of §242 BGB.[37]

[34] *RGZ* 117, 121.
[35] *BGHZ* 48, 396. This is similar to the English doctrine of part performance (e.g. *Wakeham v. MacKenzie* [1968] 2 All E.R. 783).
[36] *BGHZ* 25, 47, 51; *BGHZ* 43, 292.
[37] For example, *RGZ* 153, 59.

The instances we have given of the effect of the principle of good faith constitute only a small part of a very copious supply.[38] It expresses itself in many different rules, but quite unsystematically. These rules constitute a realm of law of their own, whose function is typically one of control. In functional terms it is quite comparable with the role of equity *vis-à-vis* common law in the Anglo-American systems, as a number of individual parallels would demonstrate.

---

[38]Staudinger/Weber, *Kommentar zum BGB* §242 (12th ed. Berlin 1978) (with 1,546 pages devoted to this provision alone).

9

The Law of Tort

Every day a very large number of people suffer harm. Often it is a matter of *accident*, as when a pedestrian is injured and a vehicle damaged in a collision, or when a workman is killed in an explosion in a factory, or a child is hurt in a playground scuffle, or a housewife is injured by a defective appliance in the kitchen. But harm does not always occur by accident, and it does not always take the form of death or injury to a person or of damage to a thing. Thus an entrepreneur may suffer economic harm if someone infringes his patent or if a consumer magazine disparages his product.

In all these cases the victim will want to know whether he can claim compensation for his harm and from whom. If he is in a contractual relationship with the party from whom he seeks compensation, the contract may provide a legal basis for a damages claim, but if he is not, as is likely in the cases just mentioned, the victim will have to turn to the rules of tort. The function of these rules is to determine the conditions under which, consistently with the prevailing sense of fairness and social justice, the victim may shift to someone else the cost of the harm to which he has been exposed.

The rules of tortious liability may take very different forms in different legal systems. At one extreme, there may be a large number of separate rules which apply to different kinds of harmful activity and render the tortfeasor liable if he causes harm in that way. This is the tradition of the common law,

---

*In addition to textbooks and commentaries on comparative law and the law of obligations, the following may usefully be consulted: von Caemmerer, 'Wandlungen des Deliktsrechts', in I *Gesammelte Schriften* (ed. Leser) 452ff. (Tübingen 1968); Catala and Weir, 'Delict and Tort, A Study in Parallel', 37 *Tul.L.Rev.* 573 (1963), 38 *Tul. L.Rev.* 221, 663 (1964), 39 *Tul.L.Rev.* 701 (1965); Cartwright, 'The Law of Obligations in England and Germany', 13 *I.C.L.Q.* 1316 (1964); Kötz, *Deliktsrecht* (Frankfurt 1979); Lawson, *Negligence in the Civil Law* (Oxford 1954); Opoku, 'Delictual Liability in German Law', 21 *I.C.L.Q.* 230 (1972); Handford, 'Moral Damage in Germany', 27 *I.C.L.Q.* 849 (1978).

where the law of tort consists of a bundle of relatively independent remedies (negligence, conversion, nuisance, malicious falsehood, conspiracy, and so on). The other extreme is found in French law, where the Code civil simply lays down the broad principle that everyone is liable for the damage caused by his *faute* (art. 1382). The German Civil Code strikes a middle path and provides *three* grounds of liability in tort (§§823–6 BGB). The terms of these provisions are admittedly so vague that, just as in France, one must look to the court decisions to flesh them out.

The provisions on liability for unlawful conduct (Haftung aus unerlaubter Handlung) are to be found in §§823–52 BGB, at the end of Book Two, which is devoted to the Law of Obligations (Schuldverhältnisse). Thus in the system of the BGB the rules of tort law are put on a par with the rules of contracts such as sale and loan; since contract and tort have this in common, that they both afford one party the right to claim something from the other, the various contracts and tort are dealt with in different sub-chapters of the law of obligations. Hence, much to the surprise of the common lawyer, the law of tort is not treated as an independent area of law in either the legal literature or the universities of Germany, but almost always forms just one part of books and courses devoted to the law of obligations.

I THE MAIN HEADS OF TORTIOUS LIABILITY

1 §823 par. 1 BGB

Liability in damages under §823 par. 1 BGB arises if an injury caused in an unlawful and culpable manner affects the victim in one of the legal interests enumerated in the text. These legal interests are life, body, health, freedom, ownership, and any 'other right'.

The courts and most of the writers are agreed that the requirement of *unlawfulness* is satisfied by any invasion of a legal interest specified in §823 par. 1 BGB which it is impossible to justify by a special privilege such as self-defence, necessity, and one or two others. The requirement of *culpability* or fault is satisfied if the harmful conduct is either intentional — that is, when the protected interest is invaded on purpose — or negligent. Negligence connotes a want of that degree of care which is generally regarded as necessary in social life (§276 BGB); for this purpose one must consider how a reasonable man in the same situation would have acted so as to avoid exposing others to an unreasonably high risk of injury. If the harm resulted from some special activity, such as driving an automobile, sailing a ship, or performing a surgical operation, the courts ask whether the defendant showed the care to be expected of the average reasonable driver, sailor, or surgeon. It is no excuse for the defendant to say that he was personally incapable of meeting the standard of care required of the reasonable man. As O. W. Holmes put it,

> if, for instance, a man is born hasty or awkward, is always having accidents and hurting himself or his neighbors, no doubt his congenital defects will be allowed for in the courts of Heaven, but his slips are no less troublesome for his neighbors than if they sprang from guilty neglect. His neighbors accordingly require him, at his proper peril, to come up to their standard, and the courts they establish decline to take his personal equation into account.[1]

The German courts have steadfastly adhered to this viewpoint.[2]

A further precondition of liability is that the unlawful and culpable behaviour of the defendant should have caused harm to the victim in one of the legal interests enumerated in §823 par. 1 BGB. Injury to *life* occurs where someone dies as a result of the defendant's behaviour. In such a case, claims for compensation vest not in that person but in third parties; who these claimants are and what they may claim is prescribed by §§844–6 BGB (see p. 155 below). Injury to *body or health* includes every adverse effect on corporeal well-being attributable to external factors. Mere mental upset is not enough by itself: it must have given rise to medically ascertainable consequences of a physical or psychical nature.[3] An invasion of *freedom* occurs only in those rare cases where a person's scope for bodily movement is constrained owing to the defendant's conduct, as happens when the police arrest him on the false witness of the defendant. An infringement of *ownership* occurs not only when the defendant damages or destroys someone else's tangible property, but also when, without harming the thing physically, he takes it away or

---

[1] Holmes, *The Common Law* (Boston 1881).
[2] See *RGZ* 119, 397.
[3] See *BGHZ* 56, 163.

uses it without the owner's consent. The Bundesgerichtshof has even found an invasion of ownership in a case where the plaintiff's ship, though unimpaired, was marooned in a canal whose walls had collapsed owing to the defendant's negligence.[4]

Finally, §823 par. 1 BGB protects those interests which are somewhat enigmatically styled '*other rights*'. These include all the interests that the law protects *erga omnes,* for instance, real rights such as servitudes and rent charges, the right to one's name, patent rights, and other industrial property rights that are regulated in special statutes. One does not, however, have an 'other right' in the sense of §823 par. 1 BGB in one's estate or one's finances as a whole. Economic loss is compensable under this paragraph only if it flows from injury to one of the legal interests specified therein. Certainly a person who suffers personal injury and damage to his car in a traffic accident can claim compensation for all the economic loss, such as medical expenses, which follows from the personal injury, or from the damage to the car, such as the cost of hiring a substitute; but culpable behaviour which causes the victim only 'pure' economic harm, unconnected with any personal injury or damage to property or the invasion of any 'other right', does not give rise to any claim under §823 par. 1 BGB.

If a person doing excavation work on a piece of land carelessly damages a power cable and so cuts off the power to a factory some way away, his liability to the factory under §823 par. 1 BGB depends essentially on the kind of harm that is caused by the interruption of the power supply. If all that happens is that the machinery stops and the factory's output is reduced, the factory cannot claim the lost profits since only an economic loss has been caused by the carelessness of the excavator;[5] but if the power stoppage makes the incubators in a poultry farm cool off, and the eggs spoil, the spoiling of the eggs is an injury to property and the excavator must pay not only the value of the eggs but also any other economic loss which the poultry farm suffers in consequence of their being spoilt.[6] Thus, where merely pecuniary loss results from reliance on false statements negligently made, there is no liability, at least under §823 par. 1 BGB. In general, Lord Denning's dictum in *SCM (UK) Ltd v Whittall & Son, Ltd* [1971]

[4]*BGHZ* 55, 153. [5]*BGHZ* 29, 65.
[6]*BGHZ* 41, 123.

1 *Q.B.* 337, 343, that 'in actions of negligence, when the plaintiff has suffered no damage to his person or property, but has only sustained economic loss, the law does not usually permit him to recover that loss', would be perfectly acceptable to a German court, although it would base that principle on the terms of §823 par. 1 BGB.

The provision in §823 par. 1 BGB whereby 'other rights' are protected against unlawful and culpable invasions is vague and imprecise enough to have tempted the courts into interpreting it as covering the particular interest of the plaintiff when there was no other way of achieving a reasonable result. There are two respects in which the notion of an 'other right' has been thus expanded by interpretation. Since 1954 the courts have accepted that the 'general right of personality', as it is called, is to be regarded as one of the 'other rights' that merit protection against unlawful and culpable invasions. This is now the legal basis on which claims for damages are made when one's honour is impugned, one's reputation besmirched or one's privacy invaded. These cases will be discussed in some detail later. The other important extension of the notion of 'other right' dates from the days of the Reichsgericht, which laid down that in principle the 'right to an established and operative business' ranks as an 'other right' in the sense of §823 par. 1 BGB. How to give clear conceptual shape and form to the 'right to an established and operative business' has long been a disputed question. The Bundesgerichtshof now says that in order to constitute an invasion of this right the conduct of the defendant must in some way be 'business-connected', that is, it must be 'in some way directed against the business as such . . . and not simply affect rights and interests which are separable from the business as a functioning unit'.[7] Since this formula will be as opaque to foreigners as it is to German lawyers, we must mention the most important types of case where damages are actually claimed on the ground that the 'right to an established and operative business' has been infringed.

In one group of cases the defendant has caused the plaintiff to stop some productive activity by claiming that he himself has an exclusive patent, licence, or other similar right. If it turns out that the defendant's claim was inaccurate and that it was unreasonable of him

[7] *BGHZ* 29, 65.

to make it, for example because his investigations were perfunctory, then he may be liable for the plaintiff's economic loss even if he was in good faith.[8] In other cases the plaintiff seeks protection against criticisms which hurt his business. It is true that in cases of this type liability in damages may be based on §826 BGB (see p. 156 below) or on §1 of the Law against Unfair Competition, but where for some reason this is not possible the court will discuss the case in terms of invasion of the plaintiff's right to an established and operative business. The same is true of cases where the defendant has called a boycott of the plaintiff's business, or where accurate but damaging facts about the plaintiff's business have been published by the defendant in an impermissible manner not warranted by any genuine or pressing interest.[9]

Interesting as these cases are, we must remember that in practice §823 par. 1 BGB has its main impact in cases where the plaintiff has suffered an *accident,* and seeks compensation for the consequential personal injury or property damage. In such cases it is perfectly plain that the plaintiff's body, health, and ownership have been invaded. Supposing that the defendant's conduct has been shown to be unlawful and culpable, there remains one further requirement before liability attaches under §823 par. 1 BGB: the unlawful and culpable behaviour of the defendant must have been the *cause* of the legal injury suffered by the plaintiff and of the harmful consequences that ensued. This is the perennially vexing problem of what in the common law is called 'proximate cause' or 'remoteness of damage'. The rule laid down by the Reichsgericht was that the causal connection was legally relevant whenever the conduct of the defendant was an 'adequate' cause of the harm, when 'it was apt in general to lead to the result which occurred, taking things as they normally happen and ignoring very peculiar and improbable turns of event'[10] (doctrine of the 'adequate' causal connection).

The fact is, however, that questions of causation in law call for value-judgments and cannot be answered just by applying logical and abstract standards. Recent court decisions show an increasing appreciation of this, some of them accepting, as do many writers, a teleological theory propounded by Rabel, according to which one

---

[8]*RGZ* 58, 24; *BGHZ* 2, 387; *BGHZ* 29, 65; *BGHZ* 38, 200.
[9]See *BGHZ* 3, 270; *BGHZ* 8, 142; BGH *NJW* 1963, 484.
[10]See *RGZ* 158, 34; *BGHZ* 3, 261.

must take account of the protective function of the rule that imposes the duty to make compensation: if a person causes harm by neglecting a particular rule of conduct, he should pay for such consequences as it was the meaning and purpose of that rule to guard against.

The Bundesgerichtshof has applied this doctrine. In one case the person who was to blame for a traffic accident, while liable for the victim's medical expenses, was held not liable for the legal expenses incurred by the victim in retaining an attorney to defend him, and defend him successfully, in the ensuing criminal proceedings: 'the risk of being involved in criminal proceedings and incurring the expense of one's defence therein is a risk to which everyone is exposed; it does not figure among the dangers which the law seeks to guard against by bringing the integrity of health and property under the protection of §823 par. 1 BGB.'[11] The same principle has been applied in another decision where the question was whether a motorist responsible for causing a traffic accident which blocked the highway was also responsible under §823 par. 1 BGB for the damage done to the cyclepath and footpath when impatient motorists drove over them in their eagerness to get past the scene of the accident. The Bundesgerichtshof denied liability: this was not the type of harm envisaged by the duty of care imposed upon, and infringed by, the defendant.[12]

As to proof, in principle it is for the plaintiff to establish that he has been injured in one of the legal interests protected by §823 par. 1 BGB and that the conduct of the defendant which caused this injury was unlawful and culpable. This means in particular that the plaintiff must prove that the defendant fell below the standard of care demanded of him under the circumstances and that this caused his injury. In cases where the defendant's activity was a dangerous one, his duty will be to act with the 'utmost' care, and this may help the plaintiff to establish his case. In other cases the court will say that there is prima facie evidence (Beweis des ersten Anscheins) if the accident would not normally have happened unless the defendant had been negligent, and this will be sufficient unless the defendant can explain how in this particular case the injury could have occurred despite his taking all the requisite precautions. This rule is seen as merely an aid in the evaluation of the evidence, not as shifting the burden of proof to the defendant.

[11] *BGHZ* 27, 137. [12] *BGHZ* 58, 162.

However, in cases where the plaintiff's claim for damages under §823 par. 1 BGB is for injuries caused by a defect in goods produced by the defendant, the courts have held that there is a presumption of negligence which it is for the defendant to rebut. The first products liability case in which the Bundesgerichtshof decreed such a reversal of the burden of proof was the so-called fowl pest case.[13] The owner of a chicken farm sued the manufacturer of a fowl pest virus after thousands of his chickens had died from being vaccinated with contaminated serum supplied by the defendant to the plaintiff's veterinary surgeon. The plaintiff's suit was successful. The court decided that §823 par. 1 BGB was the appropriate basis for the claim, and that the plaintiff in products liability cases like this must show that the defendant's product was defective when it left his place of business, but that he need no longer prove that the manufacturer fell below the ordinary standard of care. Instead, it is up to the latter to exonerate himself from the presumption that he was to blame for not taking sufficient care. While it is true that the court refused to introduce strict liability in tort, its reversal of the burden of proof so heightens liability for fault under §823 par. 1 BGB that in practice this is not very different from strict liability.

If the defendant can prove, and it is for him to prove it, that the plaintiff was also responsible for the accident to a greater or lesser degree, his liability will be reduced to the extent that the plaintiff's fault contributed to the harm. The rule which so provides, §254 BGB, is to be found in the General Part of the Law of Obligations of the BGB, because it applies to claims for damages in contract as well as in tort and has therefore been 'factored out' for reasons of systematic neatness. Apportionment, which is normally expressed in arithmetical fractions, need not be in equal shares and may be made even when the plaintiff's share of responsibility is greater than the defendant's.

Difficult problems of assessment of damages may arise once it has been established that the defendant is liable in whole or in part. According to §249 par. 1 BGB, the defendant must in principle 'restore the situation which would have existed but for the fact which rendered him liable'. This would mean that the *defendant* would have to find a substitute for the thing he destroyed, a doctor to look after the plaintiff he had injured. Since the plaintiff has an obvious interest in being the person to select

---

[13] *BGHZ* 51, 91.

the replacement, the doctor, or the hospital, §249 par. 2 BGB lays down for cases of personal injury or property damage that the plaintiff, instead of asking that the defendant restore the *status quo ante,* may ask for an appropriate amount of money damages. In practice this is almost always done.

A plaintiff who has suffered bodily injury may claim the *costs of cure* under §842 BGB. Nowadays this claim is usually brought not by the victim personally but by his social insurance carrier which is subrogated to his rights and sees him back to health. The victim also has a claim for compensation for the harm resulting from *impairment of earning capacity.* Here, in sharp contrast with the common law, §843 BGB provides that the defendant must make this harm good by means of an annuity or periodical payments. The court may order payment of a lump sum only in exceptional cases where there is a good reason, perhaps that the victim has had to give up his previous job because of the accident and needs a large sum of money to set himself up in business to secure his support in the future. Finally, the victim may seek a money payment in respect of his *non-pecuniary losses,* such as physical pain or mental anguish, loss of enjoyment or impairment of social life. It is true that compensation for non-pecuniary harm may only be granted in cases where this is expressly provided for by law (§253 BGB), but such express provision exists in the case of personal injury caused by a tort (§847 BGB). According to the Bundesgerichtshof, the damages awarded for pain and suffering (Schmerzensgeld) have a double function, the main one being to compensate the victim for the non-pecuniary harm caused by the injury, the other being to alleviate his sense of injustice by means of a solatium or atonement.[14] The amount of the award may therefore be influenced to some extent by penal considerations; for example, a higher sum may be awarded where the defendant's conduct was clearly reckless. In view of the fact that most tortfeasors are insured and therefore entirely unaffected by the size of the judgment, this rationale is not a very convincing one. In arriving at the final figure the judges always consult the books which list prior court awards according to the type and gravity of injury involved. The current tariff for the loss of a

[14]*BGHZ* 18, 149.

leg, for example, is between DM15,000 and 40,000 (£3,750 and £10,000) or even more, if there are other grave injuries; paraplegia rates between DM100,000 and 130,000 (£25,000 and £32,500). Schmerzensgeld is normally awarded in a lump sum, but the courts may, and sometimes do, order the defendant to pay by instalments or to pay both a lump sum and periodic sums.[15]

If a person is killed or injured, those whom he was supporting suffer harm because he ceases to support them. Since this is mere pecuniary harm, the conditions for liability under §823 par. 1 BGB are not satisfied. However, in the case where a person has been killed, §842 par. 2 BGB expressly provides that those whom he was bound by law to support may claim periodic payments from the person who killed him. The amount of these payments is based on the amount of support the decedent, given his income, would have had to provide during the period he would probably have lived. To find out who is bound by law to provide such support one must refer to the family law provisions of the BGB (§§1360, 1601ff., 1776): here spouse and children top the list. Some people have no such claim. Thus a woman has to be legally married to the man she has been living with, or she will have no claim on his death, even if he has been looking after her for years in a marital kind of way; much less does any claim vest in the person who suffers loss because the death of his contractor prevents or delays the performance of his contract.

2 §823 par. 2 BGB

Liability under §823 par. 2 BGB arises when there is a culpable contravention of a 'statute designed to protect another'. Protective statutes in this sense include all those rules of private and public law, especially criminal law, whose main purpose is to protect an individual or a group of individuals rather than the public as a whole. Here too it is of some importance what form the harm takes, whether injury to health, damage to property or merely economic harm; for harm is compensable only if it results from the very danger that it was the purpose of the protective statute to guard against. If the protective statute is directed to the prevention of personal injury and property

[15] BGH *NJW* 1973, 1653.

damage only, no claim for damages for pure economic loss will lie under §823 par. 2 BGB. But if a director of a limited company culpably fails to start liquidation proceedings when the company becomes insolvent, and so commits an offence under §64 of the Companies Act (GmbHG), he will be personally liable under §823 par. 2 BGB in connection with §64 GmbHG to the company creditors who thereby suffer mere economic loss: the very purpose of §64 of the Companies Act is to protect creditors from harm of this type which is caused by delay in bringing the liquidation proceedings.[16]

The rules on contributory negligence and assessment of damages which have already been discussed are also applied to claims brought under §823 par. 2 BGB, but the rule as to burden of proof is different: once the plaintiff has established that the defendant was in breach of a protective statute, the defendant must rebut the ensuing presumption that this was negligent.[17]

3 §826 BGB

The third head of general tort liability in the BGB is §826. A person is liable under this section if he 'intentionally causes harm to another in a way which offends *contra bonos mores*'. The courts have used this provision to impose liability in a wide variety of cases where one party has caused harm to another by behaviour so offensive and improper as to shock the average person in the relevant sector of society. It is not necessary to show that the defendant actually intended to cause the harm: it is enough if he consciously acquiesced in the possibility that harm might occur. Liability is thus incurred by a person who knowingly procures a vendor to make over to him goods which the vendor had already sold but not delivered to another,[18] and by a person who goes behind his partner's back and indulges on his own account in transactions that should have benefited the partnership. If the defendant's conduct is incompatible with good morals and if he intended that the plaintiff suffer harm or must be taken to have known that he might, liability under §826 BGB attaches even if the harm is purely economic. Thus a

[16] *BGHZ* 29, 100.   [17] *BGHZ* 51, 91, 103ff.
[18] RG *JW* 1931, 2228.

public accountant who is reckless in drawing up an inaccurate report on the financial position of a company is liable under §826 BGB if he knew that the report would come into the hands of third parties and might well cause them economic loss.[19] In such cases the German courts, unhampered by the doctrine of consideration, are also quite ready to find that a contractual relationship existed between the party giving the information and the party relying on it.[20]

II VICARIOUS LIABILITY

In both the civil law and the common law the nineteenth century saw the triumph of the view that a tortfeasor must be to blame before he could be made to pay: no liability without fault. While this principle did not deter either the common law or French law from making a master liable for the torts committed by his servants even if he personally was not at fault, the draftsmen of the BGB applied the principle of no liability without fault even in the area of liability for servants. §831 BGB provides that the master is liable for the harm caused by his servants only if he was at fault in their selection or supervision, that is, if he failed to come up to the standard of care called for in the circumstances. It is true that §831 BGB contains a presumption of such fault in selection or supervision, but this presumption is rebuttable, and if the master rebuts it he is free from liability.

The first precondition for making a defendant liable under §831 BGB for harm caused by a third party is that the third party be his servant (Verrichtungsgehilfe). In the view of German courts a person qualifies as a servant if he is subject to the defendant's supervision and control or, in the case of a doctor, captain, or other professional or expert, if he forms part of the defendant's organization. Accordingly, independent craftsmen, carriers, and building contractors are not, as a rule, 'servants'.

A further requirement is that the harm be caused by the servant 'in the exercise of the function assigned to him'. The

---

[19]BGH *WM* 1956, 1229.
[20]See, for example, BGH *NJW* 1973, 321.

servant need not have been employed to do the very act that triggered the harm; if his act fell within the steps normally incidental to the execution of his functions, that is enough. For example, suppose that a carpenter working on a scaffold throws a piece of wood into the highway and injures someone; he may have thrown it in order to stop children playing with the dangerous scaffolding materials on the sidewalk, or he may have thrown it to make a passing acquaintance jump. Only in the former case is there a sufficient internal connection between the harmful act and the duties of the workman.[21] Of course, there may be borderline cases where it may be difficult to say whether or not the servant acted 'in the course of his employment', as the common lawyer would put it.

If §831 BGB does not make it sufficient that the servant himself be at fault in causing the plaintiff's harm, nor does that section make it necessary. Nevertheless, the courts recognize that there is no need to make the master exculpate himself when his servant has conformed to the requisite standard of care.[22]

Once these preconditions have been established, the defendant principal will be liable unless he can adduce the exculpatory proof (Entlastungsbeweis) by showing that he selected, trained, and supervised the assistant with all the precautions generally required. The courts are extremely strict with regard to this proof, especially where the plaintiff has been injured by the driver of a train, tram, or truck, or by some other kind of employee in the transport business, such as a level-crossing keeper or a station attendant. The courts have gone so far as to say that large transport undertakings must conduct regular and secret tests to supervise their drivers, the examiner following the driver and checking how safely he is driving. This applies to all employers when the driver is young, or when the driver of a truck is inexperienced or newly appointed.

One might have thought that the circumstances of the accident itself would be the central issue, but in suits based on §831 BGB they are rather lost sight of. If a person is injured in a train crash, the fault of the train driver is usually quite clear, and the suit under §831 BGB turns instead on whether the defendant

---

[21] BGH *VersR* 1955, 205.   [22] See BGH *NJW* 1954, 913.

railway exercised sufficient supervision and control over him; the requisite degree of control and supervision depends on how well the driver did in his driving test, how much experience he had at the time of the accident and whether he had, to the knowledge or constructive knowledge of the employer, been guilty of similar errors before. It is true that there are special statutes which impose strict liability for injuries caused by the operation of railways or motor vehicles (see p. 161 below), so that in these cases the defendant railway or carrier is also answerable under this head; but since these statutes limit the recoverable damages and do not provide damages for pain and suffering at all, a plaintiff who wishes to obtain full compensation will even here have to rely on the rules of general tort law, including §831 BGB.

There is general agreement among German lawyers today that §831 BGB is thoroughly unsound in policy. The only reason it has remained so long in the BGB is that the judges have done much to undermine its effects. The courts' power to grant contractual claims for damages has proved especially useful. If the plaintiff's claim can somehow be based on a contract, the defendant can no longer exculpate himself under §831 BGB and will be unconditionally liable to the plaintiff 'contractor' for the fault of his personnel (§278 BGB). Thus, if a car salesman omits to display a warning notice in a showroom whose floor has just been polished, a customer who slips and falls on it can always claim damages from the firm on a 'contractual' basis. If the purchase had already been concluded and the customer was there to discuss the details, the German courts construe the contract as obliging the dealer not only to deliver the automobile but also to see to the safety of his sales premises; his failure to observe this duty of care, which would in any case constitute a tort, is here concurrently a 'positive breach of contract' (see p. 105 above).

Even if there is no contract between the parties, and the customer is just visiting the premises in order to look at the new models, the same is true. The courts give a contractual flavour to this situation by holding that, as soon as negotiations start and before any contract is ever concluded, there arises a mutual duty of care, breach of which renders the dealer contractually liable for *culpa in contrahendo*. And should the customer's wife happen to accompany him and be injured, she also has a contractual claim, since it is constant German practice to give her the benefit of the protective effects of her husband's contractual relations. The courts do this by using the concept of the contract for the benefit of third parties.

160 *The Law of Tort*

In contrast to the English common law, §328 BGB makes it possible for a contract to confer on a third party the right to demand performance from the contractor. The courts have extended this so as to permit the third party not to demand performance of the contract but rather to claim damages in proper cases where the contractor has breached a contractual duty which he owed not only to his fellow-contractor but to certain other people as well. Thus in one case decided by the Reichsgericht a tenant had contracted with a repair firm for the repair of a gas water heater in his house. The gas-fitter was careless and the heater exploded. It was not the tenant who was injured, however, but the cleaning-woman he employed. The cleaning-woman was allowed to base her claim for damages on the repair contract. Thus the repair firm became unconditionally liable for the fault of their fitter under §278 BGB, and §831 BGB was outflanked.[23] For other techniques used by the courts to evade the misconceived policy of §831 BGB, see Kötz pp. 131 *et seq.*

In the circumstances it is not surprising that §831 BGB has been heavily criticized in the Bundesrepublik. The efforts to reform it have culminated in a bill drafted by the Ministry of Justice. The proposed new text is as follows: 'A person who appoints another to perform a function is bound along with that other to indemnify a third party for the harm caused by an intentional or negligent tort committed by that other in the execution of his task.' If this is enacted, the employer will no longer be able to exculpate himself, and the law of Germany will be in harmony with that of the other legal systems in Europe.

III STRICT LIABILITY

When a person has to pay for harm simply because he had control of the particular dangerous activity from which it arose, German lawyers talk of Gefährdungshaftung, or strict liability. In such cases neither the defendant nor any of his servants need be to blame for causing the accident, and it is of no use to the defendant to show that his selection and supervision of his staff were beyond criticism.

1 Instances

Strict liability exists in German law only where the legislator

---
[23] *RGZ* 127, 218.

has seen fit to provide for it by special enactment. The most important of these cases are as follows.

(a) Strict liability for death or personal injury caused *through the operation of railways* was introduced by §1 of a statute of 1871; enacted shortly after the reconstruction of the Empire, it is called the Imperial Law of Liability. For property damage caused by railways strict liability was introduced in 1940. Harm is caused 'through the operation' of a railway if it results from any incident of its technical operation such as sudden braking, collapse of the track, sparks from trains, signal failure, and so on. People getting on and off trains or in railway stations are also covered if the accident is due to a danger specific to railway travel, such as high running-boards, passengers in a hurry, or crowded platforms; on the other hand, if someone in a station falls down stairs simply because the steps are defective, he will have to rely on §823 par. 1 BGB. The railway is not liable for accidents caused by *force majeure* (höhere Gewalt), which the courts take to mean external and elemental forces of nature or the conduct of third parties whose effects could not have been prevented even by the most extreme precautions.

(b) In 1943 the Imperial Law of Liability was extended by special statute to cover *installations for the transmission or supply of electricity or gas,* including high tension cables, gasholders and gas-pipes. Those in charge of such installations must pay for all harm to person or property attributable to emanations of electricity or gas from them. A further amendment of the Law enacted in 1978 extended strict liability to harm caused by *piped fluids, fumes, and gases.* Here, too, the only defence is *force majeure.*

(c) The liability of the custodian of a *motor vehicle* is less strict. It was introduced in 1908 and is now contained in the Road Traffic Act of 1952 (StVG). §7 StVG makes the *Halter,* or custodian, of a motor vehicle (normally, but not necessarily, the owner) liable for any damage to person or property that arises 'through the operation' of the vehicle. Liability is excluded

> 'if the accident is caused by an unavoidable event attributable neither to a defect in the construction of the vehicle nor to a failure of any of its functional parts. In particular an event is unavoidable when it is attributable to the behaviour of the victim, or of a third party not involved in the operation, or of an animal, and both the custodian and the driver of the vehicle have taken all the care called for in the circumstances of the case.'

This leaves the custodian liable for failure of the components of the vehicle, even if unforeseeable and unavoidable, such as a tyre defect, axle fracture through metal fatigue, brake failure, or the steering seizing up; but if the accident is due to an 'external' event, such as bad weather, road conditions, an animal running in front of the vehicle, or bad driving by other motorists, the custodian is not liable if he can prove that the driver and he himself displayed 'all the care called for in the circumstances'. Such care is defined by the courts as 'care going beyond what is usually required, extreme and thoughtful concentration and circumspection'.[24] Only in very few cases has the accident been held to have been 'unavoidable', for example, where a child darts into the road from behind a parked car so suddenly that even an 'ideal' driver using extreme care could not avert the accident. Passengers in the vehicle can use the Road Traffic Act to sue its custodian only if they were being carried by way of business and for reward, as in a taxi or bus; in other cases injured passengers must use the general provisions of the law of delict (§§823ff. BGB).

(d) The Air Traffic Law of 1936 lays an especially strict liability on the *operators of airplanes* for harm caused to persons and things except those on board pursuant to contract. In practice this means harm caused on the surface of the earth, whether by flight noise, by crashing, or by crash-landing: the custodian of an airplane is liable even if he can show that the accident was caused by *force majeure* (§§33ff. Air Traffic Law).

(e) Operators of *installations for the production and fission of nuclear materials* are rendered strictly liable by the Atomic Energy Act of 1959.

(f) Another important instance is the liability imposed by the Water Maintenance Act of 1957 (§22) for *pollution of water*. Water for this purpose includes the water table as well as all ponds, lakes, rivers, and streams; and anyone who introduces into such water substances that alter its composition is liable to pay for the harm, including pure economic loss, which others thereby suffer. The same liability, with a defence of *force majeure*, is imposed on the operator of an installation for the manufacture, storage, or carriage of materials. If any materials are

---

[24]BGH *VersR* 1962, 164 and repeatedly thereafter.

introduced into water as defined, the operator of the installation is liable in damages. 'Installation' for the purpose of the statute includes not only storage tanks for oil and paraffin, tanker vessels and oil pipelines, but also petrol tankers of the kind that supply filling stations.[25]

(g) A recent statute which introduces strict liability is the *Drug Act of 1976*. §84 lays down that a person who puts a drug into commerce is liable if death or serious personal injury results from the use of the drug as prescribed. The harmful effects of the drug so used must, however, go beyond those which current medical opinion would regard as acceptable, and they must be attributable to the process of development or production. Similar liability attaches if the drug is mislabelled or if its directions are, in the light of medical knowledge, ill-advised.

(h) In all the cases so far mentioned, strict liability is based on special statutes outside the BGB. In the BGB itself there is only §833 par. 1, which imposes strict liability on the *keeper of an animal* for the harm it causes. This applies only to what are called 'luxury animals': if the animal is a guard dog, a draught horse, a milk cow, or other creature used for its keeper's profession or business, the keeper is not liable unless he is at fault. Such fault is, however, presumed under §833 par. 2 BGB, so the keeper escapes liability for the animal only if he can prove that he exercised due care in looking after it.

2 *Characteristics*

If in any of the aforementioned cases the victim himself was at fault, the court may apportion the damages after evaluating the respective contributions to the harm of the fault of the victim on the one hand and the dangerousness of the railway, motor vehicle, installation, or animal on the other.

A characteristic feature of all these strict liability statutes (except those in (f) and (h) above) is that they put a monetary ceiling on the defendant's liability and that claims in respect of pain, suffering, and other immaterial harm are wholly excluded except under §833 BGB. For instance, the liability of the custodian of a motor vehicle is limited to DM500,000 (£125,000) in

---

[25] *BGHZ* 47, 1.

the case of death or personal injury to a single victim, and to DM750,000 (£187,500) if there is more than one. Should the damages be payable in the form of an annuity, the annual payments may not exceed DM30,000 (£7,500) and DM45,000 (£11,250) respectively. These limits are periodically increased by statutory amendment in order to take account of inflation and other factors, but their existence and the impossibility of obtaining damages for pain and suffering explain why persons injured in traffic accidents almost always sue under §§823ff. BGB as well as under the Road Traffic Act. In the result, there is often a long and tiresome dispute about fault even though liability under the Road Traffic Act is perfectly clear. To alleviate this difficulty many writers have proposed that damages for pain and suffering should be available under the Road Traffic Act as well.

Finally, it is worth mentioning that a duty to insure against liability up to a certain amount has been imposed in some, though not all, of these cases. Insurance must be taken out against the risks mentioned in (c), (d), (e), and (g) above, but no duty to insure has been imposed on the enterprises rendered liable by the Imperial Law of Liability, nor on persons operating an installation from which materials might escape and pollute water. Of course, railways, power companies, and like enterprises are usually big enough to be self-insurers, and cover against strict liability under the Water Maintenance Act, though not required by law, seems to be fairly common in practice, even among private householders with an oil tank in the basement.

Given the large number of detailed statutes providing for strict liability, it is not surprising that the German courts have always held that the imposition of strict liability is a matter for the legislature and not for the judiciary. For this reason both the Reichsgericht and the Bundesgerichtshof have held on a number of occasions that it is impossible to apply the existing statutes by analogy to other installations or activities, even though they may be at least as dangerous as driving automobiles or keeping animals.[26] Consequently, it is for the legislature to extend strict liability to new areas on a case-by-case

---

[26] See *RGZ* 78, 171; *BGHZ* 51, 91; *BGHZ* 55, 229.

basis; this means, of course, that the law will always trail some way behind advances in technology. Various writers have suggested that the existing statutes should be supplemented by a provision of the BGB which imposed strict liability on the operators of 'dangerous' installations and the possessors of 'dangerous' substances. These proposals have proved unavailing.

IV PROTECTION OF HONOUR, REPUTATION, AND PRIVACY

German law has several different ways of protecting a person's honour and reputation. It is true that until recently §823 par. 1 BGB was of no assistance, because, while it protected a person's life, body, and health against culpable invasions, it did not purport to protect his honour and reputation, but §823 par. 2 BGB gave the victim a claim for damages when the defendant's statements impugning his honour and reputation amounted to the crime of insult or slander (§185 Criminal Code), for then the defendant would be in breach of a protective statute in the sense of that paragraph. Again, §824 BGB can render a person liable in damages if he publishes facts which he knew, or should have known, to be false and apt to impair another person's credit or cause him harm in his trade or profession in some other way. §826 BGB is also of some importance here, since to attack a person's honour or reputation may be seen as behaviour *contra bonos mores*. Another instance arises from §12 BGB, which provides that one may stop someone else using one's name against one's will: as this right to one's name is treated as an 'other right' under §823 par. 1 BGB, the person whose name it is can also claim damages for any loss he suffers from anyone who culpably uses it. The 'right to one's own picture' is protected by §§22ff. of the Law of Artistic Creations of 1907: 'pictures' may be circulated or displayed only with the consent of their subject, if alive, or of his relatives. A special exception was made for figures of contemporary history, publication of whose picture is generally permitted without consent unless it infringes a 'justified interest' of the subject, as would be the case if he were photographed against his will in his private surroundings and the public had no proper interest therein. Finally, one may invoke the laws protecting literary creations and copyrights. It was on this basis that the publication of confidential letters by Nietzsche was enjoined, since the Reichsgericht found that in these letters Nietzsche had shown himself to be an 'artist in correspondence'. By contrast, some letters of Wagner's remained unprotected since they were simply 'business letters'.[27] It is undeniable, however, that the publication of business letters may constitute an invasion of the writer's privacy.

Many writers found this state of the law unsatisfactory. They argued that all these cases involved different forms of attack on one and the same legal interest, namely the human personality in all its manifestations, and that one should therefore protect the general right of personality just as much as the right to one's name or picture or artistic or literary creations. This view was never accepted by the Reichsgericht because the judges felt that such a right would be 'a concept unamenable to precise definition'.

Only after the Second World War did German courts take the crucial step. The need for effective protection of human dignity and personal freedom had become abundantly manifest during the Nazi dictatorship, and the Basic Law, unlike any previous German constitution, gave these values a prominent and important position in arts. 1 and 2. In addition, technological advances had made it much easier to invade the private sphere: secret surveillance devices were no longer a matter of science fiction. It had also been realized that sensational press reports on the private life of individuals might constitute an invasion of their personality even if the circumstances reported were true in fact and not defamatory in effect. The ground was therefore prepared for the path-breaking decision of the Bundesgerichtshof in 1954, when for the first time it recognized the general right of personality as an 'other right' under §823 par 1. BGB.[28]

The suit was brought by the attorney of Dr Schacht, who had been Economics Minister under Hitler, against a weekly periodical which had published an article objecting to Dr Schacht's founding a bank. Dr Schacht had instructed the plaintiff to write to the periodical and demand that certain corrections be made in the article. The plaintiff's letter made it perfectly clear that he was writing in his capacity as Dr Schacht's attorney, but the defendant left out parts of the letter and

[27]*RGZ* 69, 401; *RGZ* 41, 43.   [28]*BGHZ* 13, 334.

published the rest in its column of 'Readers' Letters', which made it look as if the plaintiff was supporting Dr Schacht of his own accord and as a private individual. The plaintiff claimed that the defendant be made to correct this misrepresentation by a suitable retraction, and his claim succeeded at first instance.

The reasons given by the trial court were in line with the decisions of the Reichsgericht: by the manner in which the defendant had published the plaintiff's letter it had brought him into disrepute and lowered him in the public estimation; the defendant was therefore guilty of defamation (§187 Criminal Code) and accordingly liable under §823 par. 2 BGB. The court of appeal, however, felt unable to hold that the publication of the letter defamed the plaintiff, so it dismissed the claim. The Bundesgerichtshof allowed the plaintiff's appeal and said that it was immaterial whether the defendant was guilty of a crime or not. Founding on the Basic Law, arts. 1 and 2, it held that the general right of personality as such was protected by private law: to publish the letter in mangled form as was done here, and thus to present a false picture of the author's personality, constituted an invasion of 'the private sphere of the author as protected by the right of personality'.

Other decisions in quick succession used the law of tort to protect other aspects of the right of personality. The courts held it actionable to use the name of a famous artiste in an advertisement, to publish a picture under circumstances which suggested that the person portrayed was a murderer, and to publish a factitious interview with Princess Soraya, a figure well known in international society. It was also held tortious to make a secret recording and to communicate a confidential medical certificate to a third party, subject to the qualification that such behaviour might be justified if it served an overwhelming interest such as the discovery of a serious crime.

In the case of inaccurate, incomplete, biased, or defamatory newspaper reports, balancing the values of the right of personality on the one hand and the freedom of the press on the other is particularly delicate. Here the courts have always recognized that, in relation to politicians, publicists, and other persons in the public eye or 'questions of public importance', the press is entitled to go in for very severe and even one-sided criticism, whereas much sterner standards will be applied to the gratuitous publication of private or family affairs with names attached.

A plaintiff who has established that his right of personality has been invaded is entitled to claim that the invasion be counteracted, as by the retraction of a defamatory statement, and that it not be repeated. Furthermore, if the defendant was at fault, the plaintiff may claim damages for any material harm he may have suffered. For a long time, however, the plaintiff could not obtain damages for harm to his feelings or for any other non-economic harm, since §253 BGB unequivocally lays down that damages in respect of such harm may be ordained only in the cases prescribed by law, principally for corporeal injuries. Since 1958, however, the Bundesgerichtshof has disregarded the terms of this provision.

This happened in the famous case of the 'gentleman rider'.[29] A photograph of the plaintiff, a well-to-do businessman, which had been taken when he was riding in an equestrian tournament, was used without his consent to advertise a product of the defendant which was supposed to improve sexual potency and was popularly believed to do so. The court of appeal based its award of DM10,000 on the theory that he had suffered *pecuniary* harm: he could have charged a sum of money for his consent to the publication, which had not been paid for. The Bundesgerichtshof rejected this theory, since the plaintiff, given his social position, would not have consented to the publication under any circumstances. Nevertheless, it upheld the decision of the court of appeal and awarded the plaintiff the money as damages for moral harm, notwithstanding §253 BGB. The underlying reasoning has varied somewhat in subsequent opinions, but in essence it is that the protection of the right of personality by private law would be 'patchy and inadequate' unless the defendant had to reckon on a sanction that properly reflected the seriousness of his behaviour and its consequences.[30] It is now standard practice for the court to award damages for moral harm if the invasion of the plaintiff's right of personality is 'grave'; here the motive of the defendant, the seriousness of his fault, and the mode and extent of the harmful invasion are relevant considerations.

---

[29]*BGHZ* 26, 349.
[30]*BGHZ* 35, 363; BGH *NJW* 1965, 685; BGH *NJW* 1971, 698.

10

The Law of Property

I FUNDAMENTALS

1 Property Law in the BGB

The rules on ownership and property rights are contained in Book Three of the BGB, with the same treatment being given to moveables and immoveables, so far as possible, and the provisions applicable to both being 'factored out'.[1] Land is defined as a portion of the earth's surface as delimited in the cadastral register, rights and dealings in which must be entered in the Grundbuch. Apart from land, the kinds of property are chattels and rights. Note that the rules applicable to land are extended so as to apply to buildings and all fixtures as well.

Pride of place is given to the rules on ownership of land and chattels, its protection, and its acquisition. Then there are real rights, such as usufruct and servitudes, including rights of way, rights to take timber, and so on. Security rights form the third large group, pledges of moveables and the land securities of mortgage (Hypothek) and rent-charge (Grundschuld).

Book Three of the BGB is largely self-contained, supplemented only by the Grundbuchordnung (Ordinance on the Land Register) and a few ancillary laws (see Section IV 1 below). Special enactments make it possible to create security rights in ships and airplanes analogous to those on land.

2 The Basic Principles of Property Law

(a) In German law real rights, such as ownership, and the lesser real rights, such as pledges and servitudes, have quite different characteristics from the personal rights that arise from obligations. In essence, an obligational contract is made by two people and they alone are affected by the rights and duties it engenders; the obligational bond is relative. By contrast, the

[1] For this notion, see Chapter 4 Section IV 3 above.

owner has no direct relationship with any other person, and is thought of as having a right *in* the thing, a right of dominion, a right protected against everyone, and therefore called an absolute right; as such, it is protected by the law of tort and by special rules in the law of property. The same is true of other real rights. The distinction between such a right and a personal right such as we find in the law of obligation should thus be clear.

(b) Herrschaftsrechte, as absolute rights are sometimes called, must be respected by the whole world, especially the leading example, ownership. But such respect cannot be exacted unless there is some means of knowing the standard content of property rights, so the law of property cannot afford the same profusion of forms as exists in the law of obligations, where the parties are free to fashion their contracts as they choose. There is therefore a *numerus clausus*, a limited number of possible kinds of property right. The different kinds of right that the legislator has made available to the citizen are exhaustive, and individuals cannot add to them by creating a new one. Two exceptions will be considered later: the Sicherungseigentum, or security title, and the Anwartschaftsrecht, or inchoate ownership.[2]

(c) The relationship between real rights and the rights and duties that arise from obligations was indicated in our earlier analysis of the legal transaction of sale:[3] a duty is created by the law of obligations, here contract, and is performed by the transfer of real rights. The obligational contract, created by offer and acceptance, generates duties to deliver and pass ownership in the property, on the one hand, and to pay the purchase price on the other. These duties are fulfilled by means of real transactions, conveyances of the property and of the money. The legal transactions of transfer belong to property law and not to the law of obligations, from which they are kept distinct by the principle of abstraction.

(d) In contrast with rights arising from obligations, real rights are *divisible*. This means that their attributes can be distributed between various holders; this is done by imposing one real right on another, the leading instance being the encumbrance of ownership with a security right. When several

---

[2] On this see Sections V 1 and 3 below.
[3] See Chapter 4 Section IV 4 above.

persons have rights in the same thing, the law of property has the function of determining their respective entitlements. Such encumbrances are found principally in land law, where protection is guaranteed by entry in the Grundbuch.

3 Possession and Ownership

(a) Book Three of the BGB treats of possession as well as ownership and real rights. Possession is regarded not as a right in itself, but as a physical relation to a thing, namely the exercise of factual control over it, and it is protected by the BGB against invasion and infringement by third parties, regardless of the possessor's title. If a person's possession is groundlessly challenged, or if he is dispossessed without justification, he may retain or retake the property by force, if need be, or else by legal process (§859 BGB).

The practical importance of these provisions lies mainly in the protection they afford the possessor against noxious intrusions (Immissionen). If the tenant of a dwelling is badly bothered by gas, smoke, noise, or heat, this is treated as a disturbance of his possession, and he may demand the cessation of the disturbance simply as possessor, without having to have recourse to the owner.[4]

(b) With regard to moveables, possession also has a publicity function: it is supposed to be an outward manifestation of ownership such that third parties can rely on it. If a possessor of a thing sells it and says he is owner when he is not, the buyer may still become owner by *acquisition in good faith* (§932 BGB), provided that the thing has not been stolen from the true owner. More on this fundamental principle of German law will be found in Section III 2 below.

The application and flexibility of the rules on possession have been greatly enhanced by the statutory concept of the mediate or indirect possessor. If the person exercising factual control over a thing is doing so on behalf of another under a contract such as lease or deposit, he is the immediate possessor, but his contractor is also treated as possessor — the mediate or indirect possessor (§868 BGB). It must be said that this extension has virtually destroyed the usefulness of possession as a publicity

---

[4]On the owner's claim for an injunction, see Section II 3 below.

device, for nowadays it is almost impossible to infer from the appearance of possession where ownership may lie.

II OWNERSHIP

1 Freedom and Duty: The Two Poles Of Ownership

(a) Ownership is much more that a form of property right: its economic and political significance is vast. The words of Blackstone are as true as ever in Germany today:

> There is nothing which so generally strikes the imagination, and engages the affections of mankind, as the right of property; or that sole and despotic dominion which one man claims and exercises over the external things of the world, in total exclusion of the right of any other individual in the universe.[5]

The BGB starts from the same viewpoint; according to §903 BGB, the owner of a thing may act with it as he chooses and exclude others from all effect on it. But this is immediately qualified by the proviso of legality, or Gesetzesvorbehalt, as it is called: the owner's freedom must yield to any conflicting law and to the rights of third parties. There has hitherto been an unbroken tendency to tighten the legal constraints on the owner's freedom to deal with his property as he thinks fit.

(b) The Basic Law guarantees ownership, and the right of succession to it (art. 14 GG).[6] But it contains not only the proviso whereby statutory restraints may be imposed but also the words 'Eigentum verpflichtet' (ownership entails duties): ownership should be exercised for the general good as well. There is thus a certain tension between the legal protection of ownership and its social obligations.

Ownership is constitutionally guaranteed because it provides the physical basis and material means for the free development of the individual and the responsible conduct of his life. [7] It therefore needs to be constantly redefined and delimited, even today.

---

[5]Blackstone II *Commentaries on the Laws of England* 2 (London 1765–8).
[6]Badura, 'Eigentum im Verfassungsrecht der Gegenwart', *Sitzungsbericht T zum 49. Deutschen Juristentag, 1972 (Closing Address)*; Sendler, *'Zum Wandel der Auffassung vom Eigentum'*, DöV 1974, 73f. See also the commentaries on the Basic Law, art. 14.
[7]*BVerfGE* 24, 367, 389.

The fact that ownership is a constitutional right means that it is protected against the state. This is especially important with regard to expropriations [8] for such purposes as building highways and constructing electric grids. The rules are partly contained in public law enactments, providing for fair compensation, and partly taken from general principles. Quite apart from these invasions, which are permissible only against compensation, there are social constraints on property which do not amount to expropriation. Here the courts, in determining whether compensation is payable, use the idea of 'special sacrifice' (Sonderopfer)[9] exacted from the individual, though they sometimes base their judgment on the extent and seriousness of the invasion.[10]

Discussion continues over the principle and extent of freedom of private property, especially in the area of the means of production, regarding which the Bundesrepublik, unlike the German Democratic Republic, has not accepted the idea of special public ownership.[11]

[8]Ossenbühl, *Staatshaftungsrecht* (2d ed. Munich 1978); Bender, *Staatshaftungsrecht* (2d ed. Karlsruhe 1974).
[9]*BGHZ* 23, 30, 32 (the inclusion of a plot of land in the green belt constitutes a prohibition of building, but no compensation is payable if the prior use was agricultural); *BGHZ* 30, 338, 341 (roadblock of normal duration does not amount to expropriation; it is otherwise if it is unduly prolonged). On this see Bender, 'Sozialbindung des Eigentums und Enteignung', *NJW* 1965, 1297f.
[10]*BVerwG* 7, 297 (duty to rebuild house destroyed in war).
[11]See the Civil Code of the German Democratic Republic of 19 June 1975 (Gesetzblatt der Deutschen Demokratischen Republik 1975, 465). Section 18 is entitled 'Socialist Property' and reads:
(1) Socialist property includes the property of the people, the property of socialist co-operatives, and the property of social organizations of citizens.
(2) The property of the people forms the basis of socialist relations of production, and is to be used and enhanced in accordance with the needs of society and the principles of the socialist planned economy. The socialist state controls the use and enhancement of the property of the people through the agency, *inter alia*, of people's enterprises, combined operations, dirigist organs, state departments and agencies, socialist co-operatives, and social organizations as well as citizens.
(3) As directed and planned by the state, the property of socialist co-operatives is used to fulfil their economic roles, to perform their duties towards socialist society and to shape the working and living conditions of the members. The rights arising from co-operative property are vested in the co-operative.
(4) The property of social organizations is used to fulfil their political, social, scientific, cultural, and other tasks. The rights arising from such property are vested in the social organizations, which must use them in accordance with their goals.
On this see Westen, *Das neue Zivilrecht der DDR* 70f. (Berlin 1977). On the discussion in West Germany see Pittner, 'Öffentlich-rechtliche Elemente der Unternehmensverfassung', in V *Planung* 59f. (Baden-Baden 1971).

2 The Protection of Ownership against Dispossession

(a) The BGB is concerned only with protecting ownership as a matter of private law. As an absolute right, ownership is very well protected by the law of tort, which gives a claim against any third party who adversely affects it (§823 par. 1 BGB).

The owner can demand the return of his thing from the person in possession of it (§985 BGB), unless the possessor can show some right to possess emanating from the owner himself (§986 BGB). This applies equally to moveables and to land. This *real action*, the *rei vindicatio* of Roman Law, is an invariable attribute of ownership. It reaches only the specific property, not any substitute or surrogate.

(b) After providing the owner with a claim for possession, the BGB deals in detail with the relationship between owner and unjustified possessor (§§987ff. BGB), allocating the risk of loss or damage, and regulating entitlement to profits and liability for expenditures.[12] The rules for the owner–possessor relationship are self-contained, and with few exceptions exclude those of tort and unjustified enrichment.

3 Protection against other Invasions

The real action conferred by §985 BGB is of use only in cases of dispossession; other invasions call for different remedies. Such invasions may take various forms: people may trespass; steam or gas or noise may be a nuisance (§906 BGB); and the construction of a building on the neighbouring plot may cut off light or interfere with television reception. If such invasions cannot be justified, the owner can demand that they cease (§1004 BGB), and an injunction may lie if future invasions are feared. This provison, which protects all absolute rights, has been applied very extensively by the courts;[13] suit may be

---

[12] For example, suppose that G sells and transfers a motor-car to K, and that both legal transactions are void because G is mad, but not noticeably so. An owner–possessor relationship will arise between G and K. G can claim back the car under §985 BGB. If K has damaged the car by negligence, he will be liable if he knew of his defect in title (§§989, 990 BGB). He must disgorge any profits he has made from the use of the car (§§987, 988 BGB), but may claim compensation if he replaces worn tyres with new ones (§994 BGB).

[13] von Caemmerer, 'Wandlungen des Deliktsrechts', in II *Hundert Jahre deutsches Rechtsleben, Festschrift zum hundertjährigen Bestehen des Deutschen Juristentages* (ed. von Caemmerer, Friesenhahn, Lange) 49f. (Karlsruhe 1960).

brought at the first threat of invasion, and there is no need to prove fault.

But here, too, the owner's power to exclude must be qualified by his duty to forbear; especially between neighbours there arise duties to act with mutual consideration. In this area there is also much regulation by public law, notably planning and environmental law.

III MOVEABLES

1 Acquisition by Agreement and Delivery

(a) The rules on the transfer of ownership in moveables have a prominent place in the BGB and still play an important role in practice, especially when title is transferred or reserved for security purposes, as will be seen in Section V below.

In order to transfer a moveable there must be, first, agreement between owner and transferee on the transfer of ownership and, second, delivery of the thing (§929 BGB). The conveyance is thus *bipartite*. The agreement, being a legal transaction modelled on contract, is also called a real contract, and in general it causes fewer problems that delivery.

(b) The second requirement for transfer of ownership — actual delivery of the thing — is in line with the needs of daily intercourse. The Code makes things easier by permitting substitutes for delivery. Thus there is one form of transfer in which the party effecting the transfer can retain possession of the thing, a device that has proved astonishingly useful for chattel mortgages (see Section V 3 below): the act of delivery is replaced by an agreement that a relationship of indirect possession should exist (§868 BGB). In such a relationship, which may arise under a deposit, lease, or loan, the acquirer becomes indirect possessor and the transferor retains direct possession. But to allow actual delivery to be replaced in this way by the creation of such indirect possession is to lose all outward indication that a transfer has taken place. The device has therefore been criticized as inconsistent with the publicity function of possession, and it must be admitted that it comes close to circumventing the bipartite nature of conveyance.

2 Good Faith Acquisition

Acquisition in good faith is one of the important legal institutions that the BGB took from Germanic, as opposed to Roman, law. A transferee may acquire ownership of a thing even if the transferor did not own it. For this, the acquirer must believe the transferor to be owner: his 'good faith' is his ignorance of the transferor's want of title. In addition, the acquirer must obtain possession of the thing from the transferor (§932 BGB). Finally, the thing must not have been stolen from its true owner (§935 BGB). If these preconditions are satisfied, the good faith acquirer from a non-owner acquires a fully valid and unimpeachable title. This involves the extinction of the true owner's title, so the rules of unjustified enrichment are applied as between him and the non-owner who effected the transfer, with the result of requiring the latter to give up to the former the proceeds of the sale (§816 BGB).

This accords with the old German saying 'Hand wahre Hand', which means that the original owner can look only to his bailee, the person who made the unauthorized transfer. The protection of commerce requires that priority be given to the acquirer in this way. Many problems arising out of unauthorized transfers are simply and swiftly solved by this principle of acquisition in good faith, but it must be noted that the true owner remains protected if he has been the victim of theft or similar form of dispossession, since in such cases no good faith acquisition is possible, even at third or fourth remove (§935 BGB).

3 Other Modes of Acquisition

(a) There are other ways to acquire ownership in moveables than acquisition by the legal transaction of agreement and delivery. Some of these rest on the view that a single thing should, so far as possible, have only one owner. If the oil delivery man, instead of putting the oil into his customer's oil tank, puts it into that of a third party, ownership in the oil is not transferred to the third party for want of agreement to that effect. But once the oil put into the tank is inseparably mixed with the oil that was already there, separate ownership is impossible. Either the parties own the oil jointly, or the third party owns it all, including what was delivered (§§947, 948 BGB). The rights of the parties can be adjusted by the rules of unjustified enrichment (§951 BGB). The same reasoning applies when coins or notes are put in a money box and can no longer be distinguished. Similar questions are raised by consolidation, when different objects are blended together so as to form a new thing.

(b) A change in ownership may also arise when a thing is *processed*, where the value of the work exceeds the value of the material (§950 BGB). The artist will own the painting even if the canvas and paints belong to others, but he remains under a restitutionary obligation to pay them for divesting their title. Improvements effected by industrial or craft processes are normally made under a contract that regulates these matters.

(c) The acquisition of fruits, such as standing timber or corn (§§953ff. BGB), is another matter covered in this context. This is the converse of consolidation. Unless there is a contract that vests them in someone else, the owner of the matrix becomes owner of the fruits when they are separated. *Prescription* is another mode of acquisition: the possessor who believes he is owner becomes owner in fact after the lapse of ten years (§937 BGB). Here, too, are the rules about finding, for if the finder notifies the police of his find and the true owner does not claim it within six months, he acquires 'new' ownership in the thing (§§965ff. BGB).

IV LAND

1 Basic Changes since the BGB

(a) Ownership of land was the source of wealth and prestige in the days when the economy was primarily agricultural, and even in the industrial age it is still held in high esteem. Land is normally indestructible, and its supply is limited and incapable of extension. The law relating to land, along with succession and the family, is the oldest part of all legal systems, but it is not only among lawyers that transactions in land continue to excite attention.

(b) Book Three of the BGB contains the technical rules for the acquisition and encumbrance of land. The assumption is that if one wants land to build or live on, one really must own it.

The BGB had no truck with different levels of ownership or long-term leases, but social developments and increased demand for private property made it necessary to supplement the Code in certain important respects. The Ordinance on Heritable Building Rights (Erbbaurecht),[14] enacted in 1919 after the First World War, made it possible for a landowner to burden his land by granting someone else a real right to build on it. The holder of such a heritable building right is considered as owner of the land for the period of its validity, and the true owner, as 'naked owner', receives an annual payment (Erbbauzins), which may be indexed. A heritable building right cannot last for more than 99 years, and in practice they are often granted for shorter periods such as 66, 50 or 33 years. On the expiry of this period, the land reverts to its owner and he must pay for the improvements.

(c) A further step in the same direction was taken after the Second World War with the Wohnungseigentumsgesetz (Law on Home Ownership).[15] This enactment made it possible to own a storey or part of a storey in a building, a concept which the draftsmen of the BGB had expressly rejected for fear of confusion and conflict. The owners of the individual apartments or divisions of the building are treated as co-owners of the whole plot, with co-ownership of the portions in common use (entrance hall, gardens) and special ownership in their own dwellings. The owners meet together and elect administrators to look after the common conveniences. The initial forebodings about the drawbacks of this new form of ownership have proved groundless. Both additions to the BGB — the heritable building right and the ownership of apartments — have been welcomed in practice and widely employed, especially in urban areas since the Second World War. They have contributed to the restructuring of the provision of dwellings and have made it much easier for people to acquire a home of their own.[16]

---

[14]Verordnung of 15 January 1919. This repealed §§1012–17 BGB, which had contained a hereditary building right which operated only as a matter of obligation and which was of no importance.

[15]Gesetz über das Wohnungseigentum und das Dauerwohnrecht of 15 March 1951, *BGBl* I, 175.

[16]Law of 23 June 1960, *BGBl* I, 341. Schneider-Kornemann, *Soziale Wohnungsmarktwirtschaft* (Bonn 1977).

2 Acquisition of Ownership in Land

(a) We have already seen how German law distinguishes between the real transaction by means of which the contract of sale is performed and the obligation to which the contract gives rise (principle of abstraction). In practice, however, contract and conveyance are closely allied. Even for the obligational contract there are requirements of form: every sale of land must be fully notarized (§313 BGB). In the case of building land or land already built on, the local authorities have a public law right of pre-emption under the Bundesbaugesetz (Federal Building Law),[17] and a declaration must be made before ownership is transferred. For agricultural and forest land, approval must be obtained under the Grundstücksverkehrsgesetz of 1961 (Law on Dealings on Land),[18] which seeks to control speculation in productive land. Furthermore, if none of the many exemptions applies, a tax of 7 per cent of the purchase price (Grunderwerbssteuer) is payable to the treasury before the transfer takes place.

(b) If these requirements are satisfied, the real transaction may proceed. Ownership is transferred by *agreement* and by *entry in the Grundbuch* (§873 BGB). This principle applies generally to all rights in land, and is analogous to the rule for moveables where the two components of the conveyance are agreement and delivery. There is an old German term, Auflassung, for the agreement to transfer land. This also must be fully notarized (§925 BGB), a more solemn formality than is required for the transfer of other rights in land, where it is enough that the notary certify the signature. Here we see a trace of the importance that used to be attributed to the transfer of ownership of land as a matter vitally affecting family property.

The second requirement, entry in the Grundbuch, leads us into a new area.

---

[17]Law of 23 June 1960, *BGBl* I, 341. It deals with planning permission and the construction of buildings, and is supplemented by the building laws of the several Länder.

[18]Gesetz über Massnahmen zur Verbesserung der Agrarstruktur und zur Sicherung land- und forstwirtschaftlicher Betriebe of 28 July 1961, *BGBl*, I, 1091, 1652, 2000. (Law on Provisions for the Improvement of the Agronomy and for the Safeguarding of farming and forestry enterprises).

3 The Grundbuch

(a) The Grundbuch is a land register with special features which render it remarkably efficient. The relevant part of the earth's surface being limited, and thus susceptible of exact measurement, it is in fact measured and mapped and divided into portions which are numbered as separate lots. All these individual plots into which the territory of the Bundesrepublik is thus divided are entered in the *cadaster*, as it is called, which is kept current whenever plots are affected, for example, by road construction, division, or consolidation. The boundaries of each plot of land are clearly marked on the maps. In each Land the survey is entrusted to the Landesvermessungsamt, which has subordinate surveying groups in the localities.

(b) The Grundbuch builds on the cadaster and shows the legal status of every numbered plot of land, whose boundaries are precisely described in the cadaster and marked *in situ* by numerous boundary stones placed there by the official survey. In addition to the area and location of each plot, the Grundbuch, in Part I, lists current and previous owners. It also lists any third parties who have rights which burden the land. These encumbrances are listed in Part II, save for land securities, mortgage, and rent-charge, which are listed separately in Part III. This permits one to see clearly which overlapping encumbrances have priority (§1179 BGB), a matter of great importance. Priority depends on date of entry, subject to modification if all parties make a contract to that effect. A first-charge mortgage is very much more valuable than one in third or fourth place, since a first mortgage is much more likely to be satisfied if the property has to be sold.

(c) The Grundbuch is operated by the Amtsgericht for each jurisdictional area. The procedure is laid down by the Grundbuchordnung (Land Register Ordinance),[19] which specifies the procedural and substantive preconditions of registration, and their proper form. Entries are made only on application, any alteration or cancellation requiring the consent of the person whose name is registered. The necessary documents must be notarized, or at least bear signatures witnessed by a notary.

---

[19]Grundbuchordnung of 24 March 1897, *RGBl* 139, as promulgated on 5 August 1935, *RGBl* I, 1073.

This may make the proceedings rather cumbrous, but it makes the Grundbuch more reliable.

4 The Grundbuch in Operation

(a) The crucial feature of the Grundbuch as a register is the principle that any proposed change in the legal position *must* be entered in it in order to produce the desired effect: registration has constitutive effect. Without registration a registrable legal transaction is ineffectual: the acquirer does not become owner until he is registered as such in the Grundbuch. What one loses by such dependence on the Grundbuch one gains in its dependability: in the result the Grundbuch gives an almost perfect reflection of the current legal status of any plot of land.

Divergence arises only when a registered owner's title comes to an end otherwise that through a legal transaction, for example, on the death of an individual or on the liquidation of a company. In these exceptional cases where a legal situation alters without alteration of the Grundbuch, the Grundbuch must be corrected. In the case of death the heir has a claim to this effect.

(b) The constitutive effect of registration applies not only to ownership of land but to all other real rights as well. Usufruct, for example, a real right which entitles the beneficiary to all the proceeds or income of the property (Niessbrauch; §§1130ff. BGB) take effect only on registration; this institution was quite commonly used by small farmers on giving up their farms, but nowadays it is rather rare. Registration is also required for the creation of *servitudes*, such as rights of way, rights to cut wood or pasture animals, rights to fetch water across another's land, way-leaves for electricity pylons or oil pipelines, and even restrictive building covenants, which are very common. The same is true for land securities, mortgage, and rent-charge, which are dealt with in detail in Section V 4 below. On the other hand, public liens for rates and taxes do not need to be registered. The wide range of registrable matters renders the Grundbuch extremely informative and reliable.

(c) This makes it possible to credit the Grundbuch with a presumption of accuracy, and so to protect those who act in reliance upon it (§891 BGB). It is a very important feature of land law that those who acquire property in good faith on the basis of entries in the Grundbuch should have this protection (§892 BGB). A person who acquires property from the person registered as owner in the Grundbuch can rely on it unless he has positive knowledge that his transferor's title is defective: he will acquire full ownership even if the registered owner's own acquisition was ineffectual. The newly registered owner benefits from the presumption that the Grundbuch is accurate, and becomes full owner with a new title (§892 BGB). If an encumbrance has been erased in error, so that no encumbrance appears in the Grundbuch, though there actually is one, the transferee takes free of it. Every real right in land, not just ownership, is susceptible of such acquisition in good faith. Thus mortgages or rent charges that have actually been paid off may be so acquired, if the Grundbuch is silent about the fact. Conversely the mortgagee is entitled, unless he knows the truth, to treat the registered owner as the right person to receive any repayments or the like (see Section V 2 below).

In all cases of good faith acquisition on the basis of the Grundbuch, the resulting economic imbalance is resolved by the rules on unjustified enrichment (§816 BGB). The person who loses out has recourse against the person who effected the disposition, never against the person relying on the Grundbuch.

V SECURITY RIGHTS

1 Security in the Law of Obligations

(a) Giving security for the performance of an obligation is not a topic that receives unitary treatment in the BGB: the Code offers no basic concept of a security right and does not regulate the interrelationship of the different kinds of security right. Nevertheless, scholars have been able to extrapolate a general concept of security right from the principles that underlie the different institutions contained in the Code. This has proved capable of effective application to the newly developed forms of these important features of everyday life.

(b) In the law of obligations in the BGB the principal form of personal security is suretyship, an old-established and proven institution (§§765ff. BGB). The undertaking to act as surety must be given in writing (§766 BGB), and it renders the surety personally liable with all his assets for the fulfilment of the principal debtor's obligation. In principle, the surety is liable only when the creditor has sued the principal debtor to judgment and has attempted to collect on the judgment, but it is normal for the surety to waive the defence that the principal debtor has not been sued first, and then he may be held liable immediately and directly. The surety's obligation depends on the legal status of the underlying debt, and is hence described as accessory.

Although personal security is still in common use, it has diminished in importance as other forms of security have developed.

(c) Another very common form of security is associated with sales on credit. This is Eigentumsvorbehalt, or reservation of title.[20] The BGB provides in §455 that the vendor of property on credit may retain ownership until the purchase price has been paid in full. As a technical matter, this is done in the form of a suspensive condition. The purchaser of the goods receives delivery right away, with ownership to follow later when the balance of the purchase price is paid; when the final payment is made, the condition is satisfied and the purchaser becomes owner without the need for any further conveyance.

This form of security, whereby ownership of a thing, separated from possession, is used as a means of securing payment of the balance of the purchase price, arises as a result of a clause in the contract of sale, which must be express. This is now extremely widespread; the clause is found in almost all credit sales.

Reservation of title has led to further developments. At first, the purchaser obtained no interest in the property until final payment of the purchase price, when he suddenly acquired full ownership. Then people began to realize that he deserved some protection at an earlier stage, when perhaps half of the purchase price or more had been paid. This led to the recognition of Anwartschaftsrecht,[21] or inchoate ownership, the buyer's expectation of acquiring title. It is treated as a real right of an independent variety, 'like ownership, only less'. Anwartschaftsrecht

[20] Serick, *Eigentumsvorbehalt und Sicherungsübertragung* (4 vols.) (Heidelberg 1963–76); Weber, *Sicherungsgeschäfte* (2d ed. Munich 1977); *BGHZ* 42, 53.
[21] *BGHZ* 20, 88.

has broken through the barrier of the *numerus clausus* of property rights, and has achieved general recognition. Inchoate ownership can be transfered or even pledged; it works just like ownership. But since it is dependent on the eventual fulfilment of the contractual condition it also depends on the continuing validity of the contract of sale; it would disappear if the contract were rescinded or cancelled. Inchoate ownership can even serve as security for credit in much the same way as a security title (see Section V 3 below).

2 Security in the Law of Property

(a) The most important security rights in the BGB are in the law of property. These real securities (Realkredit) still predominate, especially mortgage (Hypothek, §§1113ff. BGB) and rent charge (Grundschuld; §§1191ff. BGB). Together these constitute land securities (Grundpfandrechte), creating a real right capable of securing an existing or future debt. The beneficiary can ultimately have recourse against the burdened land; this is done by forced auction, the rules for which are laid down in a special statute.[22] Mortgage and rent charge differ in that the validity of mortgage depends on the legal status of the secured debt, while rent charge does not, but the two forms have grown so alike that in practice they are virtually interchangeable. Land securities must be entered in a special part of the Grundbuch in order to acquire validity (constitutive effect). In this way priority between different land securities is fixed and promulgated, which is of vital importance in practice.

Land securities also enjoy the presumption of accuracy that attaches to the Grundbuch: a registered mortgage or rent charge may be acquired in good faith from the registered holder, even if the right is actually vested in someone else or does not exist at all. The absence of registration is equally reliable: if one acquires the land in good faith, one will acquire it unencumbered by any land security that may actually exist. If the owner of property who seeks to pay off the mortgage pays the person who is registered as mortgagee, his liability is extinguished even if the mortgage has since been transferred to a third

---

[22]Gesetz über die Zwangsversteigerung und Zwangsverwaltung as promulgated on 20 May 1898, *RGBl*, 773.

party. When good faith is protected in this way, the consequences call for some adjustment; as in the case of moveables, this is done by the rules of unjustified enrichment (§816 BGB), operating exclusively between the party who has been deprived of his rights and the party who so deprived him, leaving the good faith acquirer entirely unaffected. The constitutive effect and comprehensiveness of the Grundbuch as a register have made dealing in real securities very sure and clear, so that, although there is a great deal of business in real securities, very few problems arise.

(b) The law of property also provides for security rights (Pfandrechte) in moveables and in rights. This institution is no longer very important so far as moveables are concerned: pawn shops that make cash loans against pledges are picturesque rather that prominent, and while the so-called Lombard transaction involves the creation of security rights by pledging negotiable instruments, most of the applicable rules come from the general conditions of business of the participating banks. Detailed and exhaustive though they are, the rules on pledge in the BGB (§§1205ff. BGB) have proved to be too inflexible and restrictive. In practice, therefore, pledge is replaced by other forms of security, some of them quite recent. The real problem which triggered these developments was that no pledge could be created unless the pledgee took the object into his immediate possession.

3 Ownership as Security

(a) Ownership in a moveable is more easily acquired than a pledge interest, since it is possible to transfer ownership and retain immediate possession: instead of delivering the thing, one creates a relationship of mediate possession (§868 BGB). For example, if a person's only collateral is something he needs to use in his business, such as a craftsman's tools or a hotelier's inventory, he cannot possibly create a pledge, for that would require him to hand the property over to the lender, but he can easily transfer ownership, because he can then keep the things in his possession. All that is required is to make an agreement, such as deposit or loan, whereby the transferee obtains indirect possession (§930 BGB), and the person creating the security interest can continue to use the thing. This mode of transfer has been used for the purpose of giving the transferee security for the performance of the obligation due to him. This security title (Sicherungseigentum) has exactly the same function as pledge, without any need to give the creditor direct possession. It was objected that to use ownership in this way constituted an evasion of the rules which made it impossible to create a pledge interest without giving up possession,[23] but the courts made short shrift of these scruples, and nowadays security title is fully accepted.

(b) Security title is based on a 'true' transfer, which to all appearances confers full ownership on the secured party, usually a bank. As between the parties to the transaction, however, there is a relationship of trust which limits the powers of the transferee, especially with regard to dispositions, which would be inconsistent with the security purpose. When security is no longer required, because the loan has been repaid, the security title must by reconveyed, or it may revest automatically in the transferor at the end of the contract without any further conveyance.[24] On the occurrence of the event against which security was taken, that is, if the debtor fails to fulfil his obligations, the secured party is entitled to sell the thing. The relevant rules come partly from the statutory rules on pledges, applied by analogy, and partly from the contract pursuant to which the security title was transferred. Such contracts are not regulated in the Code, but the courts have pretty well fixed their content by now.[25] With the invention of security title, a much more flexible device than the old pledge which it has ousted, ownership has gained a new function, for which it is perhaps formally too powerful.

(c) In practice, security titles are often employed in commerce (contents of warehouses), in industry (machinery and means of production), and even in the domestic sphere (chattel mortgages of automobiles, television sets, etc.). The drawback is that, as such security titles may be created without any outward evidence, creditors can no longer ascertain which of the moveables in a debtor's possession belong to him and which do not.

[23] *RGZ* 132, 186; so already *RGZ* 113, 57.
[24] See the review by Serick (above n. 20) III. 394ff.
[25] Coing, *Die Treuhand kraft privaten Geschäfts* (Munich 1973); *BGHZ* 28, 20.

4 Security Assignments

(a) Another new form of security, the assignment of rights, performs much the same function as security title. In such a security assignment the debtor assigns his business claims to the secured party, but only as security, not as an out-and-out assignment. If the debtor fails to fulfil his obligations, the secured party can collect on these rights; in other words, the assignment in such a transaction is really taken in earnest only on the occurrence of the event secured against, for only then can the secured party collect on the claims assigned.

Both the claim assigned and the assignment that is made of it fall within the law of obligations, but there is a parallel with security title in the law of property in that, instead of being transferred outright, the right is transferred only as security. Here, too, the right is to all appearances fully assigned, but is subject to limitations arising from the inner relationship created by the security agreement.

(b) A right can be assigned by mere agreement (§398 BGB). There is no need to give notice to the debtor, who is protected against any adverse effects the assignment might have for him; for example, he will be released by payment to the original creditor, although the claim has meanwhile been assigned to a new creditor (§407 BGB). *Future* rights which have not yet arisen are also capable of assignment. The right need not be specified or described in great detail, not even by reference to its value or to the date when it is to arise. It is enough to indicate the general class of business from which the claims are to arise. Thus it is possible, for example, to assign all the rights that will result from the sale of a specified article or class of articles, or those that will arise from sales to customers with names falling in the first half of the alphabet. Nor need the total of the claims be fixed. The secured party can include in the contract a duty to notify the debtors, but the validity of the assignment does not depend on this: in fact, the assignment is commonly kept secret so as not to impair the assignor's credit. The easy flexibility with which future claims may be assigned is not without its perils. The transferor who assigns his rights to all sums to fall due in the future loses his freedom of action. Such block agreements have therefore been regarded with some suspicion by the courts, who have introduced some constraints.[26]

5 Conflicts between Security Rights

Since the Code does not treat securities as an independent area, and since the different forms of security rights have evolved separately as the needs of business required, there are few rules to apply when they come into conflict.

Such conflicts can arise through successive assignments of the same claims or successive transfers of security title in the same property. A typical pattern in practice is as follows. Goods are supplied to a trader or processor with a reservation of title, frequently supplemented by an assignment to the vendor of any claims the purchaser may acquire in the future from subsales of the goods in question. This 'extended reservation of ownership' is very common.[27] The trader or processor then gets credit from a bank, offering as his only collateral the very goods that have been delivered under reservation of title or the claims that are to arise on the resale. The bank thus obtains a security title or a security assignment. If the event secured against now occurs, and the transferor cannot fulfil his obligations, the security right of the supplier comes into headlong collision with the security right of the bank. As to the assignment of future rights, no satisfactory solution to the problem of priority has yet emerged. A proposal that the security rights should be divided between the participants is at present under discussion.[28]

---

[26]*BGHZ* 26, 190. See also Chapter 5 Section IV 2 above.
[27]*BGHZ* 7, 365; *BGHZ* 27, 306.
[28]Erman, *Die Globalzession in ihrem Verhältnis zum verlängerten Eigentumsvorbehalt* (Karlsruhe 1960).

11

Family Law and the Law of Succession

I FAMILY LAW

Despite radical changes in social and economic conditions since the BGB came into force, much of it is still textually the same as it was in 1900. There is one significant exception: Book Four, which is devoted to family law, has been profoundly altered by various enactments in the period since the Second World War.

The family law of the BGB, as originally enacted, bore all the marks of the age of the conservative and patriarchal bourgeoisie. Decisions during the marriage were for the husband to make, and it was he who exercised parental power. The rules of matrimonial property were based on the practice then prevailing among the class of officers and officials, that the spouses brought to the marriage an interest-bearing capital sum which was to be administered by the husband. The rules of divorce and illegitimacy were influenced by Christian morality. Accordingly, divorce was possible under the BGB only when the breakdown of the marriage was due to the fault or insanity of the defendant spouse, and illegitimate children were deliberately disfavoured, as compared with legitimate children, for fear that extramarital adventures, which the law discountenanced, might be legalized, and immorality and undesirable concubinage encouraged. The law treated an illegitimate child as unrelated to his father: he was fobbed off with a claim for maintenance whose amount depended on his mother's social position and which came to an end when he was sixteen.

Since the Second World War family law has been fundamentally reformed. This legislation was primarily triggered by structural changes in the family and altered views on the roles of spouses and parents, but there was a special stimulus to this development in Germany: the Basic Law of 1949 not only laid down in categorical terms that 'men and women have equal rights' (art. 3 par. 2 GG), but also provided that any legal rules which conflicted with this principle should cease to have effect as from 31 March 1953 (art. 117 GG). This date, however, passed before the legislature had made any alternative provision. There was thus a lacuna in the law, and the judges had to fill it in the meanwhile, holding, for example, that the husband had lost his right to manage and enjoy his wife's property. The judges faced special problems where the implementation of the principle of equal rights called for brand new positive rules of law, but they performed the task with success, and there are some people who find the judge-made law of this period superior to the Law on the Equal Rights of Man and Woman which finally came into force in 1957, that is, four years late.[1]

(a) One of the principal aims of this enactment was to procure equal rights for the wife in relation to *household management and outside employment*. The original rule was that it was the wife's duty to 'conduct' the household, but here, as elsewhere, the husband had the final word in case of dispute. The wife was entitled to take a job, but the husband could put an end to it if he was of the view that it was causing her to neglect the housekeeping. The Law of 1957 declared that the wife had 'full responsibility' for the household, and that, so far as consistent with her matrimonial and family duties, she was also entitled to take employment (§1356 BGB). Here, again, the 'wife as housewife' is taken as the norm, and while, as a matter of mere statistics, this is doubtless true even today, the legislator has since realized that it is inappropriate to have a statutory endorsement of such an allocation of roles. §1356 BGB was accordingly redrafted in a 'sexually neutral' version and enacted in 1976: questions of home management and outside employment are to be determined by agreement between the spouses.

(b) With regard to *parental power*, the Law of 1947 was rather unprogressive, for while it provided that parental power should be exercised for the benefit of the child as mutually agreed by the parents, and that in case of difference of opinion the parents must try to reach agreement, it gave the final decision to the father if no such agreement could be reached. It was the father, too, who had the right to represent the child in law (§§1628–9 BGB). These rules indubitably favoured the father, so the Bundesverfassungsgericht declared them void as contrary to the principle of art. 3 par. 2 GG.[2] Subsequent decisions have

---

[1] Gleichberechtigungsgesetz of 18 June 1957 (*BGBl* 1957 I 609).
[2] *BVerfGE* 10, 60; *NJW* 1959, 1482.

held that parental power and the power to act as a child's legal representative belong to the spouses jointly, and that if they cannot reach agreement on any important issue, the decision will be made by the court. This solution to the problem has worked quite well, and has been incorporated in the new Law on Parental Care.[3]

(c) The most important reform in the Law of 1957 was the introduction of a new *statutory property regime*, that of 'community of acquisitions' (§§1363ff. BGB), whose objective is to divide equally between the spouses any property that has accrued to either of them during the subsistence of the marriage. Division takes place not while the marriage subsists, but only when it comes to an end, by divorce or the death of either party: until this occurs, the parties retain their separate property. If the marriage ends in *divorce*, the spouse with the larger increase in his property must hand over half the difference. In practice this balance may not be easy to compute, since there may be disagreements over the valuation of individual items or doubts concerning what property originally belonged to whom. Such complications are avoided if the marriage ends by *death*, for there is a rule, rather a rule of thumb, to the effect that an extra quarter of the decedent's estate may be added to the statutory portion of the surviving spouse in the succession (§1371 BGB), regardless of any increase in the property of either spouse during the marriage.

Here is an example. If there is a *divorce* after ten years of marriage, when the husband's property is worth DM100,000 and the wife's worth DM20,000, neither party having had any assets at the outset, the husband must pay the wife DM40,000. If the marriage ends by the *death* of one of the spouses, intestate but not childless, the surviving spouse will receive half of the estate: because there are children, the spouse would normally receive a quarter of the estate under §1931 BGB, but a further quarter is now added under §1371 BGB.

The purpose behind this property regime of 'community of acquisitions' is to help the wife whose devotion to home and children has prevented her from earning any income of her own

---

[3]Gesetz zur Neuregelung des Rechts der elterlichen Sorge of 18 July 1979 (*BGBl* 1979 I 1061). The term 'parental care' is used in the new Law in place of the term 'parental power', which seemed too authoritarian. See §§1626ff. BGB, as amended by the new Law.

or making any savings during the marriage. There can be no doubt that in the normal case the property acquired by one of the spouses during the marriage is due to the efforts of both. But take the case where the wealth of one party has appreciated handsomely only because it was very substantial to begin with; here one may well ask whether the other party should, on divorce, be able to claim half of the increase. Again, if one member of a childless couple has worked extremely hard and profitably while the other has idled away, one may ask whether the grasshopper, on divorce, should be able to claim half the ant's remaining earnings. In both cases the legislator has given a positive answer. In practice, indeed, it is almost impossible to decide how far one party's acquisition of property has been rendered possible by the other's housekeeping, bringing up the children, outside employment, or emotional support. If an engaged couple dislikes the statutory regime of community of acquisitions, which is admittedly arbitrary to some extent, they can always agree to exclude it and keep their property strictly separate.

(d) The next important reform in family law was made by the *Illegitimacy Act of 1969*,[4] enacted pursuant to the constitutional mandate in art. 6 par. 5 GG: 'Illegitimate children shall be provided by legislation with the same opportunities for their physical and spiritual development and their place in society as are enjoyed by legitimate children' (official translation). Under the new provisions of the BGB, the illegitimate child is regarded as related to his father in law as well as in fact. The consequence is that the illegitimate child has much the same claim to maintenance as a legitimate child, and his rights of succession are also very similar: he has an equal statutory right of succession with the difference that, if his father leaves a surviving spouse or legitimate children, he has no claim to any part of the succession in kind, but only a money claim to the value of his portion of the succession (§1934a BGB).

(e) The Law of Marriage of 1976[5] deals principally with the

---

[4]Gesetz über die rechtliche Stellung der nichtehelichen Kinder of 19 August 1969 (*BGBl* 1969 I 1243).

[5]Erstes Gesetz zur Reform des Ehe- und Familenrechts of 14 June 1976 (*BGBl* 1976 I 1421).

preconditions and consequences of *divorce*. As to the *conditions* of divorce, prior law rested on the principle of fault.

Under the old rules, divorce was in principle possible only if the defendant spouse had broken the marriage bond or had committed some other grave matrimonial offence which brought the matrimonial partnership to an end. Of course, if both spouses wanted a divorce, they used in practice to agree that the complainant should allege that the defendant had behaved cruelly and injuriously, and that the defendant should not contradict these allegations; then the judge would accept what was said and end the marriage. This method was not possible if the defendant did contest the divorce; then it had to be established that the defendant had in fact been at fault in causing the breakdown of the marriage. This procedure was subjected to increasing criticism from the end of the 1960s onwards. It was said, for example, that to make the spouses fight to establish fault simply increased their bitterness, that there were no binding standards for what constituted a matrimonial offence, and that the reasons for the actual breakdown of marriages had nothing to do with 'fault' as currently understood.

Like many other European countries, Germany has now abandoned the principle of fault in its divorce law; it has adopted the principle of 'breakdown of the marriage', and seems to have carried it to its logical limits. Breakdown of marriage is the only ground of divorce under the new law. A breakdown of the marriage is taken to have occurred if the spouses are no longer on living terms and it is not to be expected that they can re-establish the society that they have lost (§1565 BGB). It is not necessary for the judges to inquire in every case whether the marriage has actually broken down, because there is an irrebuttable presumption in §1566 par. 1 BGB that the marriage has broken down if spouses, both of whom want a divorce, have lived apart for *one* year, or in other cases have lived apart for *three* years (§1566 par. 2 BGB).

There are only two cases under the new law where it is relevant whether a spouse was at fault with regard to the breakdown of the marriage. First, spouses who have not lived apart for as much as a year can be divorced only if it would be an 'unbearable hardship' on the complainant, in view of some personal characteristic of the defendant, to have the marriage continue (§1565 par. 2 BGB). This provision is designed to prevent hasty divorces, and to stop people relying on their own misconduct as a ground for immediate divorce before living apart for a year. Of course, if both parties want a divorce, all they need do is to assert that they have already lived apart for a year, for many family courts accept such an assertion at its face value and proceed forthwith to hold that the marriage has broken down.

Once a marriage has broken down, a divorce must be granted, save in exceptional cases: if the marriage, for special reasons, must be kept afoot in the interests of the children, or if divorce would be a 'severe hardship' for the defendant by reason of unusual circumstances (§1568 BGB). In either case the divorce will be granted when these conditions cease to be satisfied. The question whether the breakdown of the marriage is attributable to the *fault* of the complainant can play a role in the application of the 'hardship clause', but one must bear in mind that it is only in very exceptional circumstances that it falls to be applied at all,[6] and that it has no application if the spouses have lived apart for *five* years (§1568 par. 2 BGB).

If fault has virtually no role to play any longer in the availability of divorce, the same is true as to the *consequences* of divorce. Under prior law, for example, a woman who was predominantly or exclusively responsible for the divorce could claim no maintenance from her husband. This is no longer a factor. Nor do the present rules make any distinction depending on whether the claimant for maintenance is the husband or the wife, though it is true that very few cases have arisen in which a man who has lived off his wealthy or active wife has claimed that she should continue after the divorce to keep him at his accustomed standard of living. The basic principle of the present rules is that divorced spouses are responsible for their own maintenance. Only in so far as a divorced spouse is prevented from earning a living by the need to bring up the children is there a maintenance claim, or if he or she is too old or infirm to earn (§§1569–73 BGB). In either case the claimant need engage only in 'appropriate activity' suitable to his education, age, health, and prior life-style (§1574 BGB).

---

[6]See BGH *NJW* 1979, 1042.

The property that has accrued to the spouses during the marriage is divided on divorce (see p. 191 above). So far as such property consists of immoveables, shares, and other items that can be realized at the moment of divorce, there is no problem, but nowadays one of the most important earned assets of many people is their right to a *pension on retirement or prior invalidity*. It is usually only the earning partner, normally the husband, who has such a right, and it matures — that is, the social insurer or the private pension fund starts paying out — only on his reaching a certain age or being incapacitated before then. The prior law was that, if divorce was decreed before the pension was payable, the divorced wife had only her general claim for alimony against her former husband, which, after his retirement, would be paid by him out of his pension. This meant that she often received much less than she needed to live on. Such rights to a pension on retirement or prior invalidity are now treated on the same footing as assets acquired during marriage, i.e., they must be divided, on the ground that their accrual to the employed spouse was due to the joint efforts of them both. The splitting of such entitlements on divorce — half and half (§§1587ff. BGB) — is normally achieved by granting the wife an independent right against the social insurer or private pension fund for herself and by diminishing the husband's right to that extent. The rules laid down in the statute are much too complicated and detailed to be explained here, but while serious difficulties and anomalies keep cropping up in practice, it is surely undeniable that the new rule is sound in principle.

II THE LAW OF SUCCESSION

1 Principles

The law of succession, in Book Five of the BGB, is regarded as technical and complicated. To some extent this is true, but it nevertheless forms an important part of the law relating to property, and succession, like ownership, is given constitutional protection by art. 14 GG.

Two principles underlie the German law of succession: universal succession (Gesamtrechtsnachfolge) and automatic inheritance (Vonselbsterwerb).

Universal succession means that, instead of having the various objects and rights of the decedent pass individually or severally to the heir or heirs, all the decedent's assets and rights are aggregated with his obligations and liabilities so as to make of all his legal relations a single complex, which is treated as a unit, including property, tangible or intangible, such as contract rights, patents, and shares, and also liabilities of all kinds arising from contract or tort or even taxes. All these pass together to the heir as a single unit, or to the heirs jointly, if there is more than one. For this transfer on death no conveyance is required even for the totality, let alone the individual component rights or duties.

Nor is any co-operation on the part of the heir required when these rights and duties pass to him at the moment of inheritance. The heir inherits on the death of the decedent even if he knows nothing whatever about it, even if he cannot yet be identified. This is why one speaks of 'automatic inheritance' (Vonselbsterwerb) or says that 'the dead vest the living' (der Tote erbt den Lebendigen), meaning that the inheritance attaches to the heir regardless of his own intentions in the matter. It is possible for the heir to disclaim the inheritance that has fallen to him (Ausschlagung), but he has nevertheless become heir in the meanwhile. Since all the decedent's obligations pass directly to the heir on inheritance, disclaimer is often the only way the heir can disburden himself of an overindebted estate. Disclaimer must take place within a period of six weeks, starting when the heir learns of the inheritance. An heir who disclaims ceases retrospectively to be heir, and the person next in line becomes heir with retrospective effect. The Amtsgericht, the Succession Court, and the Ortsgericht, a local court especially for inheritance matters, look after the estate in the meanwhile.

2 *Liability of the Heirs, and the Devolution of the Estate*

On the passing of the estate as a whole to the heir, two different patrimonies coalesce: the heir has both his own property and the estate he has inherited. Similarly, he now has the decedent's obligations as well as his own. In principle, he is liable for all these obligations with both properties; he can limit his liability by starting bankruptcy proceedings in respect of the estate he has inherited or making a composition with its creditors, but it is usually much easier for him to disclaim the inheritance.

The *devolution of the estate* is by and large the responsibility of the heir or heirs; there is no executor such as exists in Anglo-American law. The will itself, if there is one, may appoint a person (Testamentsvollstrecker) to look after the devolution and administration of all or part of the estate, but this is uncommon; in practice it occurs in only a small proportion of succession cases.

On the request of the heir, the Succession Court will issue him a certificate of heirship (Erbschein); once the period for disclaimer has run, this certificate becomes conclusive of his entitlement as against third parties, and facilitates his dealings with banks, insurance companies, and so on. The certificate of heirship is usually necessary before the heir can be registered in the Grundbuch as the new owner of any immoveable property contained in the estate.

3 Wills

In the BGB there are two methods by which a person may inherit: he may inherit in accordance with the rules of succession laid down by statute, or he may be appointed heir by a will made by the decedent. Freedom of testation is entailed in private autonomy, and it is protected as such. There are several different forms of valid *will*. One may write one's will out by hand, giving the date and place of writing, or one may have it drawn up in an official document by a notary. Only exceptionally does a will need to be witnessed. The will itself may be kept by the testator or deposited with the court; on the testator's death the will is officially opened by the Succession Court and is kept under its control.

There are also various special forms of will (emergency will, or Nottestament, will made by a sick person, and so on). Here we may mention the *gemeinschaftliches Testament,* or Berlin will, as it is called: a will drawn up by both spouses simultaneously may provide, with binding effect for the survivor, that the children shall inherit. The *inheritance contract,* or Erbvertrag, concluded between the testator and his heirs is also recognized; if it is in notarial form it restricts the testator's freedom to alter his will.

4 Statutory Intestate Succession

In the absence of a valid will, succession takes place in accordance with the statute, which puts relatives in five mutually exclusive classes. The first class includes children and their issue, the closer excluding the more remote: for example, if the decedent is a grandfather whose son is alive, his grandchildren by that son are excluded from inheriting. Issue inherit equally *per stirpes*. The second class includes the parents of the decedent and their issue (normally the decedent's siblings and their children). The third class includes the grandparents of the decedent and their issue, and the fourth class the great-grandparents and their issue. All other relatives together constitute the fifth class. If the decedent had no relatives at all, the state is the statutory heir.

A surviving spouse is entitled to one quarter of the estate if there are any relatives in the first class, and to one half if there are relatives in the second class but none in the first. If there are no relatives in either the first or the second class, the surviving spouse takes the whole estate. For the spouse to succeed, the marriage must have been valid at the time of the death, no proceedings for divorce must have been started, and there must be no grounds for divorce. The part of the surviving spouse is usually increased if the marriage was subject to the statutory property regime of community of acquisitions: a surviving spouse who waives the right to calculate the actual accruals during the marriage may claim an extra quarter of the estate regardless of whether any such accruals actually took place.

There is now a special rule for the succession of *illegitimate children*, dating from 1969. Theretofore they could inherit only from their mother, not from their father. Now, provided that he has been recognized by the father, an illegitimate child may inherit from the father just as legitimate children do, save that he has only a money claim up to the value of his part of the succession. The reason for denying the illegitimate child a complete right of succession in kind was to avoid making him a joint heir with the widow and legitimate children.

5 Pflichtteil, or Legitim

The succession rights of the immediate family are secured by the institution of legitim or Pflichtteil: the children, parents, and surviving spouse of the decedent have a statutory right of inheritance which cannot be curtailed by will. The legitim primes the will, save in a few exceptional cases, for example where the relative has tried to kill the testator or done him bodily injury. A person claiming legitim must have been wholly or partially excluded from the succession by the testator. He is not regarded as an heir, but has a money claim against the estate for the appropriate sum, namely half of what he would have received by intestate succession. If he has received something under the will, he can claim such amount as will make this up to the amount of the legitim. In order to calculate the amount of the legitim, the estate is valued as at the time of the death, but gifts which the testator had made to strangers in the last ten years of his life and which prejudice the legitim may be brought into account so as to increase it. The law of legitim is in practice quite important.

12

Conflict of Laws and Nationality

I NATIONALITY

While the basic principle of the German law of nationality is quite clear — as in other countries in continental Europe, it is the principle of parentage or *ius sanguinis* — the details are complex and difficult to grasp, for the law is the product of Germany's troubled history and its political situation today. The basic text, the Reichs- und Staatsangehörigkeitsgesetz of 22 July 1913 (RuStAG), has been frequently amended, most recently by a law of 29 June 1977. Originally the statute reflected the federal nature of the Reich: citizenship of the Reich depended on citizenship in one of the federal states. This dual citizenship came to an end when the federal structure was abolished by the Nazis in 1934, and unitary nationality was retained on the reintroduction of federalism after the Second World War.

After the war the problems posed by recent events were nearly insoluble by legislation. During the Nazi period political pressure or personal danger had forced many citizens, especially Jews, to leave Germany. Conversely, the citizens of Austria and the Sudetenland had been declared to be German citizens when those countries were incorporated in the Reich, and German nationality had been accorded to persons of German stock in the zones that had been militarily occupied before and during the war. When the war was over, not only did Germans get expelled from the eastern parts of the Reich, but many people of German stock from beyond the Reich fled or were driven to Germany and had to be assimilated there. After the war all those who had had German nationality thrust upon them were naturally released from it. The Basic Law regulated the legal absorption of the other groups we have mentioned: art. 116 GG laid down that anyone who had been deprived of German nationality was entitled to reclaim it on demand, as were his descendants, and it also expanded the definition of 'German' beyond that of 'German national' so as to include anyone 'who has been admitted to the territory of the German Reich within the frontiers of 31 December 1937 as a refugee or expellee of German stock or as the spouse or descendant of such a person' (official translation).

The fact that the former territory of the Reich now accommodates two states, the Bundesrepublik and the German Democratic Republic, makes for a new problem. Although its sovereignty is limited to its present territory, the Bundesrepublik sees itself as the legal successor of the German Reich, and since it regards German nationality as unitary, it does not treat inhabitants of the German Democratic Republic as aliens: although the German Democratic Republic has its own nationality, any of its citizens who comes to the Bundesrepublik is entitled to be treated just like any citizen of the Bundesrepublik.[1]

The present law relating to German nationality, which may be acquired by birth, legitimation, adoption, or naturalization (§3 RuStAG), still adheres to the principle of parentage, accepts the principle of the equal rights of man and woman, and seeks to avoid both dual nationality and statelessness.

If either the father or the mother of a legitimate child is a German national, the child acquires German nationality on birth, as does an illegitimate child if its mother is German (§4 I RuStAG). In neither case is place of birth relevant: the rules apply equally to children born abroad. The illegitimate child of a German father becomes a German national on being legitimated under German law, that is, by the marriage of the parents, or on declaration of legitimacy by the court of wards (§§1719, 1723 BGB) (§5 RuStAG). So, too, does a minor on being adopted by a German (§6 RuStAG).

Naturalization (Einbürgerung) is the process whereby a foreigner can acquire German nationality. This is a sovereign act for which application must be made. German nationality is acquired on delivery of the certificate of naturalization (§16 I

---

[1]The Bundesverfassungsgericht confirmed these principles in a decision of 31 July 1973 in which it examined and accepted the constitutionality of the Basic Treaty between the Bundesrepublik and the German Democratic Republic of 21 December 1972; *BVerfGE* 36, 1.

RuStAG), which normally makes it clear that it does not confer citizenship on members of the applicant's family. A legal right to naturalization is enjoyed by only certain classes of person in cases clearly specified by law, such as the illegitimate child of a German father and those we have mentioned who are German by origin but not yet by citizenship. In all other cases naturalization is at the complete discretion of the authorities and will be granted only when it is in the public interest. The applicant must normally have settled in Germany, be capable of doing business, and have no criminal record (§8 RuStAG). But the satisfaction of all these conditions does not entitle one to naturalization, although there are certain relaxations for those who have previously been German citizens, for homeless aliens, for refugees from abroad, and for stateless persons.

Like other Western countries, the Bundesrepublik faces the problem that many people want to become citizens although its cultural traditions and dense population make it an inappropriate place for them to settle in. First, there are the numerous immigrant workers, especially from other members of the European community, who came to Germany when the economy was booming. Then there are those who claim a right of asylum as political refugees in order to take up residence in Germany and perhaps acquire German citizenship; the authorities have to examine all such cases and reject those who are claiming asylum as a pretext for looking for work in Germany. Finally, many students from developing countries wish to stay in Germany, contrary to the interests of their homelands.

A person who is about to acquire citizenship in another state is released from German nationality on demand (§18 RuStAG); German nationality is also lost by a person who applies for and acquires foreign nationality (§25 RuStAG), by a German with dual nationality who makes a declaration of waiver (§26 RuStAG), and by a minor who, on adoption by an alien, acquires the nationality of his adoptive parent (§27 RuStAG).

II PRIVATE INTERNATIONAL LAW

The Introductoy Law to the Civil Code (Einführungsgesetz zum Bürgerlichen Gesetzbuch — EGBGB) contains in arts.

7–31 some of the rules of German private international law (IPR), but it is no more than a partial collection of provisions, no attempt having been made at a comprehensive codification. Important parts of German conflicts law have thus been developed by the courts and legal writers.

German private international law as it is today has been profoundly marked by the thinking of Friedrich Carl von Savigny (1779–1861). Himself much influenced by the American lawyer Joseph Story (1779–1845), Savigny helped to overcome the old theory of statutes which for centuries had dominated European conflict of laws. According to this theory, the spatial range of application of any law, when there was a conflict, depended on its content: distinctions drawn between *statuta personalia, statuta realia,* and *statuta mixta* depended on their subject matter and determined their range of application. Savigny put forward the converse viewpoint: 'for every legal relationship there is an appropriate system of law to which it naturally belongs, a home system, and that is the system one must discover'.[2] This was a new and more flexible method for linking the facts of a case to a legal system. Mancini (1817–88) also said much that is important for modern European conflicts law: while he emphasized that the application of a legal system very often turns on the nationality of a party, and said that everyone was primarily to be judged by his own law, he also stressed that there were rules that the public interest of a state requires to be applied to all persons on its territory, an idea which is central to the modern doctrine of *ordre public.*

A school of American lawyers has recently sought to resolve conflicts of law with reference to the content of the rules in relation to the policy interests of the legal systems involved; these views are discussed with interest, but not generally adopted, by German lawyers today. They are more inclined to accept the new theory of the law merchant (*lex mercatoria*), according to which international commerce is in the process of generating substantive rules which apply regardless of location.

It is still the generally accepted view that the function of private international law is to produce a 'conflictual justice', regardless of the content of the substantive rules to be applied,

---

[2]von Savigny, VIII *System des heutigen römischen Rechts* §§348, 360, pp. 28, 108 (Berlin 1849).

that is, to apply the rules of whichever system is closest to the real facts of the case. If the rules of German private international law invoke the private law of a foreign system, the judge applies that private law, and if the conflicts rules of that foreign system refer back to German law (*renvoi*), the German judge applies German law. Should the conflicts rules of the foreign system refer to a third system, the German judge will apply the rules of that third system. The foreign rule will be applied whatever its content or whatever the result it produces, unless 'its application conflicts with good morals or with the purpose of a German enactment' (art. 30 EGBGB). This entrenchment of *ordre public* is, however, rather narrowly construed in Germany, and it is generally agreed that the limit 'purpose of a German enactment' is too broad: foreign law is excluded only when it conflicts with the basic principles and values of the German legal system.

The heartland of *ordre public* in Germany is the system of values expressed in the chapter of the Basic Law devoted to basic rights. A decision of the Bundesverfassungsgericht indicates the role it may play. A Spaniard who had left the Church met a Protestant German woman whose prior marriage had been validly terminated by divorce under German law, and he wished to marry her. For this purpose he required a certificate of marriageability from the President of the Oberlandesgericht, but the President refused to issue one. According to art. 13 I EGBGB, the question whether there is any obstacle, such as bigamy, to a marriage falls to be determined separately for the man and the woman by the national law of each. Spanish law would not recognize the divorce obtained by the German bride-to-be, so the Spaniard faced the obstacle that his marriage to her would be bigamous. The Bundesgerichtshof endorsed this view, but the Bundesverfassungsgericht decided otherwise. Like all other cases, those involving private international law must be decided consistently with the Basic Law, which is in principle well disposed to international law (art. 25 GG). Here there was no question of disregarding foreign law as such, only of operating the German legal order without the intrusion of values contained in a foreign law which were in conflict with it. Under German law marriage is terminable by divorce, and the right of a divorced person to remarry is constitutionally guaranteed (art. 6 I GG). Accordingly, a restrictive construction must be given to art. 13 I EGBGB, and the question whether the status of the German wife was an obstacle to the marriage by reason of bigamy should be determined exclusively by German law, even in relation to the Spaniard's freedom to marry. Alternatively, one could directly invoke the entrenched German *ordre public* (art. 30 EGBGB).[3]

We shall now give a brief summary of a few of the important rules of German private international law. A natural person's legal capacity depends on his personal status, that is, on the law of the state of which he is a citizen. The same is true of capacity to do business (art. 7 I EGBGB), subject to this, that once capacity to do business has been acquired, it is never lost by a change of status (art. 7 II EGBGB); if a foreigner enters a legal transaction in Germany, he has the capacity to do business which he would enjoy under German law if he were a German (art. 7 III EGBGB). Personal status is also determinative for declarations of majority and for incapacitation (Entmündigung). Legal persons fall under the law of the place of their head office, according to the majority view,[4] but there is a minority view that would apply the law of incorporation.

Legal transactions are generally controlled by a status of their own rather than by the status of the parties. The aim of this is to procure that, so far as possible, each legal transaction falls under only one legal system. As to questions of form, the law of the transaction generally applies (*lex causae*: art.11 I 1 EGBGB), but it is sufficient if the parties follow the form prescribed in the place where they conclude the transaction (art. 11 I 2 EGBGB; *locus regit actum*). Whether a person has power to represent another in transactions is controlled by the law of the place where he purports to use such power, save that, if the transaction concerns real property, the law of the transaction applies. As to rights in property, moveable and immoveable, the principle is that one applies the law of the place where the property is situate (*lex rei sitae*); if moveables are sold and sent abroad (*res in transitu*), German lawyers apply the law of the destination as soon as the goods have left the country of origin.

Torts are controlled by the law of the place of the tort (*lex loci delicti commissi*);[5] if the act was done in one place and the damage occurs in another, the victim can choose whichever law is more

---

[3]*BVerfGE* 31, 58, disapproving *BGHZ* 41, 136.
[4]*BGHZ* 53, 181.   [5]*BGHZ* 23, 65, 67.

favourable to him. With regard to contracts, German lawyers, like those in other countries, accept the principle of the autonomy of the parties and leave the parties free to choose their law: the law chosen by the parties applies, and the German judge does not inquire whether that foreign law would itself permit such a choice. The choice of law may be express or implied; if the parties have made no choice, one must try to discover the hypothetical will of the parties, that is, the 'centre of gravity' of the transaction, in the hope of finding a single law to apply to the contract.[6] If this method fails, one applies the law of the place of performance, although in the case of synallagmatic contracts this often has the unfortunate result that two different legal systems may have to be applied.[7]

German conflicts rules regarding family law are based on the principle of nationality. For questions of affiancement and marriage, the law of both parties is to be applied, the rights of each being judged by the law of his own state (art. 13 I EGBGB); this applies to questions of the preconditions of marriage (see the example on p. 204 above) and to the question whether the marriage is valid, voidable, or void. If the marriage takes place in Germany, its form is controlled by German law alone (art. 13 III EGBGB). For the rest, German conflicts rules are designed so far as possible to subject a marriage to one legal system only. This is usually the law of the husband, and while this rule may seem to be merely a technical expedient, it is permissible to ask whether in some cases the rule may not amount to a violation of the basic principle of the equal rights of man and woman. Claims for maintenance, rules relating to names, and other personal effects of matrimony are subject to the common national law of the spouses; thus German law applies to German spouses wherever they live (art. 14 I EGBGB). If there is no common national law, the wife may choose, as regards her name, between the law of her homeland and the law of the couple's normal domicile.[8] Questions of matrimonial property are determined by the national law of the husband at the time of the marriage (art. 15 EGBGB), which continues to apply even if his nationality changes. Divorce falls under the national law of

---

[6] *BGHZ* 19, 110, 112; BGH *Betrieb* 1969, 1053.
[7] *BGHZ* 43, 21.   [8] *BGHZ* 56, 193.

the husband at the time the suit is raised (art. 17 EGBGB), but no German court may grant a decree of divorce on the basis of foreign law unless it could have granted it under German law (art. 17 IV EGBGB); German law is invariably applied in divorce suits brought by German wives (art. 17 III EGBGB).

The question whether or not a child is legitimate is determined by the law of the mother's husband at the time of the birth (*arguendo*, from art. 18 I EGBGB). If the child is legitimate, its legal relations with its parents, such as the duty of maintenance or the right of education, fall under the national law of the father or, if he is dead, under that of the mother (art. 19 sentence 1 EGBGB).

Children normally domiciled in the Bundesrepublik are covered by the Hague Convention for the Protection of Minors of 1961, which provides for the application of the protective rules of German family law; thus if such a child is endangered or neglected, the court of wards can put it in a foster family or in a home (§ 1666 I BGB). If a child is illegitimate, its relations with its mother are controlled by the mother's national law (art. 20 EGBGB). The father's duty to maintain the child and to reimburse the mother are also controlled by the national law of the mother at the time of birth, subject to this, that a German father cannot be exposed to more extensive obligations than would arise under German law (art. 21 EGBGB). In other respects the relationship between the illegitimate child and its father, including legitimation, depends on the national law of the father. Adoption is covered by the national law of the adoptive parent (art. 22 I EGBGB). Before a foreigner can legitimate or adopt a German child, however, German law requires the agreement of the child or his family and the approval of the court of wards (art. 22 II EGBGB). The legal incidents of any other relationship are determined by the national law of the person whose duty is in issue. Questions of guardianship and care fall within the jurisdiction of German courts only if the persons involved are German, but it is possible to apply the German rules of guardianship and care to a foreigner if his own state has failed to take care of him and if, by his own law, he needs care or is subject to a German interdiction order (art. 23 I EGBGB).

In matters of succession the rules applied are those of the

state of which the decedent was a national at the time of his death (the principle of arts. 24–5 EGBGB), though the heirs may invoke the law of his domicile in order to reduce their liability for the debts of the succession (art. 24 I EGBGB).

III JURISDICTION AND PROBLEMS OF THE LAW OF PROCEDURE

The statutory provisions regarding the international jurisdiction (Zuständigkeit) of German courts are incomplete and dispersed, but it is sometimes possible to invoke the rules of the Code of Civil Procedure (ZPO) regarding venue or internal competence. The rules as to jurisdiction and venue coincide in the following cases. Legal persons may be sued at the place of their head office (§17 ZPO); for claims to ownership of real property, the court where the property is situate is competent (§24 ZPO), for contractual disputes the court of the place of performance (§29 ZPO), for torts the court of the place of the tort (*locus delicti commissi*: §32 ZPO). §606b ZPO lays down a special rule: no German court may entertain a matrimonial dispute between aliens except under special circumstances.[9]

It is very often in uncontentious matters (freiwillige Gerichtsbarkeit), where the court's role is that of guardian or supervisor of private rights, that German courts have to apply substantive foreign law, for example, if part of a succession which is subject to English law is in Germany, or if a foreign father is bound to maintain a child who is living in Germany.[10] Here German law differs from English and American law in holding that a German judgment is valid even in the absence of international jurisdiction.[11]

Even where the German conflicts rules require the application of foreign private law, a German court naturally applies its own law of procedure and no other. Different legal systems have different views about what constitutes procedural law: in

[9] (1) At least one of the spouses must be regularly domiciled in Germany and the national law of the husband must recognize the validity of the German decree, or else one of the spouses must be stateless; (2) the wife must have been German at the time of the marriage and be bringing suit for a declaration of nullity, for rescission, or for a declaration of the existence or non-existence of the marriage.

[10] *BayObLGZ* 1959, 8; Dölle, *Internationales Privatrecht. Eine Einführung in seine Grundlagen* §2 VIII (Karlsruhe 1972).

[11] Kegel, *Internationales Privatrecht. Ein Studienbuch* §22 II (4th ed. Munich 1977).

Jurisdiction and Problems of the Law of Procedure 209

Germany the statute of limitations is regarded as substantive law, not procedure. Should foreign substantive law fall to be applied under German conflicts rules, the court itself must find out what the content of that foreign law is; it will then try to interpret and apply it just as the foreign judge would do. Foreign law is thus not a fact to be proved by the parties, except in a case where the court is wholly unfamiliar with it and unable to discover what it is (§293 ZPO); and even here the court is not bound by the evidence adduced by the parties.

German courts recognize foreign judgments if the statutory conditions are met: the state whose court rendered the judgment must have had international jurisdiction; the judgment must not conflict with German *ordre public*; and there must be reciprocity, that is, the state in question must recognize German judgments (§328 ZPO). Before one may execute on a judgment in Germany, one must have a German judgment of execution (*exequatur*: §§772, 773 ZPO); here too the court checks that the preconditions for recognition are satisfied, and in matrimonial causes there is a special investigation by the Ministry of Justice in the relevant Land.

Germany is a signatory of many multilateral and bilateral treaties on legal and judicial co-operation. Especially important is the Hague Convention concerning Civil Procedure of 1954, which in most countries replaced the old Convention of 1905. Then there is the Hague Convention of 1965 on the Service Abroad of Judicial and Extrajudicial Documents in Civil or Commercial Matters, and on the Taking of Evidence Abroad in Civil or Commercial Matters of 1970. There is also the 1958 Convention on the Recognition and Enforcement of Judgments regarding the Maintenance of Children. Germany is a signatory of the EEC Convention on the Jurisdiction of Courts and the Enforcement of Judgments in Civil and Commercial Matters of 1968, which came into force in 1973. Germany is also a signatory of the European Convention on International Commercial Arbitration of 1961, and the UN Convention on the Recognition and Enforcement of Foreign Arbitral Awards of 1958. There are bilateral treaties between Germany and many other states, including one with Great Britain in 1960.

Awards of private arbitrators can be executed only on the order of a court (§1042 ZPO), which may check that the arbitration was properly conducted and that the basic principles of procedure were respected (§1042–1042d, 1041 ZPO). This applies to foreign and internal arbitrations alike.

Commercial Law

In its narrow sense Handelsrecht, or commercial law, designates the rules laid down by the German Commercial Code as applicable to merchants. In a broader sense, it includes other areas of law relating to mercantile activities, such as the law of negotiable instruments, and a few important types of transaction which are found in general civil law but are much used by bankers and other people in commerce (Section IV below). In addition there are distinct areas of law affecting traders and trading, such as company law, the law of connected enterprises (Chapter 14 below) and competition law (Chapter 15 below).

I THE MERCHANT AND HIS LAW: THE COMMERCIAL CODE (HGB)

1 Survey

The German Commercial Code (Handelsgesetzbuch: HGB) contains provisions regarding the status of merchants (§§1–7 HGB), the commercial register (Handelsregister: §§8–16 HGB), the form of business names and their registration (Firma: §§17–37 HGB), and the duty to keep proper accounts (§§38–47b HGB). The HGB also covers special types of commercial agency (Section III below) and lays down some general rules regarding commercial transactions (§§343–72 HGB) as well as some special rules for the contracts made by commercial vendors (§§373–82 HGB), commission agents (§§383–406 HGB: see Section III below), forwarding agents (§§407–15 HGB), warehousemen (§§416–24 HGB), carriers (§§425–52 HGB), brokers (§§93–104 HGB), and maritime traders (§§476ff. HGB). The code also has special rules on commercial partnerships (§§105–77a HGB) and on silent partnerships (§§335–42 HGB: see Chapter 14 below).

2 Scope of Application: Who is a Merchant?

The provisions of the HGB apply only to merchants and their

commercial transactions, the applicability of the code turning on the key concept of 'Kaufmann' (merchant) and, subsidiarily, on the concept of 'Handelsgeschäft' (commercial transaction). In adopting the concept of Kaufmann as his point of departure, the draftsman of the HGB was following the rather old-fashioned idea that the different professions within a society constitute separate estates, each with its special law: indeed, the title of Book I of the HGB is 'Handelsstand', or 'The Estate of Merchants'. Other codes have used the notion of business transaction as the main determinant of applicability, e.g. the French Code de commerce (art. 1: 'Acte de commerce') and, to some extent, the old German ADHGB (see Chapter I Section II above). The Uniform Commercial Code in the United States regulates selected commercial transactions, while the modern Turkish Commercial Code, which was drafted under German influence, starts out from the concept of the 'business enterprise'.[1] The idea of the business enterprise is not, however, completely ignored by the HGB, for it defines merchants in relation to typical business activities (§1 HGB) and also refers to the size and organization of businesses (§§2–3 HGB).

A Kaufmann in German law may be an individual, a commercial partnership, or a company. A company (AG; GmbH) invariably has merchant status, whatever its business or activities, by reason of the legal form of its organization; here the Commercial Code makes reference to the special legislation on companies (§6 HGB: 'Formkaufmann'; compare §3 AktG; §13 III GmbH; §17 II GenG). Individuals and partnerships must meet the requirements of §§1–3 HGB before they can have the status of merchants.

The HGB, in §1 II, enumerates nine categories of business activity (Grundhandelsgewerbe: basic commercial business) which invariably confer the status of merchant.

1. Purchasing and reselling goods or securities, whether or not any processing of the goods is involved. This covers, for example, the manufacturer of cars as well as the car dealer, but not persons who sell land or houses, or those who exploit natural resources, such as miners

---

[1]The concept of commercial law as a distinct and separate body of law within a legal system is discussed generally and from the point of view of comparative law in Schmitthoff (ed.), *The Sources of the Law of International Trade* 41–100 (London 1964).

or farmers, or those in the house construction business. It is possible for a construction firm to become a merchant, but only by registration (under §2 or §3 HGB) or by forming a company (§6 HGB).

2. Processing goods received from third parties (e.g. laundering, dry-cleaning, dyeing, book-binding), provided the business is run on an industrial rather than a craft basis and scale.

3. The insurance of risks against a premium. Mutual insurance is not considered to be a business at all, not being run for profit, so it is not directly covered; the provisions of the HGB have, however, been extended to mutual insurance associations by special legislation (§16 VAG).

4. Banking and money-changing.

5. The carriage of goods and persons by sea, inland waterway, or land. A person who runs a single taxi-cab would not be included because his business would be too small. Carriers by air are merchants only under §2 or §6 HGB.

6. The business of forwarding agents, warehousemen, commission agents (Kommissionäre: see Section III 4 below).

7. The business of commercial agents, other than employees, and brokers. Real estate brokerage is not covered, since it is declared to be non-commercial by the HGB (§93 HGB).

8. The business of booksellers, publishers, and art-dealers.

9. The business of printers, provided it is run on an industrial rather than a craft basis and scale.

All persons or partnerships that run a business of any of the kinds enumerated have the status of a merchant by operation of law from the moment they embark on that activity, and have all the duties and privileges of a Kaufmann as prescribed by law; thus they are bound by the rules as to business names (Firma; Section II 1 below), applications for entry in the commercial register (§29 HGB; Section II 4 below), and proper bookkeeping (§§38ff. HGB), and may use the special forms of commercial agency (Section III 2 below). But the special provisions of the HGB regarding business names (Firma), commercial bookkeeping, and commercial agency do not apply to those whose business is very small (§4 HGB: Minderkaufleute (small traders)).

The status of a merchant may also be acquired by registering in the commercial register: if a business not listed in §1 HGB is large enough to need proper commercial organization, bookkeeping, and accounting, which is admittedly a question of degree, then §2 HGB imposes a duty to register on the individual or partnership operating it. Under §3 HGB farmers and forestry enterprises are permitted, but not required, to apply for registration, either with regard to their farm or forestry business or in respect of an ancillary operation, such as a dairy or a saw-mill, provided that the business calls for the commercial organization referred to in §2 HGB. If a person whose business is either non-mercantile or very small has been registered in the commercial register, he will be treated as a merchant *vis-à-vis* third parties who rely on that registration until the register is corrected (§5 HGB: Scheinkaufmann (merchant by appearance)).

Some peculiar results follow from the manner in which the law defines the class of merchants. While the German Federal Bank (Bundesbank) is a merchant under §1 II no. 4 HGB, other public bodies with huge turnovers, such as the Federal Railways or Federal Post Office, are not, because the services they render are regarded as being in the public interest (§452 HGB; §7 PostG; §41 I BundesbahnG). Those who exercise the so-called liberal professions, such as doctors, lawyers, architects, or private teachers, are not merchants, but the doctor who runs a large sanitorium can become a merchant either by registration under §2 HGB or by choosing to run the business in a certain legal form, such as a limited liability company (GmbH; §6 HGB). The same applies to mining enterprises, which are not included in §1 II HGB. Farming and forestry were traditionally regarded as non-mercantile, so until recently such enterprises could achieve mercantile status by registration only for an ancillary business such as a saw-mill; now, however, they may achieve mercantile status for their main activities by registration, for the law was amended in 1976 in order to take account of the development of large-scale forestry and farming enterprises.

Even the law regarding small traders is not free from difficulty. A person who runs a newspaper kiosk is a merchant under §1 II no. 1 HGB, usually a 'small trader' under §4 HGB, since his business is small 'in nature and extent'. A man who runs a taxi is not a merchant at all, for although the carriage of persons is included under §1 II no. 5 HGB, it is mercantile only if it is conducted by larger units (Anstalten). The person who runs a hotel is a merchant only if he provides food and drink (§1 II no. 1 HGB), though even so he is often a small trader under §4 HGB. When the courts have to decide whether the 'nature and extent' of a business is such that it ought to be commercially organized (which is important for §2 and §4 HGB), they consider the complexity of the business and the degree of organization it needs, while the chambers of commerce and industry tend to differentiate on the basis of turnover figures.

Merchants are like everyone else in having private lives, and they may make contracts in their private capacity to which the HGB does not apply. The HGB applies only to their 'commercial transactions' (Handelsgeschäfte), that is, all the transactions relating to their commercial business (§343 HGB). But since it is not always easy for third parties to distinguish between private and commercial acts, §344 HGB establishes a presumption that a merchant's transactions relate to his business. If only one party to a contract is a merchant, it will be governed by the HGB, with the exception of those provisions which expressly require that both parties be merchants (§345 HGB).

For example, when a car-dealer buys furniture for his home, the purchase is not a commercial act so far as he is concerned, but the contract will nevertheless be a commercial sale if the vendor is a merchant under §1 II no. 1 HGB (§§343, 345 HGB). The provisions on commercial sales would consequently apply (§§373ff. HGB). Under §377 HGB the buyer must inspect the goods supplied and give immediate notice of any defect in the goods if he wishes to hold the vendor liable for it. But §377 HGB is one of those provisions that apply only if the transaction is a commercial act for both parties, so in this example the buyer would not be subject to this particular duty.

3 The Philosophy of the Commercial Code Exemplified: Some Rules on Commercial Acts

In laying down special rules within the general framework of private law, the HGB is performing two functions. First, it contains juridical forms for transactions that are purely commercial, such as the commission business (see Section III 4 below) or the special form of agency known as the Prokura. Second, it affords protection to those involved in commerce, where important affairs are often concluded very swiftly: the protection results in part from the rules regarding business names and the commercial register (Section II below), and in part from making the merchant more strictly answerable to third parties for his actions and words than a private individual would be.

Under the BGB, a guarantee (Bürgschaft) or the recognition

of a debt is valid only if it is in writing (§§766, 780, 781 BGB), but an oral guarantee or recognition of a debt will bind a merchant (§350 HGB). Again, a merchant who promises to pay a contractual penalty is bound to pay it even if it is very high, whereas under the BGB such a penalty promised by an individual could be reduced by the courts to a 'reasonable amount' (§343 BGB). On the other hand, a merchant who performs a service or effects a transaction for a customer at the customer's request may claim the standard remuneration even if there is no contractual provision for it (§354 HGB).

The HGB modifies the general principles of the law of contracts in some respects. As a general rule, the formation of a contract requires offer and acceptance, and if the offeree does not accept the offer expressly or by implication, no contract is concluded. Under §362 HGB, however, if the offeree is a merchant whose business is the management of the affairs of others, he is bound to make it clear if he is unwilling to accept: silence on his part constitutes acceptance. Accordingly, if a customer instructs his broker to buy certain securities, and the broker neither replies nor effects the purchase, the broker will be contractually liable. §362 HGB applies only when the offeree's business involves the 'management of the affairs of others'. Even so, the merchant will be bound to answer the order or offer only if there was already some business connection between the parties or if he offered his services spontaneously. Thus §362 has no application to simple offers for sale. A tailor need not reply to a supplier who offers to sell him more cloth.

But the idea that silence can exceptionally constitute the acceptance of a contract has been extended to other cases by the courts. Under the general law, a contract is not formed by the late acceptance of an offer: the late acceptance constitutes a new offer. But a merchant may be under a duty to answer this new offer, and to make it clear that he does not propose to accept it: his silence is tantamount to acceptance of this new offer.[2] A corresponding rule has been developed by the courts with respect to what is called the commercial letter of comfirmation (kaufmännisches Bestätigungsschreiben). This is a letter which puts in writing what was previously agreed, often orally, between the parties. A merchant who receives such a letter from the other party must answer it if he thinks that it does not reflect

[2] *BGHZ* 18, 212; BGH *LM* §346 HGB (D) no. 7b.

their common understanding, and his silence will constitute acceptance of the letter as it stands, unless it differs so widely from the prior negotiations that the sender could not reasonably expect the other party to accept its contents.[3] Under the rule regarding letters of confirmation it is not always clear whether the letter is simply evidence of the prior agreement or constitutes an offer that turns into a contract by silent acceptance. Whichever it is, it makes no difference to the practical result.

The provisions of the HGB on commercial sales (§§373–82 HGB) are designed to protect the vendor of goods against the silence or sloth of the buyer. One such rule has already been mentioned (Section 1 above): the buyer must inspect the goods supplied for defects or disconformity from the normal or agreed quality. He must inform the vendor immediately if he wishes to complain of a defect in the goods, or he loses all the rights which would accrue from their defectiveness (§377 HGB). The same rule applies if the goods supplied are different in kind or quantity from what was ordered in the contract, unless the difference or deviation was obvious (§378 HGB).

The HGB extends the protection enjoyed by the bona fide purchaser of moveable property under the BGB. §932 BGB requires that the acquirer believe the vendor to be the owner of the goods, but in commerce merchants are often entitled to sell goods belonging to someone else, for example, when acting as a Kommissionär (see Section III below). Thus §366 HGB extends protection to the good faith purchaser of goods which he knew the vendor did not own, provided that he bona fide believed that the vendor was entitled to dispose of the goods on behalf of their owner. The same rule applies when merchants pledge goods or commercial documents.

Another special rule of commercial law worth mentioning in conclusion is that a merchant has a particular lien in respect of sums due over any goods or instruments belonging to his debtor which came into his possession in the course of normal business relations.

4 Commercial Code and Civil Code

As will be clear from the examples we have given, many provisions of the HGB can be understood only in the light of the

[3] *BGHZ* 11, 1; *BGHZ* 40, 42; *BGHZ* 54, 238.

general rules laid down by the BGB, which it is their function to modify, amplify, or exclude. Thus the HGB contains very few provisions about contracts of sale: the reciprocal duties and rights of buyer and seller are regulated in principle by the BGB (§§433ff. BGB). The rule that the merchant who buys goods must inform the vendor immediately if wrong or defective goods have been supplied (§§377, 378 HGB) must be set in the context of the rules of the BGB concerning the vendor's liability for defects (§§459ff. BGB). So, too, the special protection afforded by the HGB to the good faith purchaser (§366 HGB) makes sense only when one knows what is laid down in the property section of the BGB about acquiring a moveable from a non-owner (§§932ff. BGB). And the provisions of the HGB regarding commercial brokers are only modifications of the rules contained in the BGB about the brokerage contract (§§652 ff. BGB).

But other types of contract, such as the commission contract, although founded on the general contract law contained in the BGB, are so fully regulated in the HGB that we must regard them as separate institutions altogether (Section III below). The provisions of the HGB on commercial partnerships (Chapter 14 Section I below) contain rules additional to, and sometimes deviant from, the provisions of the BGB relating to civil partnerships, a large number of which nevertheless remain applicable to commercial partnerships (Chapter 14 below).

It should now be clear that commercial law is only a special part of general private law and cannot be understood or applied except in the light of its provisions.

5 Other Sources of Commercial Law

A number of areas of commercial law in the wider sense are the subject of special enactments, either because they have been removed from the Commercial Code, e.g. the law of stock companies (special legislation in 1937 and 1965), or because they have always been the subject of distinct enactments, e.g. the law of bills of exchange. The special legislation on companies will be described in Chapter 15 below. The law of bills of exchange (1933) and the law of cheques (1933) were enacted in accordance with the respective international Geneva Conventions of 1933 which unified the continental European law on these subjects. There are laws against unfair competition (1909) and against restraint of competition (1957) (see Chapter 15 below).

According to §346 HGB the customs and usages of merchants 'determine the meaning and effect of acts and omissions' of merchants; they are an important guide for the interpretation of a merchant's contracts and other acts. Although they are not customary law, commercial customs and usages often have a similar function, since the interpretation of contracts by the German courts is very flexible and the courts do not hesitate to supplement the gaps in a contract. The courts sometimes consult the chambers of industry and commerce in order to ascertain what the commercial custom or usage may be. With respect to international business transactions, the German courts would take account of international usages, customs and terms of commerce, especially those found in the semi-official texts of the International Chamber of Commerce (e.g. inco-terms, rules on letters of credit).

In business practice, contracts with standard terms and form contracts are widely used among merchants themselves as well as between merchants and private customers. Only if such terms have been agreed, i.e. if they have been incorporated in the contract, do the courts apply them, but then they interpret them in much the same way as an enactment. Standard contract terms and form contracts are subject to special legislation designed to protect the customer (see Chapter 5 above).

II THE BUSINESS NAME (FIRMA) AND THE COMMERCIAL REGISTER

1 The Firma

In colloquial language the term 'Firma' sometimes means a business entity or factory, just like the English term 'firm'. In legal language, however, 'Firma' denotes the name used by a merchant (individual, partnership, or company) in business life (§17 HGB). It is under his Firma that a merchant concludes and signs contracts (§17 I HGB), and sues and can be sued (§17 II HGB). The Firma has to be registered in the commercial register. Only registered or registrable merchants have a Firma, that is, individuals or partnerships that run a basic mercantile business (Grundhandelsgewerbe) in the meaning of

§1 HGB, unless their business is quite small (Minderkaufleute under §4 HGB), or those who register under §2 or §3 HGB. Companies have a Firma because they are always merchants (Section I 2 above).

Two conflicting principles underlie the provisions of the HGB on the Firma (§§17ff. HGB): when a business is started up the rule of 'truth of the firm name' (Firmenwahrheit) must be observed, but it is limited by the principle of 'continuity of the firm name' (Firmenbeständigkeit) if there are subsequent changes in the business or its proprietors. The Firma must be true when it is first formed and registered, and its contents are prescribed by law.

Individuals must use their true surname and first name (§18 I HGB). They may add an indication of the nature of their business, but it must not be calculated to deceive people (§18 II HGB); thus Müller can have the Firma 'Hans Müller Möbelgrosshandlung', but not 'Hans Müller & Co', for that indicates the existence of partners; nor, if he has a smallish local business, may he add 'internationale Möbelgrosshandlung'. The Firma of a general partnership (OHG; see Chapter 14 Section I below) is composed either of the surnames of all general partners ('Müller & Schmidt') or of the surname of at least one general partner with an indication of a partnership ('Müller OHG'; 'Müller & Co'; 'Möbelhaus Müller & Co'). The Firma of a limited partnership (KG) contains the surname of at least one general partner with an indication of a partnership ('Möbelhaus Müller KG'; 'Müller & Co KG') (§18 II, IV HGB). The Firma of a stock company (AG) indicates the subject or nature of the business and its legal form as a company ('Farbwerke Höchst Aktiengesellschaft') (§4 AktG); the same rule applies to a private limited company (GmbH), which, however, may use the surnames of partners instead of the subject of the business in its Firma ('Duisburger Stahlhandels-GmbH'; 'Müller, Mayer, & Schmidt GmbH'; 'Müller & Co GmbH') (§4 GmbHG).

Every Firma must be clearly distinguishable from every other Firma in use in the area; if necessary, distinctive words must be added (§30 HGB). A person who buys or inherits a merchant's business (see Section II 2 below), or operates one under a contract of lease, can keep the old Firma unchanged or add an indication of successorship, provided that the previous holder or his heirs give their consent (§22 HGB). In this way the new holder can benefit from the goodwill of the original Firma, and the principle of 'truth of the Firma' gives way to the principle of 'continuity of the Firma' in order to maintain the value attaching to the Firma.

Thus a stock-company can continue the Firma of the prior sole proprietor or partnership ('Daimler Benz AG'; 'Siemens AG'); a partnership that takes over the Firma of a sole proprietor is not bound to add the indication 'OHG' or 'KG', but may do so if it pleases. But if an individual, Müller, takes over the Firma 'Meyer OHG', he must strike out the indication 'OHG' in the Firma so as to avoid giving the false impression that it is still a general partnership. The Firma 'Meyer & Schmidt' also indicates a general partnership, but views differ on the question whether Müller as an individual can continue the business under this Firma. Probably he may not;[4] at least he must add the words 'Alleininhaber Müller' (Müller sole proprietor).

If the sole proprietor of a business takes on a partner, the Firma can remain unchanged (§24 I HGB), but if a partner whose name is included in the Firma leaves the business, the old Firma can be kept only if he gives his consent (§24 II HGB).

2 Changes of Ownership

One must draw a very clear distinction between the Firma, the name under which the business is carried on, and the Handelsgeschäft, or the business enterprise itself, as an economic unit with all its assets and debts. The distinction becomes clear on a change of ownership. The rule given above (Section I 1) relates to the continuation of the Firma (abgeleitete Firma), not to questions of ownership and liability, to which we now turn.

A Firma can be transferred only in conjunction with the business to which it relates (§23 HGB). Private law does not recognize a business, or even its assets, as a unit of property. It is true that a single transaction suffices for an agreement to sell a business or its assets (§§433ff. BGB), but this contract of sale can be performed only by effecting separate conveyances of the individual items comprised in it, such as goods, land, rights of action, patents, and so on. In a partnership or company, on the other hand, changes in the ownership of the enterprise can be effected indirectly by altering the number of partners or share-

---

[4] OLG Hamm, *Betrieb* 1973, 2034. On the converse case of a partnership continuing the Firma of an individual merchant, see *BGHZ* 62, 224 (no obligation to add an indication of partnership).

holders or the size of their interest: the property of the partnership is in the joint co-ownership (Gesamthandseigentum) of the partners, and the property of a company belongs to it as a legal person (see Chapter 14 Section I below).

Since third parties may be unaware of a change in the ownership of a firm where the Firma continues in use, and may suffer harm if they have extended credit to the prior owners or have other rights against them, the law provides for the liability of the new proprietors. Anyone who acquires a commercial business and runs it under the old Firma is liable to the business creditors, but not the private creditors, of the previous proprietor (§25 I HGB), and a contractual exclusion of such liability has to be entered in the commercial register and published or immediately communicated to the creditors before it affects them (§25 II HGB).[5] If a merchant dies and his heir continues the business under the old Firma for more than three months, he will be liable to the decedent's creditors in full (§27 HGB) without being able to limit his liability to the value of the estate as he might do under the law of succession.

If a person joins the sole proprietor of a business and forms a partnership in the form of an OHG or KG, the new partnership is liable for the debts previously incurred by the proprietor in this business even if the Firma is changed (§28 I HGB); this liability does not arise from the legal impressions caused by the Firma, but rather from the fact that the same business continues to exist with all its property and debts.[6] The liability can be excluded by special agreement, provided it is entered in the commercial register and published or communicated without delay to third-party creditors (§28 II HGB). A person who joins an existing commercial partnership as a partner is invariably liable for the existing debts of the partnership and cannot exclude his liability at all (§130 HGB).

3 Legal Protection of the Business and the Firma

The 'right to an established and operative business' is now included by the courts among the interests that are protected against invasion by the law of tort (§823 I BGB; see Chapter 9 above). Further protection is afforded to the enterprise by the

---

[5] See *BGHZ* 29, 1.   [6] BGH *NJW* 1966, 1917.

law of unfair competition and restrictions on competition (see Chapter 15 below).

A person who is using a Firma which is irregular or belongs to someone else can be required to stop using it by the court in charge of the register (§37 I HGB; see Section II 4 below). If anyone is injured in his rights by another person's unauthorized use of his Firma, he can demand the discontinuance of this unauthorized use (§37 II HGB), and the erasure of the Firma from the commercial register if it has been improperly entered there. A claim for damages under the law of tort (§823 BGB) arises if the person making the unauthorized use was at fault, and a claim for damages or an injunction may also arise from invasion of the right to one's name (§12 BGB) or under the provisions of the law of unfair competition which render it unlawful to profit from the business reputation of another by using a Firma or other description of one's business that is apt to mislead (§1, 3, 16 UWG; see Chapter 15 Section I below).

4 The Commercial Register

The commercial register is designed to inform other businessmen and members of the public about the vital legal and economic facts relating to merchants. The commercial register is maintained by the Amtsgericht and is in two parts, Part A for individual merchants and partnerships, and Part B for companies. A distinction is drawn between making actual entries in the register and attaching documents and forms to it. For example, an increase in the capital of a stock company must be entered in the register (§188 I AktG), but its annual statement of accounts only has to be attached (§177 I AktG). Anyone may make an inspection and demand an authorized abstract of the commercial register and the documents attached to it (§9 HGB). Most entries are made on the application of the merchant. Sometimes he is under a duty to apply; for example a person who is a merchant under §1 HGB must apply to have his Firma included in the commercial register (§29 HGB) and the court can require him to do this on pain of a fine (§14 HGB). But certain facts, such as the opening of bankruptcy proceedings against the merchant, are entered in the register on the court's own motion (§32 HGB). Before making any entry the court checks it for legality and formal accuracy, but it goes into the underlying facts only when it has reason to doubt their veracity. In this it usually invokes the assistance of the chamber of commerce and industry. All entries must be promulgated in the *Bundesanzeiger* and at least one other newspaper (§10 HGB).

The matters where there is a duty to apply for registration include the following: the Firma on starting a basic commercial business (§1 HGB) or a commercial business under §2 HGB (§29 HGB), the change of location of a business (§13c HGB), the creation of a subsidiary branch (§13 HGB), the alteration of the Firma or its proprietors (§31 HGB), the granting or withdrawal of a Prokura, a special commercial power of attorney (§53 I, III HGB; see Section III 2 below), and the incorporation of a stock company or limited liability company (§36 AktG, §7 GmbHG).

All entries in the register carry a presumption of legality and accuracy, which may be very important procedurally, but otherwise their effects may differ. Many entries have constitutive effect, that is, they actually alter the legal position; thus a person who is running a commercial business under §2 HGB or an agricultural business under §3 HGB becomes a merchant only on the making of the entry; and entry is necessary in order to give legal personality to a nascent stock company or limited liability company (§41 I AktG; §11 I GmbHG). Other entries have declaratory effect, that is, they simply testify to a legal situation already in existence, such as the status of merchant under §1 HGB, or the creation or withdrawal of a Prokura (see Section III 2 below). The law protects third parties who rely on the contents of the commercial register: a fact requiring registration affects third parties only fifteen days after registration and publication (§15 II HGB).

For example, if Dörr, a merchant, withdraws the Prokura (power of attorney) which he had granted to his man Laspe, and Laspe nevertheless borrows money from a bank in Dörr's name, Dörr will be contractually liable to the bank if the withdrawal of Laspe's Prokura was not registered and published (§15 I HGB). This applies even if the fact that Laspe had been granted a Prokura was never registered, as it should have been, for even in that case the withdrawal of the Prokura must be registered (§53 III HGB); the question is, however, disputed.

A third party is thus entitled to rely on the absence of an entry in the commercial register; his negative reliance (negative

Publizität) is protected. Further protection of positive reliance was introduced by an amendment to the code in 1969 which was designed to harmonize the law within the European Community: reliance on an official publication to the effect that a certain entry has been made in the commercial register is protected, even though the publication or the entry be false, and though the merchant in question neither caused nor knew of it (§15 III HGB).

Suppose, for example, that Dörr, a merchant, has granted a Prokura to Laspe and caused this fact to be entered in the register, but that it is published that the grantee of the Prokura is Laaff. If Laaff borrows money from the bank in Dörr's name, Dörr will be liable to the bank. Whether this applies when it was not Dörr at all, but someone else entirely, who gave the Prokura and asked for it to be registered is doubtful.

III AGENCY IN CIVIL AND COMMERCIAL LAW

1 The German Concept of Agency

The basic concept of agency in German law (Stellvertretung) is much the same as in English law: the agent (Stellvertreter) is a person whose conduct in transactions, by making or receiving declarations of will, has direct legal effect in the person of the principal (Vertretener) (§164 I BGB). Unlike English law, German law recognizes only disclosed agency (offene Stellvertretung): the agent must act in the name of the principal (§164 II BGB). In accordance with the so-called 'principle of abstraction', German law makes a sharp distinction between the external relationship (Aussenverhältnis) that arises between agent and third party, and the inner relationship (Innenverhältnis) that exists between agent and principal. In the external relationship what is vital is the power of the attorney (Vertretungsmacht), his authority as agent to act with effect for the principal. In the internal relationship between principal and agent a contract of mandate or other legal duty will be controlling.

The power of an agent may arise in various ways: it may be granted by legal act (in which case it is called Vollmacht) or conferred by law; thus parents act as the statutory agents of their children (§1626 II BGB), and a guardian is the statutory agent for his ward (§1793 BGB). In the case of a legal person, those through whom it acts, such as the president and managers (Vorstand) of a registered association or Verein (§§21–53 BGB), or the board of management of an Aktiengesellschaft (§§76–94 AktG), are called its 'organs', and are designated by law as its agents (§26 II BGB; §78 I AktG). The provisions of the BGB relating to Vollmacht or the power of agency which is granted by legal act (§§164–81 BGB) also apply, so far as they can, to cases of agency arising by law.

There are no formal requirements for the creation of an agency (Vollmacht; §167 II BGB): the principal may create it by declaration to the agent or to the third party (§167 I BGB), either orally or in writing, and may revoke it in the same manner (§168 sentence 3 BGB).

According to the principle of abstraction already mentioned, such revocation is possible even though the underlying relationship between principal and agent remains in being (§168 sentence 2 BGB). On the other hand, if the underlying relationship is brought to an end (for example, because the principal's instructions have expired through lapse of time), the power itself is taken to end, unless there are contrary indications (§168 sentence 1 BGB). Unless it is a general power, an agency may be made irrevocable; such irrevocability, indeed, in exceptional derogation from the principal of abstraction, may arise from the basic relationship of principal and agent (§168 sentence 2 BGB), and is assumed to do so in the case of powers coupled with an interest.[7]

The law seeks to protect third parties who rely on the continuance of the agency. Thus an agency must be revoked in the same manner as it was created, whether by special communication to the third party, by general publication, or in a written document. Until, for example, the document is returned or declared inoperative, a third party who acts on the assumption that the agency still exists receives protection (§§170–3 BGB).

The scope of the agency, that is, what transactions the agent may conclude, depends on the terms of the declaration of agency, for example, whether his agency is general or delegable (Untervollmacht). The power of an agent does not normally include a self-serving power to effect a contract between himself and his principal, but he may do so if he has special authority to

---

[7] BGH *WM* 1965, 107; *RGZ* 53, 416, 419.

this effect or if it is necessary in order to fulfil a legal obligation (§181 BGB; see below on commission agents). If the question arises whether a contract made by the agent is voidable for mistake or deceit, or whether title has passed to the principal as bona fide purchaser, it is the agent's mistake or good faith that is relevant, not that of the principal; it is different, however, if the agent was acting on the specific instructions of his principal (§166 BGB).

If an agent acts without authority *(falsus procurator)* he is personally liable to perform the contract or pay damages (§179 BGB), but the principal may ratify the contract with retroactive effect (§§177, 184 BGB), and in the meantime the innocent third party may withdraw (§§177 II, 178 BGB). Responding to the idea which runs throughout the German law of agency, that protection should be afforded to third parties who rely on express declarations or specific conduct, the courts in a whole series of cases have found a valid power of agency where in reality no such power had been granted by the principal. One can perhaps make a comparison here with the English doctrine of authority by estoppel. The cases fall into two groups: first, where the principal's conduct has created the impression that he has granted the supposed agent power to represent him (Anscheinsvollmacht; see also §56 HGB, and Section III 2 below); second, where a person acts in commerce as if he had authority from the principal, and the principal culpably fails to make it clear that no such power exists (Duldungsvollmacht).

Anscheinsvollmacht, or ostensible agency, may arise when managerial employees make declarations in the normal course of business on the firm's notepaper: for example, the employee of an insurance firm who issues a cover note may be treated as empowered to do so.[8] Repeated prior authority to act may serve as a basis for ostensible agency in a subsequent matter.[9] And the display of the main architect's name on the board of a construction site has been held to give rise to ostensible agency.[10]

2 *Commercial Power of Attorney: Prokura and Other Forms*

The HGB defines the powers of agents under the typical forms of commercial agency. Here again the idea of protecting third

[8]OLG Köln, *VersR* 1965, 54.   [9]BGH *LM* §167 BGB no. 4.
[10]OLG Stuttgart, *NJW* 1966, 1461.

parties is predominant: they should be able to tell the extent of a person's power from the type of commercial agency he has been granted.

The widest form of commercial agency is the Prokura. 'A Prokura empowers the procurator to undertake in and out of court all manner of transactions appertaining to the management of a commercial business' (§49 I HGB). The Prokurist or procurator is thus the alter ego of the merchant who empowers him to act. A Prokura is granted to only the most trustworthy senior colleagues. The merchant principal can certainly lay down in his agreement with the procurator that the procurator shall enter only transactions of a certain type, but such a limitation on the scope of a Prokura applies only in the inner relationship of principal and procurator and is ineffective in relation to third parties (§50 I, II HGB). Only for dispositions or encumbrances of land does the procurator require express extensions of his Prokura or agency (§49 II HGB). The power of a procurator can be limited to matters concerning a particular establishment, provided that the establishment has its own Firma or business name (§50 III HGB), and a limitation can also be achieved indirectly by granting the Prokura to several persons jointly (Gesamtprokura, §48 II HGB).

Only the proprietor of a business or his statutory agent can grant a Prokura, and the grant must be express (§48 I HGB). It is not transferable, but it is unaffected by the death of the proprietor of the business. It is revocable at any time (§52 HGB). Both the grant and its revocation must be entered in the commercial register (§§53 I, III HGB), and while the efficacy of the grant or revocation does not depend on registration, third parties who rely on the commercial register are protected in the manner already described (Section II 4). The procurator should sign documents with the name of the firm, his own name, and an indication that he is acting in his capacity as procurator (§51 HGB; e.g., 'Werkzeugmaschinenfabrik Schoppa & Wagner, Gerhard Etzel p.pa.'), but nowadays it is regarded as sufficient if he just gives the name of the firm and signs his own name.

There are powers of representation falling short of a Prokura that a merchant can also grant. The Code uses the term Handlungsvollmacht, or commercial power of attorney, when a person, without being made a procurator, is empowered to run a business or to enter specified transactions or types of transaction (§54 I HGB). Such a power may cover all the transactions appertaining to the business or the usual incidents of the transactions specified in the authorization. Since there may thus be quite wide variations in the power to act, the Code indicates the typical scope of such a power, and protects third parties by providing that, if the principal has conferred a lesser power, he must show that the third party was aware of the restriction (§54 III HGB). The typical power never extends to transactions in land, signing negotiable instruments, giving credit, or acts in litigation (§54 II HGB). The agent can delegate his power when this is normal (§54 I HGB), and he can transfer it to another with the principal's special permission (§58 HGB). The agency is not registrable in the commercial register.

A person with Handlungsvollmacht signs with an indication of his capacity, making it clear that he has no Prokura (§57 HGB; e.g., 'Möbelfabrik Ostmeyer & Co, i.V. Allroggen'). Such a power of agency is commonly granted to responsible senior employees such as chiefs of division, or to those who are in constant contact with the public such as cashiers or tellers in a bank. But independent merchants, such as commercial agents (Section III 3 below), may also receive such a power. A Person employed in a shop or open warehouse to which the public has access is treated as empowered to effect sales or receive money or goods in a manner typical of such a shop or warehouse (§56 HGB). This is a statutory example of Anscheinsvollmacht or ostensible agency (Section III 1 above).

3 The Commercial Agent (Handelsvertreter)

The commercial agent is an independent businessman (§§1 II no. 7, 84 I HGB) who is retained by an entrepreneur to negotiate business for him or enter transactions in his name, and who is paid a commission for doing so. A Handelsvertreter, or commercial agent, must be free to fix how and when he works; if he is subject to instructions on these matters, the Code treats him as a commercial manager (Handlungsgehilfe in the sense of §§59–82a HGB) rather than as a commercial agent (§84 II HGB). The commercial agent usually acts as a go-between, and the principal concludes with customers the contracts that he has negotiated. If a commercial agent without a power of attorney does conclude a contract on his principal's behalf, the principal is taken to have ratified the contract if he does not reject it expressly as soon as he hears of it (§91a I HGB). This is another instance of the idea of Anscheinsvollmacht or ostensible authority.

It is possible, however, for the commercial agent to be empowered to conclude contracts in the name of the entrepreneur. In this case his authority is of the Handlungsvollmacht type (Section III 2 above), but though he can conclude contracts on behalf of his principal he cannot subsequently amend their terms or extend the period of payment (§55 II HGB), and he needs special authority to accept payment for his principal (§55 III HGB). It is, however, within the normal scope of his authority to receive notice from his principal's customers of defects in goods or contractual performance, and in such a case he may ask the court in his principal's name (§55 IV HGB) to ensure that the evidence be recorded or secured (Beweissicherung: §485 ZPO).

The internal relationship (see Section III 1 above) between commercial agent and principal entrepreneur is constituted by a contract of services involving the management of affairs (§§611–30, 675 BGB), amplified by the rules laid down in §§84–92c HGB, whose principal purpose is to protect the interests of the commercial agent, often economically the weaker party and as dependent on the entrepreneur as a managerial employee. In a contract with a foreign agent, a contrary clause in the contract may render these provisions inapplicable (§92c HGB).

The commercial agent must bestir himself to obtain business for his principal, the entrepreneur, and must protect his interests in so doing (§86 I HGB). He must provide the entrepreneur with all appropriate information, especially regarding new business. He must act with all the care a proper merchant would exercise (§86 III HGB), as in checking new customers for creditworthiness. He must treat his principal's trade and business secrets as confidential, even after the end of the contract (§90 HGB). Unless the contract provides otherwise, a commercial agent is free to act for several principals, even if they are in the same line of business, unless their interests would be seriously prejudiced thereby. Often, however, a commercial agent is bound by a restrictive covenant. Such a clause must be in writing if it is to bind the agent after he has ceased working for the principal; it may not bind him for more that two years; and it must contain an engagement by the principal to pay adequate compensation to the agent during its currency (§90a I HGB). The agent is not bound by the clause if he terminates the contract by giving notice for an important reason, e.g., on the grounds of unfair conduct by the principal. On the other hand, if misconduct by the agent entitles the principal to terminate the contract, the agent remains bound by the clause but loses his claim for compensation (§90a II HGB).

The entrepreneur principal is obliged to supply the materials required by the commercial agent for his activites (drawings, designs, prospectuses, price lists, other advertising materials, business contacts), and to furnish all necessary information (§86a HGB). When the agent finds him customers, the principal is free to do business with them or not, but he must inform the agent of his decision without delay (§86a II HGB).

The Handelsvertreter is entitled to a commission fee with respect to all contracts concluded during the term of his contract with the principal which are the fruit of his present or former endeavours. This also applies to repeat orders by customers (§87 I HGB). If a commercial agent is appointed to a specified area or group of customers, he is entitled to commission on all contracts entered into with persons in this area or group, even if they do not result from his efforts (§87 II HGB). If business attributable to an agent's efforts is concluded after the end of his contract with the principal, but reasonably soon thereafter, he may claim commission (§87 III HGB), and his successor may not.

As soon as the principal has completed his side of the transaction, the agent's claim to commission falls due, but it expires if it becomes clear that the third party is not going to perform (§87a I, II HGB). A commercial agent may claim special commission if he collects payment from customers (§87 IV HGB), or guarantees their performance *(del credere* commission; §86b HGB). The commercial agent must bear his own normal business overheads unless some contribution from the principal is customary in that branch of trade (§87d HGB).

The contract of commission agency comes to an end on the expiry of the agreed term, on the death of the agent, by six weeks' notice to expire at the end of any calendar quarter (three months' notice if the contract has run for more than three years), or by extraordinary termination for an important reason (§§89, 89a HGB: on extraordinary termination see Chapter 16 below). On the termination of the contract, the commercial agent is entitled to an adjustment payment (Ausgleichszahlung) as compensation for loss of commissions and as consideration for his efforts to enhance the principal's business and goodwill (§89b HGB). The payment may amount to a year's commission income, averaged out over the past five years. The right to an adjustment payment cannot be excluded in advance (§89b IV HGB), but does not arise where the principal entrepreneur has terminated the contract for grave cause or where the commercial agent has terminated the contract without good reason such as illness, old age, or unfair conduct by the principal.

Müller acted as commercial agent for the Möbelfabrik Fa Rainer Göbel for many years, and built up an extensive network of faithful customers. He was then killed in a traffic accident caused by his own carelessness, and his widow claimed the adjustment payment under §89b HGB from Göbel. She was held entitled to succeed, since the claim was not barred by the carelessness of the decedent.[11]

4 The Kommissionär

The Kommissionär is an independent merchant under §1 II no. 6 HGB. His business consists in buying and selling goods or securities in his own name on the account of another — the Kommittent — and thereby earning commissions (§383 HGB). Since the Kommissionär acts in his own name, the agency is undisclosed and therefore not recognized as agency by German law (see Section III 1 above): the Kommittent does not become party to the contracts of sale which the Kommissionär enters on his account.

We must therefore distinguish at least three different contracts in the Kommissionär's business. First, there is the contract of commission whereby the Kommittent engages the

---

[11]BGHZ 24, 214; on the negligence of a commercial agent, see BGHZ 41, 122; according to BGHZ 45, 385 a claim for adjustment payment may still lie even though the agent has committed suicide.

Kommissionär to buy or sell for a commission. This is a contract for the management of an affair in the sense of §675 BGB, for which the HGB has special rules (§§383–405 HGB). Second, there is the contract of sale which the Kommissionär then concludes with a third party in pursuance of his commission. This contract is then performed by the transfer of the object of the sale, so that a Kommissionär who is engaged to buy often becomes owner of the purchased goods. Third, there is the transfer by the Kommissionär to his Kommittent of the economic benefit of the business so transacted; thus, where he is engaged to buy, the Kommissionär must transfer the ownership in what he has bought or assign his right to claim delivery of it from the third party, and when he is engaged to sell, he must transfer the proceeds of the sale or assign the right to claim the price.

Commission business used to figure largely in the export and import business, and is still important in the art world, the wine trade, and, above all, in dealings in securities. An example may help. Suppose that A engages an art dealer, K, to buy a valuable painting for him, and gives him a large sum of money on account towards the purchase price. Until A obtains the picture or gets his money back, he runs the risk of K's insolvency. If K can buy the picture from its owner, V, K will be able to transfer to A his claim against V for the delivery of the picture. If, before he does so, one of K's creditors tries to execute on this claim, A is nevertheless protected (§392 II HGB): as against K's creditors, K's claim against V is treated as already belonging to A.[12] For the same reason K cannot make a valid transfer of this claim, say as security to one of his own creditors, even if the transferee is in good faith.[13] K's creditors may, however, execute on the picture itself once V has transferred it to K. The only way A can protect himself against this is by procuring a rapid transfer to himself, which requires an agreement with the owner over both the transfer of ownership and the transfer of possession (§§929ff. BGB). It is possible for A to make a prior agreement with K that K should possess the picture for A: in such a case, A would become the indirect possessor through the so-called anticipatory *constitutum possessorium*. The requisite transfer agreement between A and K can also be made beforehand. Alternatively, A can give K authority to effect the transfer agreement between them: this instance of contracting with oneself is permitted under §181 BGB (see Section III 1 above). Finally, there could be a direct

[12] *RGZ* 148, 190. [13] BGH *BB* 1959, 975.

transfer of ownership from V to A, but for this A would have to make K an open agent, and K would have to conclude the agreement with V in A's name.

If A asks his bank to buy quoted securities for him, the bank can accept this commission and itself sell these shares to A (Kommission mit Selbsteintritt (commission with personal intervention) under §400 HGB); the bank must inform A of the fact (§405 HGB) and charge him the going price on the stock-exchange or in the market. But the bank can also bypass the commission contract altogether and act as simple seller (or buyer) *vis-à-vis* A (Eigengeschäft). Transfer of ownership in the securities can be effected by the bank's sending the customer a numbered list of the securities on deposit in the bank (§18 III DepotG); if the securities are in general deposit in a collecting bank, the bank can make the customer joint owner of part of the holding.

The commission contract determines the duties of Kommittent and Kommissionär in much the same way as in the case of a commercial agent. The Kommissionär must execute the business with care and must safeguard the interests of his Kommittent, submit accounts, and transfer any proceeds (§384 HGB). He must not go beyond the instructions of the Kommittent, such as a price limit (§§395, 386 HGB), but if he can do the business more favourably, this benefits the Kommittent (§387 HGB). In a case where the Kommissionär intervenes personally, his duty to account and safeguard the principal's interests is effectively satisfied by proving that he kept to the market or exchange price (§400 II HGB). The Kommittent must indemnify the Kommissionär for expenses incurred in negotiating the business, and must pay the normal or agreed commission (§396 HGB). A Kommissionär who guarantees the third party's performance of the transaction in question can claim a special commission as a *del credere* agent (§394 HGB). The Kommissionär has a security interest in the goods on commission in respect of moneys due to him from the Kommittent (§397 HGB).

5 Agency and Sales Oranization

An enterprise that wishes to set up a distributive network has a wide range of legal devices at its disposition, many of which are not even mentioned in the HGB. Let us first take the case that an enterprise organizes its sales on the basis that it is itself to become party to the contract with its customers, as can often be observed in sales of gasoline or new cars. These contracts may be effected by employees of the enterprise or by independent salesmen. In the former case, the enterprise runs sales outlets or has an 'Aussendienst', very common in the insurance business, where the contracts are concluded by employees with a commercial power of attorney. Otherwise the distribution can be entrusted to independent salesmen, such as commercial agents, with or without power to conclude contracts under §55 HGB, who also have an important role in insurance marketing.

Distribution may also be effected through independent salesmen who contract with consumers in their own name. Here the Kommissionär appears. The special position of the Kommissionär who works exclusively for one enterprise, which remains in the background, is not regulated by the HGB. In practice he is called a Kommissionsagent, and the courts apply the rules relating to the commission contract reinforced by the provisions that protect the commercial agent (e.g. §§87 II, 89b HGB). More often an enterprise uses independent Vertragshändler, or dealers under contract, who form part of the distributive organization of the enterprise but transact with the customers in their own name for their own account. They have a long-term contract for the sale of the enterprise's products in a particular area and use its trade-marks and publicity. There are no statutory provisions regarding the Vertragshändler, but the courts have held that where the Vertragshändler's contract comes to an end, he is entitled to an adjustment claim just like a commercial agent.[14]

IV CREDIT AND SECURITY

The topic of security for credit includes not only the central case where a lender wants security for the repayment of the loan, but all cases in which a person who renders a performance now but is only to receive the counterpart later (such as a supplier of goods on the terms that payment is to be made in three months) wants a present assurance that the payment will be forthcoming. Only the most important forms of security can be briefly reviewed here.

[14] *BGHZ* 68, 340.

Personal security is exemplified by the contract of suretyship (Bürgschaftsvertrag: §§765–77BGB). The surety agrees to be liable to the creditor for performance to him by a third party (§765 BGB); unless the surety is a merchant (§350 HGB), his undertaking must be in writing (§766 BGB). The surety is basically liable only to the extent that the debt for which he has stood surety actually exists; he can use against the creditor any defences available to the third party (§§767, 768 BGB). Suretyship therefore does not always enable the creditor to execute swiftly on the security, and for this reason security by means of guarantee has become more common. The guarantor is liable for the result specified in the contract of guarantee, for example, that the debtor actually perform. Unlike the surety, however, the guarantor is under a duty which is independent of the existence or amount of the underlying debt. The BGB does not regulate the contract of guarantee, although it recognizes it. In addition to suretyship and guarantee, there is the device of cumulative assumption of debt: if the principal debtor finds a person who is willing to be liable to the creditor along with him, they become debtors in common (Gesamtschuldner). This form of security is also unregulated, though recognized, by the BGB. Finally, a common form of personal security in commerce occurs when a person accepts, endorses, or even draws a bill of exchange in order to help someone else obtain credit: banks frequently offer their customers such an Avalkredit.

Real security in assets may be preferred to personal security. Since German property law respects the principle of specificity, such a security can be created only in precisely identified individual things or rights, not in a collection or stock of things: the floating charge is unknown to German law, although in practice forms of security have been developed that are not unlike it in their result. Real security, or Pfandrecht, entitles the person for whose benefit it is created to obtain satisfaction of a claim or money debt by executing on a specified thing (§§1113, 1191, 1204, 1273 BGB). Security rights in land differ from those in goods and choses in action. The most important security rights in land are the mortgage or Hypothek (§§1113ff. BGB), created for a particular debt and dependent on the existence and extent of the debt just like suretyship (Akzessorietät, the accessory security: §§1137, 1163 BGB), and the land charge or Grundschuld, which, like guarantee, is not dependent on the existence of the debt secured (§§1191 BGB). Securities in land must be entered in the Grundbuch or land register, and are frequently used to raise credit for individuals, farmers, and, above all, land developers. Security in land extends to any buildings on it and their accessories, such as the tools, equipment, and machines appertaining to the business being run there, provided that they belong to the owner of the land(§1120 BGB).

The principle of publicity runs through the security law of the BGB. Publicity for mortgages and land charges is achieved by requiring them to be registered in the Grundbuch. In the case of pledges of moveables, possession must be transferred to the creditor, and the pledgor must give up possession entirely (§1205 BGB); notification to the debtor is required for the pledge of a chose in action (§1280 BGB). These two requirements were found to be tiresome in practice: a merchant cannot give up the possession of his goods in order to raise credit since he needs them to trade with, and people were reluctant to publish the amount of security they were giving. Today, therefore, the pledge is hardly ever used in order to raise credit, except in the form of the pledge of negotiable instruments, almost always with a bank (Lombardkredit). The provisions on pledges also apply to pledges which arise by operation of law under the HGB and BGB, such as the Kommissionär's right to secure his claims against his Kommittent by looking to the goods he was commissioned to buy (§397 HGB; see Section III 4 above).

In recent years businessmen have been using a quite different type of security in moveables and choses in action: instead of pledging such assets to the creditor, they make him full owner of them, subject to a fiduciary duty to exercise his rights as owner only so far as his security interest requires. Such a transfer of ownership in goods is called 'Sicherungsübereignung' (transfer by way of security): the ownership in the goods is transferred to the creditor, but the possession remains with the debtor, who now 'possesses for the owner'. This method of transfer suffices for the conveyance of ownership (§930 BGB), though not for the creation of a pledge. In Sicherungszession (assignment by way of security) the chose in action is fully assigned to the creditor instead of being pledged. In both cases the person providing the credit acquires complete ownership, but he is in a fiduciary position and may exercise his rights only for the purposes of his security; once those purposes have been answered, by payment or otherwise, the transferor may stop a third party executing on the asset. As to specification, it is sufficient if the goods so transferred are always identifiable, as by reference to their location: security can thus be given in the revolving stock in warehouses. Choses in action, which are commonly assigned in bulk, must be determinable at the time of the assignment.

Financiers such as banks often use both these devices to secure their advances. So, too, do suppliers in respect of the price due to them. Their general conditions of business usually provide that ownership in the goods is retained until they are paid for in full (Eigentumsvorbehalt: §455 BGB), but this is a somewhat ephemeral security: the purchaser will likely resell, in which case ownership may pass to the subpurchaser by good faith acquisition (§932 BGB), and if he processes them first, title may pass to him by operation of law (§950 BGB).Consequently, a special arrangement is made which gives the supplier further security: he takes a present transfer of the title which the purchaser may acquire by processing the goods and a present assignment of the rights which the purchaser will acquire on reselling to his customers. The rights resulting from such a verlängerter Eigentumsvorbehalt (extended reservation of title) often come into conflict with similar security rights acquired by banks. In such a case the courts always give priority to the supplier.

In the result, the principle of publicity which the legislator of the BGB had in mind has been completely reversed: the fact that a merchant has huge stocks of goods or an enormous turnover tells one nothing about his financial position, for all his goods and claims may already have been made over to third parties, and the unsecured creditor may, as so often happens, leave the bankruptcy proceedings empty-handed.

14

Partnerships and Companies: Business Organization*

I COMPARATIVE SURVEY

1 The Models in the BGB and their Variants in Business Law

Despite the differences in their legal traditions, all the industrialized countries of the West that have market economies in which business is run by private owners and investors have developed very similar legal forms of business organization. Germany is no exception. Here, too, we find the common legal forms of business organization, partnerships and companies.

The Civil Code itself provides the model for the partnership. According to §705 BGB, partnership is a contract whereby the partners agree to promote a common purpose and to make the stipulated contributions. In commercial life syndicates of banks for issuing loans or debentures may take the form of civil partnerships, as may consortia of construction firms for individual building projects. Two variants of the civil partnership are provided by the Commercial Code: the general commercial partnership (offene Handelsgesellschaft, OHG; §§105–60 HGB), and the limited partnership (Kommanditgesellschaft, KG; §§161–77a HGB). Both of these are widely used for the organization of businesses of small or medium size, while the KG is now also used for investment projects (Publikums-KG; see Section III 2 below). For the person who simply wants to invest in a mercantile business, the option of the silent partnership exists (stille Gesellschaft; §§335–42 HGB), and there is a

*Apart from works on the law of business associations and translations of the German legislation, one may usefully consult: Fitting, Wlotzke and Wissmann, *Kommentar zum Mitbestimmungsgesetz* (2nd ed. Munich 1978); Horn and Kocka, *Law and the Formation of the Big Enterprises in the Nineteenth and Early Twentieth Centuries* (Göttingen 1979); Kraft and Kreutz, *Gesellschaftsrecht*, (2nd ed. Frankfurt 1977); Reinhardt, *Gesellschaftsrecht. Ein Lehrbuch* (Tübingen 1973).

special form of partnership if the business is a maritime one (Partenreederei; §§489–508 HGB).

When we come to incorporated associations, the Civil Code also provides a model: the registered association (eingetragener Verein; §§21–53 BGB). Although the registered association is rarely used for business purposes, its rules are not without relevance in business law, since some of them also apply to associations of other types; for example, the provision on vicarious liability (§31 BGB) has been extended to render a company liable in tort for its directors, and it has even been applied to commercial partnerships by the courts.[1] Large enterprises are usually organized as a stock company or public limited company (Aktiengesellschaft, AG). The AG used to be regulated in the Commercial Code, but since 1937 it has been the subject of special legislation, and is now regulated by the Aktiengesetz of 1965 ('Law of Shares' or Company Act, AktG). Ever since 1884 company law has imposed strict rules on public limited companies so as to protect creditors and investors against fraudulent and incompetent promoters and directors, and to make them suitable investments for the public. While the shares of many AGs are quoted on the stock exchange, others are unquoted and still closely held, for example by the members of a family. Privately held companies may use a special legal form, the Kommanditgesellschaft of Aktien (KGaA), which has one or more general partners in addition to shareholders (§§278–90 AktG). The Gesellschaft mit beschränkter Haftung (GmbH), or private limited company, was introduced by an Act of 1892, amended in 1898. It has a simpler legal structure than the AG, being tailored for a small number of shareholders, and it has some of the features of a partnership; its shares cannot be traded on the stock exchange. The GmbH is very popular for small- and medium-sized businesses, and even some very big enterprises use it. A foreigner who was setting up a German subsidiary or investing in a joint venture would normally choose the GmbH.

If an association is to be set up not to make a profit but to enhance the business of its members, it may incorporate as a registered co-operative (eingetragene Genossenschaft) under

[1] *RGZ* 76, 35, 48.

special legislation (Genossenschaftsgesetz of 1889, GenG). Its purposes might include the common purchase of goods or raw materials, the common marketing of products, or the creation of a credit system based on mutual help (Genossenschaftsbanken). Some mining companies, under old regulations relating to mines, are still in the form of a Gewerkschaft rather than an AG or GmbH. Insurance companies may be organized as an AG or as a mutual insurance association (Versicherungsverein auf Gegenseitigkeit, VVaG) under the law for the control of insurance (Versicherungsaufsichtsgesetz of 1931, VAG). There is also legislation for specialized banks (mortgage banks, investment funds).

2 *Ownership, Liability, and Legal Personality*

The distinctive feature of a company as compared with a partnership is that it has legal personality. The partnership is not a persona in law: the partners are the proprietors of the enterprise, the joint co-owners of the assets (Gesamthänder). They are jointly and severally liable for the debts incurred by the partnership (Gesamtschuldner; see §§421ff. BGB), though in a KG the liability of the limited partner or Kommanditist is limited to the unpaid amount of his partnership contribution (§171 HGB).

As personae in law the AG and GmbH are themselves the proprietors of the enterprise they run and the owners of its assets. The shareholders are owners only in an intermediate way, in their capacity as members of the incorporated association. Nor are the shareholders personally liable for the debts of the company. For continental lawyers limited liability is a natural consequence of the company's having juristic personality, but it is not a necessary one. In the KGaA, the general partners are personally liable for the debts of the company, while shareholders are liable only to pay in the unpaid amounts of their shares. In the GmbH, a sort of joint liability exists in this respect, for all members are liable for the unpaid calls of fellow members (§§24, 31 GmbHG); in addition, members may have to make further contributions if the statute of the GmbH so provides (§§26–8 GmbHG).

3 Lifting the Corporate Veil

In German law there are only a few cases where a shareholder is personally liable to the creditors of his company. Anyone, including a shareholder, who uses his influence on a company to its detriment is liable in damages to the company, its shareholders, and its creditors (§117 AktG); a similar liability is imposed on a shareholder who receives money or other assets from a company otherwise than as provided in the Act (§62 AktG); and the statutory rules on combines (Konzernrecht; see Section VII below) make a parent company liable for its subsidiary under certain conditions. But apart from these special statutory provisions there are only a few cases where the courts have 'lifted the corporate veil' and held a shareholder personally liable to the creditors of his company.[2]

Even where a company is in the control of a sole member and has an inadequate capital base, the courts are not always ready to go behind the company and make the sole member liable,[3] though if special circumstances exist, liability may be imposed in tort for causing financial loss to the creditors in an immoral manner,[4] and the sole member of a company has been held liable to corporate creditors where he has not kept his own affairs clearly separate from those of the company he controls.[5]

4 Management and Representation

Partnerships and companies have to act through natural persons. They are called 'Organe'. The partners themselves act for the partnership (Selbstorganschaft), while in the case of a company (AG, GmbH), people are appointed to act as Organe; they need not be shareholders, and usually are not (Fremdorganschaft; for details, see Section VI 2 below).

We have already seen how the German law of agency generally makes a clear distinction between the external relationship, that is, the power of the agent to represent the principal, and the internal relationship (see Chapter 13 Section III 1 above). It is the same with Organe. We must distinguish between their power of representation and their power of management, that

---

[2]Wiedemann, I *Gesellschaftsrecht* §4 III 2 (Munich 1980).
[3]*BGHZ* 68, 312, 315.
[4]BGH *WM* 1979, 229.
[5]BGH *WM* 1958, 463.

is, their ability to act and make contracts with third parties in the name of the partnership or company on the one hand, and their right as against partners and shareholders to deal with some particular aspect of the company or partnership on the other. Ideally, the power of representation and the right of management go hand in hand, but there may be a divergence, and certain legal consequences follow if there is.

II THE GENERAL COMMERCIAL PARTNERSHIP (OHG)

1 Formation

According to the Commercial Code, a partnership of two or more persons for the purpose of running a commercial business under a firm name is a general commercial partnership (offene Handelsgesellschaft, OHG) if the personal liability of none of the partners is limited (§105 HGB). Thus, in order to form an OHG there must be a partnership agreement (§§705 BGB; §109 HGB) which does not restrict the liability of any partner, a mercantile business, and a firm name (Firma). The OHG must be registered in the commercial register (§§106–8 HGB).

An OHG can be formed, for example, for the wholesale distribution of goods, for that is a mercantile business under §1 II no. 1 HGB; and as soon as the partners embark on their commercial activities, even before the partnership is registered, they are treated as an OHG vis-à-vis third parties (§123 II HGB). If real estate brokers join together, they normally form a civil partnership, because theirs is not a mercantile business (§§1 II no. 7, 93 II HGB), but if it requires a commercial business organization, they must apply for registration under §2 HGB, and only upon registration will they be treated as an OHG (§123 I, II HGB). A law firm can be organized as a civil partnership only (see Chapter 13 Section I 2 above).

Although the OHG is not a persona in law, it can under its Firma acquire rights, including real property, it can incur liabilities, and it can sue and be sued (§124 I HGB). The assets of the OHG are held by the partners in joint co-ownership (Gesamthand; see §§718, 719 BGB) and may be seized by the creditors of the OHG in execution of a judgment against it (§124 II HGB). The OHG is governed by the terms of the partnership agreement, the relevant provisions of the HGB

(§§105–60 HGB), and the provisions of the BGB as to civil partnerships (§§705–40 BGB) in so far as they are not excluded or modified by the HGB.

2 The Partners: Rights and Duties

Partnership in an OHG is not restricted to natural persons: a legal person such as an AG or GmbH can be a partner, as indeed can another OHG or a KG.[6] The partnership agreement determines what partners must contribute (in cash, kind, or labour) and how they are to share in the assets of the OHG. A partner's participation in the profit and loss of the partnership is normally in proportion to his share in the capital, but a special share in the profit may be fixed in the partnership agreement (§§120ff. HGB). No partner may compete with the OHG, either by running his own business or by acting as a partner in a competing partnership, without the consent of his fellow partners (§112 HGB).

Every partner may and should take part in the management of partnership affairs unless it has been put in the hands of some partners exclusively (§114 HGB), which often happens. Unless otherwise provided, each managing partner is entitled to act alone, subject to the veto of any other managing partner (§115 I HGB). Even if a partner is empowered to conduct business on his own, he must obtain the approval of his fellow partners if he wishes to go outside the normal running of the business and do something unusual: if he wants to appoint a procurator, he must ask the other managing partners (§116 II, III HGB). Usually the partnership agreement itself regulates the details of the internal working of the firm and allocates the power of decision-making. If it does not, the law itself provides that all the managing partners must be unanimous; if the agreement provides for majority decisions without specifying how the majority is to be ascertained, votes are by head rather than by share (§119 HGB).

3 Representation and Liability

Now we leave the power of management in the internal relationship and turn to the power of representation in the external

[6]Baumbach-Hueck, *Handelsgesetzbuch* §105 Anm 1 C (22nd ed. Munich 1977). See *RGZ* 105, 102 on juristic persons (AG, GmbH) as partners of an OHG or KG.

relationship. In principle, every partner is entitled to represent the OHG as against third parties, and may thus make contracts in the name of the partnership (§125 I HGB). This power of representation may, however, be excluded by the partnership agreement, which may alternatively provide that a partner may represent the partnership only if he acts in conjunction with other partners or with a procurator (Gesamtvertretung; §§125 II, III HGB). Such an exclusion or limitation must be entered in the commercial register (§125 IV HGB). If a partner does have a power of representation, it is very extensive – he may even enter land transactions or grant a Prokura — and no limitation on the scope of such a power affects third parties (§126 HGB). There is only one implied limitation on the scope of a partner's authority to represent the partnership: he cannot conclude a contract that interferes with the common partnership agreement; for example, he cannot accept a new partner. Nor can he sell the business.[7]

Example: A, B, and C, partners in an OHG, agree that no loan of over DM100,000 should be taken from a bank without the previous consent of all partners. At a time when A is extremely ill and B is on safari in Brazil, C is faced with an urgent liquidity problem in the business and borrows DM150,000 from a bank on its behalf. In addition, another bank agrees with him to become a silent partner (see Section IV below) with a capital participation of DM200,000, provided it is granted certain rights to control the management: C agrees with it that the OHG will not undertake certain financial transactions without the consent of its representative. In the result the OHG is bound by the first loan contract: C's authority to represent the OHG extends to every such contract (§126 HGB), regardless of whether there is an emergency or not, and even though, as an internal matter, it was a contract that C was not to conclude, either because of an express agreement or because it was an unusual transaction under §116 I HGB. The second contract, however, goes beyond the scope of C's authority because the grant to the silent partner of those rights of control over the management interferes with the allocation of decision-making power and responsibility under the partnership agreement. The second contract is therefore void unless A and B ratify it.

The HGB distinguishes the liabilities of the partners (§128

[7] *RGZ* 128, 172, 176.

HGB) from those 'of the partnership' (§§124, 128 HGB), but since the OHG is not a persona in law, the liability 'of the partnership' refers to the joint liability of the partners to respond with the assets of the OHG. The OHG is liable on all contracts validly concluded in its name (§124 HGB) and, by analogy with §31 BGB, for all the tortious acts of the managing partners as its Organe.[8] All partners are simultaneously liable with their personal property jointly and severally (Gesamtschuldner; §128 HGB), and anyone who enters a partnership becomes liable for its prior debts (§130 HGB). Thus a creditor of the OHG may sue the partnership and execute on the jointly owned assets of the partnership, or he may sue one or all of the partners in order to reach their private property; in practice, the suits are usually joined.

4 Dissolution

An OHG comes to an end on the expiry of the period of time for which it was to last; on resolution by the partners; on the opening of bankruptcy proceedings against the OHG or any of its partners; on the death of a partner (unless the partnership agreement provides otherwise); on a partner's giving six months' notice to expire at the end of a business year; and on a court decision at the suit of a partner, where there is sufficient reason, that the partnership be dissolved (§§131–4 HGB). If notice is given, as it may be, by the creditor of a partner who has seized his share of the partnership (§135 HGB), the other partners can opt to buy out that share and continue the partnership among themselves (§141 HGB). Where application could be made to a court to have the partnership dissolved on the ground of the misconduct of one partner or some other matter relating to him personally, application may be made instead to have him excluded from the partnership (§140 HGB); where there are only two partners, this rule gives the innocent partner the right to keep the business going by buying out the other (§142 HGB). After dissolution the partnership is wound up by liquidators, who may be partners and must act jointly: their names must be entered in the commercial register and the words 'in liquidation' must be added to the Firma (§§145–58 HGB).

[8] *BGHZ* 45, 312.

5 Changes in Membership

Since the personal link between the partners underlies the statutory model of partnership, there must in principle be an alteration of the partnership agreement whenever a new partner is taken on; and when a partner dies or goes bankrupt, as we have seen, the partnership may be dissolved (§131 nos. 4 and 5 HGB). The law itself, however, sometimes makes it possible to exclude a partner yet keep the partnership going as an economic concern with the partners that remain, and partnership agreements commonly provide that in many other cases the partnership may be carried on despite changes in its membership. Two classes of case are worth mentioning.

The partnership agreement can provide that when a partner dies his heirs, or certain of them, should become partners in his place. Since this would make the heirs liable for the full amount of all partnership debts, they are allowed by law, within three months, to become Kommanditisten, and so have their liability limited to their share of the partnership (§139 HGB). Since the effect of such a successorship clause in the partnership agreement is that the heirs automatically become members of the partnership in the place of the decedent,[9] there may be a conflict with the law of succession. For example, if only one of three children is to become a partner, and the share in the partnership is the largest asset in the estate, the other children will not be fully indemnified. In principle, the successorship clause in the partnership agreement takes precedence over the law of succession, but it is not allowed to trench upon the statutory reserved portion of any heir.[10] Many other successorship provisions are in common use, such as one that grants an heir the right to be admitted to the partnership if he chooses.

Clauses in agreements that provide for the possibility of excluding a partner from the partnership raise other problems. They may validly extend the grounds of exclusion provided by law (§§140–2 HGB; see Section 4 above), and in particular may dispense with court proceedings and permit the exclusion of a partner by mere resolution of the others.[11] If such a resolution is to be based on a serious personal objection to the partner in question, such as suspicion of improper conduct,[12] the clause is upheld by the courts. Indeed, for a time the courts even held that a partner could be excluded at the mere dis-

---

[9] *BGHZ* 22, 186, 191.
[10] Lange-Kuchinke, *Lehrbuch des Erbrechts* §5 V and §31 V (2d ed. Munich 1978).
[11] *BGHZ* 31, 295, 298.   [12] *BGHZ* 31, 304.

cretion of the other partners. Today, however, the protection of the partner whom it is sought to exclude is taken more seriously: the exclusionary provisions must be unambiguously stated in the partnership agreement; the exclusion cannot be left to the free decision of the other partners, save in an exceptional situation with special features;[13] nor may the compensation for the excluded partner in such a case be limited to the mere book value of his share.[14]

III THE LIMITED PARTNERSHIP (KG)

1 Legal Structure

The Kommanditgesellschaft, or limited partnership, is a commercial partnership created to run a commercial business under a Firma, which has two types of partner: at least one general partner (Komplementär), and one limited partner (Kommanditist). The limited partner is personally liable to the creditors of the KG only up to the amount of his unpaid partnership contribution: if he has paid his contribution and not had it paid back to him, he is not liable to them at all (§§171, 172 HGB). The entry in the commercial register relating to the KG must state the amount of the contribution to capital of the limited partners (§162 HGB), and the limited partners must not consent to the partnership's starting business before registration, or they will be personally liable without limit to partnership creditors who had no prior notice of their status as limited partners (§176 HGB). Limited partners are excluded from the management of the affairs of the KG under the statutory model (§164 HGB), and have no authority to represent it (§170 HGB); they may, however, inspect and check the annual financial statement of the partnership (§166 HGB). On the other hand, as the limited partners are joint co-owners of the partnership assets together with the general partners, they participate in any increase in the internal value of the assets. Unless the agreement provides otherwise, limited partners, unlike general partners, are free to compete with the partnership (§165 HGB).

2 The KG in Business Life

For small- and medium-sized businesses the KG is a very popular form of organization. Having a partnership with

[13] *BGHZ* 68, 212, 215.   [14] BGH *WM* 1978, 1044.

limited liability is very attractive, though for this very reason the KG tends to have a rather lower credit standing than an OHG. The statutory model may be modified in many respects by the partnership agreement: limited partners who finance the KG are often given entire internal control over the management of its affairs, and they may also represent the KG by virtue of a Prokura conferred on them.

The KG is often used nowadays when some project is being financed by a large number of small investors, rather like a capital company. The numerous Kommanditisten furnish the capital, but mostly they have very little influence, if any, on the way the business is conducted: their rights of administration are assumed by trustees. The managing partners are generally empowered to take on other limited partners, the change of membership being facilitated by the fact that the shares are administered by trustees. Attractive tax advantages for investors often attach to such a method of financing a project, but there are also certain dangers for the limited partners and the creditors of the partnership, since just a few people may have control of the management of the business and of the supervisory board which is often formed. The courts have therefore applied to these large-scale KGs some of the principles relating to stock companies.

For example, although the liability of a partner as such is rather mild (§708 BGB),[15] a partner of a KG who was also a member of its supervisory board has been put under the stricter liability of company law (§§116, 93 AktG), and he remains liable for the full period of prescription.[16] If special advantages are to accrue to individual founders or partners, the contract must make this clear.[17] The courts have set limits to what may be done by majority decision (for example, when an increase in capital is proposed, every limited partner must be given a realistic opportunity to participate in the capital increase proportionally to his present holding),[18] and they are ready to use the standard of good faith and fair dealing (Treu und Glauben) to control the content of the agreements which set up such large-scale KGs.[19]

---

[15] BGH *NJW* 1977, 2311. [16] *BGHZ* 64, 238, 245.
[17] BGH *NJW* 1978, 755.   [18] *BGHZ* 66, 82, 86.
[19] *BGHZ* 64, 238, 241.

3 The GmbH & Co. KG

As we have already seen, a legal person may become a member of a partnership (Section II 2 above). In the 1920s people started to take advantage of this fact and set up KGs with a GmbH as the sole general partner. The shareholders in the GmbH (see Section V below) are usually also limited partners in the KG. Combining the advantages of a partnership with those of legal personality, this hybrid partnership-company found many admirers. The involvement of the GmbH enabled all natural persons involved in the enterprise to limit their liability completely. Furthermore, the GmbH made it possible to put the management of the KG's business in the hands of a suitable person who was not involved in the enterprise, for the GmbH was responsible, as the general partner, for the management of the KG, and a GmbH can appoint an outsider as its own manager (see Section V 2 below). Until 1976 the KG had the advantage over a company that it was not exposed to the double burden of corporate and individual income tax, so KG agreements provided that a relatively small proportion of the profits would go to the GmbH and much the larger share be transferred directly to the limited partners, there to be taxed. The tax reforms of 1976 have largely removed this problem (see Section IX 1 below); for, although there are still slight tax advantages in the GmbH & Co. KG in some respects (mainly with regard to property tax), most of the benefits can be achieved today by a simple GmbH. Nevertheless, the popularity of the GmbH & Co. KG does not seem to be diminishing in the least.

IV SILENT PARTNERSHIP

Silent partnership (§§335–42 HGB) is a contract between a financier who acts as silent partner and the proprietor of a mercantile business (individual, partnership, or company) whereby a sum of money is provided in return for participation in the profit and loss of the business (§§335–7 HGB). The silent partner does not thereby become a merchant himself, he does not become co-proprietor of the business assets, and he takes no part in any capital appreciation of the business. He has no influence on the management of the business, and his right to obtain information is limited (§338 HGB: inspection and verification of the balance sheet). The silent partner is not personally liable to the merchant's creditors. From the economic point of view, silent partnership is very like a loan with profit-sharing, the distinction being that the loss is shared as well. The partnership agreement, however, often provides for the silent partner to have rights, such as influence on the management of the business or participation in the capital appreciation of the business property, which put him in the same position as a true proprietor ('atypical silent partnership').

V THE PRIVATE LIMITED COMPANY (GmbH)

1 Definition and Formation

The private limited company (Gesellschaft mit beschränkter Haftung, GmbH) is a business association with legal personality under a special law of 1892 (GmbHG). A GmbH can be formed for any lawful purpose, mercantile or not, by one or more persons (§1 GmbHG). The formation of a GmbH by a single person was made possible by the 1980 amendment of the GmbHG in recognition of the wide use of one-man companies in German business life. The new rule is a remarkable deviation from established legal doctrine, as the articles of association (statutes) of a company are created by contract, and every contract requires two parties at least. The company contract must be in notarial deed (§2 I 1 GmbHG). This contract must specify the Firma (see Chapter 13 Section II 1 above), and the domicile of the GmbH, its purpose, the amount of the capital stock, and the amount of the capital shares held by the several shareholders (§2 GmbHG). The capital stock must be at least DM50,000 (until the end of 1980, DM20,000), divided into shares (Geschäftsanteile) of at least DM500 (§5 GmbHG). The GmbH must be registered in the commercial register (§§7, 10 GmbHG) and becomes a persona in law only upon such registration (§13 GmbHG). Those who act on behalf of the future GmbH before it is registered are personally liable to third parties (§11 II GmbHG).

Before registration, 25 per cent of the face amount of each share, up to at least DM25,000 in total, must be paid in; the single founder, in addition, must deliver a security bond

(Sicherheit) for the unpaid balance of the capital (§7 II GmbHG). A capital contribution in kind and its capital value must be specified in the company contract (§5 IV GmbHG); such contribution must be delivered in full to the company before registration (§§7 III, 9 GmbHG). A special report on any contribution in kind and its valuation must be made and presented to the registration court (§§5 IV, 8 I nos. 4 and 5 GmbHG). Shareholders and managers are personally liable to the company for capital not paid up and also for damages in case of wrong statements to the court on the payment of capital, in case of over-valuation of contributions in kind and when any special consideration awarded to promoters is not stated in the company contract (§9a GmbHG).

2 Organization

The organs through which the GmbH operates are the management and the general meeting of the company; in addition, many GmbHs have an advisory board (Beirat) or supervisory board (Aufsichtsrat).

(a) Every GmbH must have one or more business managers (Geschäftsführer); only a natural person with a clean record (not another company) can be manager of the GmbH (§6 GmbHG). The Geschäftsführer manage the business of the GmbH (§§35–44 GmbHG) and represent it *vis-à-vis* third parties (§§35 I, 36 GmbHG). If there is more than one manager, and no other provision has been made, they must act jointly in representing the company (§35 II GmbHG). Business managers sign with the Firma of the GmbH and their own name (§35 III GmbHG). Their names must appear with the details of the GmbH on the company letterhead (§35a GmbHG), and not only must their names be entered in the commercial register, but so must any changes (§§8, 39 GmbHG). Business managers are empowered to represent the company in transactions of all kinds, and even if the company contract or a subsequent resolution limits their power of management within the company, this is effective only *vis-à-vis* the company and its members, and has no effect whatever on their power of representation *vis-à-vis* third parties (§37 GmbHG). This is another good example of the distinction so strongly made by the German law of agency between the external rela-

The Private Limited Company 253

tionship — the power of representation — and the internal relationship — the power to manage the business. Business managers are under a duty to manage the business with care, and are liable to compensate the company for any harm they may cause it by failure to do so (§43 GmbHG). Business managers may be appointed either by the contract of incorporation or by a resolution of the members (§6 II, 46 no. 5 GmbHG).

They enter a contract of employment with the GmbH, the appointment or employment frequently being made for a fixed period of time, such as five years. While the members of the company may remove a business manager at any time without reason assigned (§§38 I, 46 no. 5 GmbHG), this does not automatically bring the contract of employment to an end; if the business manager has done nothing to justify such dismissal, he has a claim for damages arising from the contract (§38 I GmbHG). The power to dismiss can, however, be confined to cases where an important reason exists, such as gross dereliction of duty (§38 II GmbHG).

(b) The supreme decision-making body of the GmbH is the totality of its members. Its position is rather like that of the general meeting of the AG, but it has distinctly more power and ability to influence the management of the business. The contract of incorporation may lay down quite freely what matters are for decision by the members (§45 GmbHG): the law simply specifies a few matters which, subject to further determination by the contract, are for the members to decide. Apart from the appointment and dismissal of business managers, already mentioned, they include the determination of the annual balance, the distribution of the profits, the calling-in of capital contributions, procedures for checking and controlling the management, and the appointment of procurators and commercial agents (§46 GmbHG). Meetings of the members at which such resolutions are taken (§48 I GmbHG) are summoned by the business managers, unless the contract of incorporation provides otherwise (§§48 II, 45 II GmbHG).

In certain cases there is a duty to call a meeting, for example, when members whose interests in the company amount to at least 10 per cent of the basic capital have so required (§50 I GmbHG), or generally when it is necessary for the good of the company (§49 II GmbHG). The rules regarding the conduct of meetings of a GmbH are much simpler than those for meetings of an AG: the notification may be sent by registered letter (§51 GmbHG); meetings may be dispensed with and resolutions made in writing if all the members agree (§48 II GmbHG), if the company contract so provides (§45 II GmbHG), or if one member has received a proxy from all the others.[20] Resolutions need only be notarized if they alter the articles of association (§53 II GmbHG).

(c) The contract of incorporation of a GmbH often provides for an advisory or supervisory board to superintend the management of the business in much the same way as the supervisory board of an AG. The rules concerning the supervisory board of an AG are applied (§52 GmbHG) unless the company contract provides otherwise. A GmbH with more than 500 workers is required under the Works Councils Act (§77 BVG) to have a supervisory board one-third of whose members represent the workers; such a supervisory board is subject to the rules applicable to the supervisory board of an AG, whatever the company contract may say. A GmbH with more than 2,000 employees is subject to the rules of the Co-determination Act of 1976 (see Section VIII below).

3 Shareholders' Rights and Duties

A notarial deed is required for the sale or transfer of any member's interest in a GmbH (§15 III, IV GmbHG). Of course this makes transfers more difficult, but that itself is consistent with the idea underlying the statutory model, namely that there is a personal link between the members of a GmbH, much as there is between the members of a partnership. Shares in a GmbH can therefore not be traded on the stock exchange: they are not items of the capital market. While the members of a GmbH have administrative and economic rights just like the shareholders of an AG, some of these rights may be profoundly modified and sometimes excluded altogether by the contract of incorporation.[21] The 1980 amendment to the GmbHG has somewhat restricted this kind of contractual freedom: the law now expressly confers on every shareholder the right to require the managers of the GmbH to give him immediate information about all relevant company affairs and to allow him to inspect the books; this right cannot be excluded

[20] BGH *LM* §46 GmbHG no. 7.   [21] *BGHZ* 14, 264.

by the contract of incorporation, and any denial of the request may be disapproved by the shareholders' meeting and, on the shareholder's application, by the court (§§51a and 51b GmbHG). The most important right of administration is the right to vote on resolutions. Subject to contrary provision in the contract of incorporation, the right to vote is proportional to the voter's share of the capital (§47 II GmbHG). The entitlement of members to the net annual profits as declared in the annual balance sheet is also in proportion to their capital contributions, but here too the contract of incorporation may provide otherwise (§29 GmbHG). Provided that the basic capital remains untouched, interim payments may be made to members even before the annual balance is struck (§30 GmbHG).

It is the duty of every member to make the requisite capital contribution in respect of his interest in the business (Stammanteil: §§19, 20 GmbHG). If he fails to do so, the company may declare him to have forfeited his membership (Kaduzierung: §21 GmbHG). All previous holders of that share in the company are liable for the shortfall (§22 GmbHG), and so are all the other members of the company (§24 GmbHG). This joint liability, which does not exist in the case of an AG, means that until all these parts of the business have been paid up in full, each member of a GmbH bears a risk of financial liability greater than the contribution entailed in his own interest in the business. The contract of incorporation may also require the members of the business to make further payments to the company over and above the initial contribution (Nachschusspflicht: §26–8 GmbHG); where such extra liability is unlimited, the member can free himself by putting his share in the company at the company's disposition (Abandon: §27 I GmbHG); if no such duty to make extra payments is contained in the contract of incorporation, only a unanimous resolution of the members can create one (§53 GmbHG). In exercising his rights of administration, the member must take account of the interests of the company, and must not exercise his rights to its detriment (see Section VII 3 below).

The individual shareholder of a GmbH is normally not personally liable to the creditors of the GmbH; special circumstances must exist before such liability will be imposed (see in general Section I 3 above). This applies also to the one-man company expressly approved by the law since the 1980 amendment (Section V 1 above). Accordingly, sole proprietorship of all shares of a GmbH as such is no ground for personal liability. It is true that, if the sole member sells all the shares in the GmbH, the courts treat him as selling the business itself and make him liable for defects in the enterprise,[22] but when this is properly analysed, it is seen to be not so much a problem of the one-man company as a more general question arising on the sale of rights. Since 1980, however, the law has established a personal liability in case of under-capitalization: loan capital provided by a shareholder in lieu of the fresh equity capital needed by the GmbH is treated as such equity capital in any bankruptcy proceedings. If a third party makes such a loan against a guarantee given by a shareholder, he is not treated as a normal creditor: he must seek repayment from that shareholder before he can claim from the GmbH in bankruptcy proceedings (§32a GmbHG). The shareholder who gave the guarantee is liable to the GmbH for any sum repaid by it to the third-party lender if bankruptcy proceedings commence within twelve months thereafter (§32b GmbHG).

4 Dissolution; Exclusion of Shareholders

A GmbH comes to an end on the expiry of the period of time provided in the contract of incorporation, on a resolution carried by a majority of the members, on a court decision, on the opening of bankruptcy proceedings, or on a final determination by the register court that the contract of incorporation is defective (§60 GmbHG). The contract of incorporation itself may specify further grounds of dissolution. It may also provide that the GmbH may buy in a share of the business even against the will of the member who owns it, but the conditions for this must be unambiguously specified (§34 GmbHG). Such amortization or calling-in has the effect of making that share of the business disappear. It is now accepted that a member of a GmbH, rather like a partner under §140 HGB (see Section II 4 and 5 above), may be excluded from the company by court decision which has been requested when there is some serious objection to him personally;[23] indeed, this may be done by a resolution of the

[22] *RGZ* 120, 283, 287.   [23] *BGHZ* 9, 157.

members if the contract of incorporation so provides. Conversely, as the Reichsgericht held, a member of a GmbH has the right to withdraw from it if there is a good reason for doing so;[24] such a right to leave does not exist in the case of an AG, but of course it is especially important when the business is such that one's share in it is not easily saleable.

VI THE STOCK COMPANY — PUBLIC LIMITED COMPANY (AG)

1 Definition and Formation

The Aktiengesellschaft (AG), as defined by the law, is a company, endowed with legal personality, and equipped with a capital stock divided into shares (Aktien), liability to whose creditors is restricted to the assets of the company (§1 AktG). An AG has the status of a merchant by reason of its legal form, regardless of its activities (§3 AktG). The constitution, or Satzung, of an AG is contained in a single company contract (Gesellschaftsvertrag; §2 AktG) and is not split into memorandum and articles of association (by-laws).

Many existing enterprises, especially sizeable ones, take the form of an AG, and although the foundation of a new AG is a rare event today, it is still worth taking a look at the relevant rules, since they make it easier to understand the AG's legal and organizational structure. The foundation of an AG goes through several stages.

First, at least five founder members who are to take up the shares must draw up the company contract in notarial form (§§2, 23 I AktG). The company contract or constitution must contain the Firma of the company, the company seat, the object of the enterprise, the amount of the basic capital, the face value of the shares, the number of shares of each face value, any different kinds of share there may be, and also the form that company notifications are to take (§23 III AktG). Freedom of contract has been severely curtailed here in order to safeguard investors and creditors: the company contract may deviate from the statutory prescriptions only where this is expressly permitted (§23 V AktG), and if any particular advantages are to accrue to individual shareholders, or if any contributions or receipts are to be in kind, the company contract must specify them precisely (§§26, 27 AktG). When all the shares have been taken up by the

[24]*RGZ* 128, 1.

founders the AG is 'constituted' ('errichtet': §29 AktG). The founders then have a notarial document drawn up in which they appoint the first supervisory board of the company and the auditors for the first year (§30 I AktG). When the supervisory board has appointed the first board of management or Vorstand (§30 IV AktG), the founders must draw up a report of the foundation with a justification of any contributions or receipts in kind (§32 AktG) and the members of the board of management and the supervisory board must check the procedure of foundation (§33 AktG). All the founders and members of the board of management and supervisory board then apply for the registration of the company in the commercial register (§36 AktG), which will occur after the court has checked that the constitution of the company and the application for registration are in proper form and consistent with the documents submitted therewith (§§38, 39 AktG).

The AG starts its existence as a legal person when it is registered (§41 AktG), and anyone who acts in the name of the AG before that moment is personally liable (§41 I AktG). Any subsidiary branch of an AG must likewise be registered in the commercial register (§42 AktG).

2 Organization: Management and Control

The procedure of foundation has already introduced us to the three organs through which the AG acts: the Vorstand, or managing board of directors; the Aufsichtsrat, or supervisory board; and the Hauptversammlung, or shareholders' meeting. The two-tier structure of board of management and supervisory board is the feature that most strikingly distinguishes German from Anglo-American company law. All legal systems have a problem with the division of power within public companies, for the management is in a very strong position in relation to the mass of small shareholders but may also be dominated by a few influential shareholders acting as a group. In German law it is accepted that the board of management should have a strong position, but certain decisions are strictly reserved for the shareholders, which is not always the case in the United States: increases in capital, alterations of the company contract, and the adoption of any so-called enterprise contracts (Unternehmensverträge: see Section VII below) must be voted in shareholders' meetings. There are also certain provisions to protect minority shareholders.

The Stock Company-Public Limited Company 259

(a) The board of management is reponsible for the conduct of the business of the company (§76 I AktG), and is not subject to any directions from the supervisory board or the shareholders' meeting about how to conduct it. The board of management may consist of a single person, but it usually has several members, as is indeed required in companies with a basic capital of over DM3 million, and under the Law of Co-determination (§76 II AktG; see Section VIII below). Decisions are taken by members of the board acting jointly or, if the rules of the board allow, by majority decision; decisions by a minority or, for example, by the chairman of the board are not permitted (§77 AktG). Restraints on what business may be conducted may arise from statutory prohibition (for example, an AG may not acquire its own shares (§71 AktG)) or from the purpose of the business as laid down in the company contract. Furthermore, the company contract may, in respect of clearly specified transactions, make it a precondition that the board of management have the supervisory board's approval (§111 IV AktG). But the supervisory board cannot require the board of management to enter any particular transaction, and it cannot itself act in place of the board of management.

The board of management represents the company in court and elsewhere (§78 I AktG). The members of the board must in principle act jointly in order to represent the company, but the company contract or a resolution of the board of management may empower individual members to represent the company, either alone or in conjunction with a procurator (§78 II, III AktG). The scope of the managing board's power of representation cannot be limited (§82 I AktG), least of all by the purpose of the company as laid down in the company contract: German law has no doctrine of *ultra vires*. In the internal relationship the members of the board of management are bound to respect limitations imposed on their conduct of the business by the company contract, the shareholders' meeting, the supervisory board, or their own rules of business (§82 AktG). The names of the members of the managing board must be given on the company letterhead (§80 AktG), and must be registered in the commercial register with an indication of their type of agency (for example, joint representation) (§39 I AktG); any alteration is also registrable (§81 I AktG).

The members of the board of management are appointed by the supervisory board for a maximum period of five years (§84 I AktG), but the appointment can be terminated before the end of this period for a serious cause, such as gross dereliction of duty (§84 III AktG). The contract of employment, which may provide for participation in the profits, is made by the supervisory board in the company's name (§86 AktG). A member of the board of management needs the agreement of the supervisory board if he wishes to engage in any business activity outside the company (§88 AktG), or receive any credit from the company (§89 AktG). The members of the board of management must keep the supervisory board regularly informed of the course of business (§90 AktG) and see to the proper keeping of the company books (§91 AktG). The board of management must call a general meeting if half of the basic capital is lost (§92 I AktG); and if the company becomes over-indebted or insolvent it must apply for the opening of bankruptcy proceedings, or of composition proceedings under court supervision, and stop making payments to third parties except under certain conditions (§§92 II, III AktG). In its conduct of the business the managing board is under a duty of care and confidence (§93 I AktG), and if the company claims damages for culpable neglect of these duties, the members of the board of management have the burden of establishing that they acted with due care (§93 II AktG).

(b) The supervisory board is to keep an eye on the conduct of business (§111 I AktG) and to call a general meeting when the interests of the company so require (§111 III AktG). The supervisory board cannot, even by purported delegation, take any steps in the actual management of the business; all that can be agreed is that the supervisory board's consent be necessary before specified steps be taken by the board of management (§111 IV AktG). The supervisory board represents the company in relation to the members of the board of management (§112 AktG). It must have at least three members (§§95, 96 AktG), the precise number in the case of larger companies depending on the Law of Co-determination (see Section VIII below). Members of the supervisory board who represent the shareholders are either chosen by the shareholders' meeting (§101 I AktG) or nominated by specified shareholders, if the company contract so provides (§102 II AktG). Representatives of the employees are chosen by the employees in accordance with the Law of Co-determination (see Section VIII below). All members of the supervisory board are under a duty of care and confidence, and are liable to the company for breach of these duties (§116 AktG).[25]

A member of the supervisory board to whom there are serious personal objections can be dismissed as a result of court proceedings, or he can be recalled by the group that appointed him to the supervisory board (§103 AktG). The membership of the supervisory board and any change in it must be notified to the commercial register, and published in the company handouts, which need not, however, be entered in the commercial register (§§40, 106 AktG). The supervisory board chooses its chairman and deputy chairman (§107 I AktG), may set up sub-committees (§107 III AktG), and decides by resolutions on which each member has a vote (§108 AktG), the chairman having a casting vote under the Law of Co-determination (see Section VIII below). The supervisory board should meet every three months, and must meet every six months; additional meetings may be called by the managing board or by a member of the supervisory board (§110 AktG).

The function and efficacy of the supervisory board, which has something in common with the outside or non-managing directors of Anglo-American company law, has been the subject of debate ever since the law of 1870, which made it a mandatory feature of the AG. Clearly, supervisory boards have varied a great deal in practice, but the continuing problem is to find the position, midway between inaction on the one hand and pre-empting the conduct of the business on the other, where a supervisory board can exercise just the right degree of control. As compared with the device of non-managing directors, the supervisory board has at least the advantage of giving clear institutional expression to the need for control. The supervisory board's best chance of exercising influence in the long term lies in its control of the personnel, its power to appoint the members of the board of management. It can also play an important part in situations of crisis or cases of mismanage-

---

[25]The extent of the duty of confidence of members of the supervisory board has become a matter of discussion by reason of the participation of representatives of workers and trade unions; on the duty of confidence, see *BGHZ* 64, 325, 330.

ment. Banks traditionally have a representative on the supervisory boards of the large industrial enterprises to which they extend credit or in which they have shares, and the fact that a person may not by law serve on more than ten supervisory boards at a time is suggestive about the practice (§100 AktG). In the interplay of the components of a combine (see Section VII below) the supervisory board can play a useful, though not a crucial, role. Most recently, the supervisory board has been chosen as the forum in which to allow the employees of big enterprises to exercise their rights of co-determination (see Section VIII below).

(c) The third decision-making body of the AG is the general meeting of shareholders (§118 I AktG). It decides the matters laid down by law and in the company constitution, notably the appointment of members of the supervisory board, the use to which the profits are to be put, the discharge of the members of the board of management and of the supervisory board, alterations in the constitution of the company, the steps to be taken to increase or decrease the company capital, the appointment of auditors, and the winding-up of the company. The general meeting of shareholders is normally called by the board of management (§121 II AktG), but it may also be called by the supervisory board (§111 III AktG). A meeting must always be called when the interests of the company so require (§121 I AktG), and is normally called once a year in order to discharge the board of management and supervisory board in respect of the year just ended (that is, to endorse their administration), to decide how the profits are to be used, and to appoint the auditors. In such an 'ordinary' general meeting the annual report and balance sheet are discussed (§§120, 163, 174 AktG). Individual shareholders at the general meeting have the right to question the board of management on matters relating to the company (§134 AktG) and may vote in proportion to the face amount of their shares (§131 AktG; for exceptions, see Section VI 3 below). The general meeting must be called with due notice, that is one month, and the notice of the meeting must contain the agenda, the motions to be put, and the recommendations of the board of management and supervisory board on these motions (§§123–7 AktG).

The voting rights of smaller shareholders are normally exercised by the banks with which the shares are deposited. The banks' right to vote in this way has been much criticized, since, rather like the device of giving proxies to the board in the United States, it leads to an accumulation of power in a few hands and to impotence on the part of small shareholders. On the other hand, unless one is to dispense with the small shareholder's right to vote altogether, there has never been any viable alternative. In 1965 the legislator was keen to popularize shareholding and create a 'shareholders' democracy', so he tried to make the depositary's exercise of its voting rights more responsive to the interests of the individual depositor-shareholder.

The shareholder's proxy to his bank must be in writing and is valid for only fifteen months at a time (§135 I, II AktG). Before every general meeting the bank must let the person whose shares it is holding know of the calling of the meeting, of the agenda, and of the motions. It must make its own recommendations as to voting thereon, and must ask the shareholder for instructions if he wishes to vote differently and has not already said so (§128 II, III AktG). What normally happens is that the bank recommends voting with the management, and that the shareholder does nothing. The bank must then vote in accordance with its own recommendations (§135 V AktG). Voting may take place secretly or openly, with the shareholder giving his name (§135 IV AktG); here it should be remembered that most shares are bearer shares.

No quorum is required at the shareholders' meeting. This means that a minority of the capital can make binding decisions, and this constitutes a further reason for trying to see that rights to vote be exercised as widely as possible, a thing that is acceptably procured by allowing depositary banks to vote for their depositors. A resolution normally requires a simple majority of the capital represented at the meeting, unless on a special matter a qualified majority is required by law (for example, changes in the constitution (§179 II AktG)) or by the company constitution (§133 AktG). Resolutions of the general meeting must be notarially recorded (§130 AktG).

A resolution passed at a general meeting may be void if it is seriously defective — for example, if the meeting was not properly called and published, if the resolution was not notarially recorded, or if its content is contrary to good morals (§241

AktG). Resolutions that are invalid in this way are without legal effect, though only shareholders, the board of management, and the supervisory board may ask a court to make a declaration of nullity (§249 AktG). A resolution of a general meeting may also be impugned if it is defective in a less serious way — for example, if it infringes a law, or serves the special interests of particular persons in a manner prejudicial to the company (§243 AktG); here the resolution is valid at the outset, but may be invalidated by proceedings in court (§241 no. 5 AktG).

3 Shares; Shareholders' Rights and Duties

An AG must have a capital stock of at least DM100,000 (§7 AktG), divided into par value shares with a face amount of not less than DM50 each; any higher nominal amount must be DM100 or a multiple thereof (§8 AktG). German law is like English law in not allowing the non-par value shares that are permitted by many corporation laws in the United States. Shares cannot be issued at a price lower than their face amount (§9 AktG), though the issue price may be higher (§54 AktG). A shareholder's only financial obligation is to pay the face amount or higher issue price; no further contribution can be demanded (§54 I AktG). The full price is normally paid at once, and at least one-quarter must be paid before the AG is registered (§§54 III, 36 II AktG). If any contribution in kind is to count towards shares, the contract of incorporation must specify the subject matter of the contribution (for example, an existing commercial business) and the nominal amount of the shares to be granted (§27 I AktG). The shareholder's right of membership is normally evidenced in an instrument (the 'share'), though it can exist and be assigned independently of any such instrument. Shares are nominative or to bearer (§§10, 24 AktG).

As moveable property, shares are transferred by contract of transfer and delivery (see §§929ff. BGB); nominative shares also require endorsement by the transferor (§68 I AktG). The contract of incorporation can require the consent of the board of management for the transfer of nominative shares ('vinkulierte Namensaktie': §68 III AktG). But bearer shares are the regular form (§24 I AktG) and in practice the commonest. The bulk of bearer shares is held in collective deposit (Sammelverwahrung) by special banks, the shareholders being co-owners *pro parte* (§6 DepotG); transfer is effected by assign-

The Stock Company-Public Limited Company 265

ment of the co-ownership rights. Thus the securities themselves are not moved about in daily trading, but only the accounts evidencing co-ownership rights in collective deposits. This 'stückeloser Wertpapierverkehr' is slowly making headway in English-speaking countries, since the daily transfer of millions of named shares is a tiresome and costly chore.

A shareholder participates in the AG in the proportion that the nominal value of his shares bears to the basic company capital. To this extent he has a right to part of the net profits (dividends) (§§58 IV, 60 AktG), to any proceeds on liquidation (§271 AktG), to take up new shares (§§186, 221 III AktG; see Section VI 5 below), and to vote in the general meeting (§§12 I, 134 AktG). The constitution of the company may, within strictly defined limits, provide for a different distribution of the profits and a different entitlement to vote (§§60 III, 12 I AktG), and thus create different kinds of shares. It is no longer possible to issue shares with extra voting rights, unless official authorization has been given in an exceptional case (§12 II AktG), for example with regard to publicly held shares in public utilities, but existing shares with extra voting rights continue to be effective. Shares that carry a preferential right to profits but no voting rights may be created (§139 AktG). Finally, the company constitution can limit the voting rights of large shareholders by fixing a maximum nominal value in respect of which voting rights may be exercised (§134 I AktG).

Several large German undertakings introduced such a limitation between 1976 and 1979 in order to deter foreign investors from acquiring shares in large numbers. The same result can be achieved by creating the tied nominative shares we have already mentioned.

A form of company exists, the NebenleistungsAG, in which the shareholders are bound to provide periodic performances otherwise than by payments in money (§55 AktG). The origin of this institution was the practice in the early days of the Industrial Revolution of setting up stock companies to run sugar refineries, and issuing shares to the growers of sugar-beet in return for their undertaking to deliver their annual crop to the company.

An AG may not acquire its own shares save in exceptional cases which are narrowly defined — for example, in order to protect the company against 'serious disadvantages', in order to compensate minority shareholders within a combine (see Section VII 2 below), or in order to have shares to reissue to employees (§71 AktG).

So far as the legal relationship between shareholder and company is concerned, the basic principle is that shareholders must receive equal treatment in proportion to their holdings; no special advantages can be given to individual shareholders, though minority shareholders are protected by special rights (see Section VI 6 below). In exercising his rights the shareholder must take account of the interests of the company, a duty that the Reichsgericht once called a duty of the most complete loyalty, which is perhaps something of an exaggeration. A shareholder will be liable to the company and other shareholders if he deliberately uses his influence on the company so as to procure a member of the board of management or the supervisory board or any senior employee to act to its detriment or theirs (§117 AktG; see Section I 4 above).

4 Accounting, Appropriation of Profits, and Disclosure

All industrial countries of the West have found that to require companies to disclose information and publish their accounts is quite an effective way of giving shareholders, creditors, and the general public in the capital market some control over them and their managements. The relevant rules in Germany were strengthened by the Company Act of 1965, which sought to render shares more attractive as a form of investment to the small investor, and to induce companies needing finance to make greater use of the capital market. The principal purpose of earlier legislation was to stop companies exaggerating their assets or filtering the capital back to the shareholders, but today the law must also guard against undervaluation of assets and the creation of unduly large secret reserves beyond the shareholders' control.

The board of management is responsible for seeing that the company's books are properly kept (§§38ff. HGB; §91 AktG). For every business year the board of management must produce an annual financial statement (Jahresabschluss), that is, a balance sheet, and a statement of profit and loss, and must amplify this by a report on the year's business (§148 AktG). The balance struck must be true in the sense of being complete, formally accurate, and based on proper principles of valuation; it must also be clear and comprehensible (§149 AktG), itemized in accordance with the detailed statutory scheme (§151 AktG). The law also lays down how the profit and loss account is to be set out (§157 AktG), and has some mandatory rules on valuation (cost of procurement, production, market price) and depreciation (§§153–5 AktG). Independent auditors must be called in to check the annual balance sheet, affix their mark of approval, and prepare the auditors' report (§§162–7 AktG); they are bound by a duty of care and confidence (§168 AktG). Then the balance sheet, the annual report, and the auditors' report are sent to the supervisory board, along with the board of management's proposals for the allocation of the profits (§170 AktG). The supervisory board in its turn checks all these documents and prepares its own report on them (§171 AktG). If the supervisory board approves the statement of accounts, it is treated as 'confirmed' (§172 AktG), though such confirmation may be left to the general meeting (§173 AktG).

A general meeting of shareholders must be called forthwith to consider the annual balance sheet, as approved by the auditors, the annual report, the proposal from the board of management for the allocation of the profits, and the report of the supervisory board (§§174, 175 AktG). It is for the general meeting to decide on the allocation of the profits — that is, on how much should be distributed as dividends, how much should go to open reserves, and how much should be carried forward on the profit and loss account (§174 AktG). The amount of profit stated in the balance sheet is conclusive for the shareholders' meeting, but the board of management's proposals as to its allocation are not binding at all.

The confirmed balance sheet, marked as approved by the auditors, the annual report, and the report of the supervisory board must all be handed in to the commercial register, and will be checked by the register court for formal accuracy (§177 AktG). At the same time the balance sheet, but not the annual report, must be published in the papers where the prescribed company communications usually appear (§177 II AktG).

German law has no effective method of requiring disclosure of business information apart from these rules of company law on accounting and the verification and publication of annual balance sheets, which were extended by a special law on disclosure in 1969 to large businesses not organized as an AG. There is much less stress on disclosure when shares are issued or companies 'go public': Germany has no body comparable to the American Securities and Exchange Commission. It is true that there have recently been improvements in the disclosure of information on the exchange, for a code of conduct voluntarily adopted by the exchanges requires businesses whose shares are quoted to submit quarterly interim reports on their state of business; but it still remains true that most of the rules requiring disclosure of business information are to be found in company law rather than in the law of securities.

5 Financing the AG

It is one of the themes of company law that the basic capital of a company must be maintained in order to protect its creditors. Hence the contributions of shareholders must never be returned to them (§57 AktG), and the basic capital is put on the debit side of the account (§151 I AktG), thereby constituting, in accounting terms, a capital liability. In addition, it is obligatory to build up a statutory reserve of 10 per cent of the basic capital, or more if the company constitution so provides (§150 AktG); this is gradually built up by annual appropriations (say, 5 per cent of the annual profits). This reserve may be used, without any alteration of the company constitution, to wipe out a loss in any particular year. Furthermore, the annual meeting, in determining the allocation of the profits, may decide to create or swell 'free' reserves by setting sums aside for this purpose.

If it is sought to increase the company's capital, four different legal methods offer themselves: each of them requires a resolution of the general meeting, with three-quarters of the represented capital voting in favour (§§182 I, 193 I, 202 II, 207 II AktG); and if any new shares are created by such capital increase, the existing shareholders, as members of the company, are in principle entitled to take them up in proportion to their existing participation (§186 AktG).

First, an AG whose capital base is small in relation to the volume of its business, but which has built up sizeable free reserves, may decide in general meeting to increase its capital out of corporate resources by turning these free reserves into basic capital and issuing the new shares to shareholders in appropriate proportions (§§207–20 AktG); such a step involves no injection of fresh capital into the company, just an alteration in its accounts.

The other three methods do result in the acquisition of new

capital. The normal form is to raise the capital by means of contributions (§§182–91 AktG). The shareholders are entitled to take up the new shares at the fixed opening price in proportion to their existing shareholding (§186 AktG); but of course they may prefer to sell their right of pre-emption on the exchange. For technical reasons this distribution of new shares to raise capital is effected by the banks, which first take up the shares themselves and then offer them to the shareholders (§186 V AktG).

Alternatively, the general meeting may decide on a 'conditional increase of capital' (§§192–201 AktG), whereby the capital will be increased only to the extent that new shares are actually bought or taken in exchange. The law allows this where convertible debentures are being created, when a merger is in prospect, and when shares are being created for the employees of an enterprise (§192 II AktG), and in these cases the shareholders' entitlement to shares is excluded. A few large businesses in Germany have used this method to encourage their employees to become shareholders, with some success: in Siemens AG, for example, one of the biggest of German companies, 20 per cent of the shareholders are employees or former employees, and in all they control 5 per cent of the basic capital.

Finally, it is possible to raise 'authorized capital' ('genehmigtes Kapital'; §§202–6 AktG), a legal form that gives the company's board of management a certain freedom of decision and the ability to satisfy its financial needs by reacting quickly to situations in the capital market, as is possible in the United States: by a resolution of the general meeting, valid for not more than five years at a time, the board of management may be empowered to issue new shares up to a specified maximum amount.

As a matter of fact, a certain under-capitalization is typical of many German enterprises today. Companies tend to finance themselves by issuing bonds or by taking direct short-term or long-term loans from banks, and despite the efforts of the legislator in the Company Act to popularize shares as a form of investment, they do not seem to be increasing their equity capital sufficiently by issuing new shares. One hindrance has been that the long-term returns on shares were much the same as those on bonds and other money obligations; the tax laws also discriminated somewhat against shares, though this effect has now been diminished by the recent reform of corporation tax (see Section IX below).

6 Legal Protection of Minority Shareholders

Special rights are accorded by the statute, as we shall see, to minority shareholders — those who command a specified but small proportion of the shares in a company — but they also obtain some protection from the rights that every shareholder enjoys as such. Thus, if the resolution of a general meeting is void, any shareholder may sue for a declaration to that effect, which will be binding on all parties (§§249, 248 AktG), and if a resolution is defective, he may impugn it in court and have it invalidated (see Section VI 2 (c) above).

An action of the latter sort will succeed if, for example, the resolution granted special privileges to certain shareholders or third parties (§243 II AktG),[26] if it attributed different dividends to similar shares or otherwise infringed the basic principle that shareholders must be treated equally,[27] or if it was passed only because the board of management unlawfully withheld information which was demanded in the general meeting (§§131, 243 IV AktG).

The shareholder also receives some protection from the fact that the company can sue the board of management or the supervisory board for damages if it has acted unlawfully, or a third party if he has abused his influence on the company (§§93, 116, 117 AktG). The shareholder himself, however, cannot sue on behalf of the company: the derivative suit with which American lawyers are familiar does not exist in German law, subject to an exception in the law of combines, whereby the shareholder in a subsidiary company may bring suit on its behalf against the parent company for the harm it has suffered (§§309 IV, 317, 318 AktG; see Section VII 2 below).

Minorities have many statutory rights. For example, a group of shareholders with 10 per cent of the basic capital may compel the company to exercise the claims for damages just mentioned (§147 AktG), may stop the general meeting voting to waive such claims (see, for example, §§50, 53, 93 IV, 116, 117 IV

[26]*RGZ* 107, 75; *RGZ* 142, 222. [27]*RGZ* 68, 210, 213; *RGZ* 120, 180.

AktG), and may insist that experts be appointed to check steps that have been taken in the foundation, management, or creation of capital of the company (§142 AktG); they may also insist that the votes on the discharge of the different members of the board of management be taken separately (§120 I AktG). A group of shareholders with 5 per cent of the basic capital also has certain rights; for example, they may insist that a general meeting be called, and may make written demand on the board of management that specified items be placed on the agenda (§§122, 124 AktG); and they may also go to court and ask to have any under-valuation in the annual balance sheet or incompleteness in the annual report subjected to special investigation (§258 AktG).

7 Dissolution and Liquidation; Merger

In the dissolution and liquidation of an AG much the same principles apply as in the case of a GmbH (see Section V 4 above). An AG may be dissolved by the expiry of the time specified in the company constitution, by a decision of the general meeting with a majority of three-quarters, by the opening of bankruptcy proceedings, and by a final determination by the register court that there is some defect in the constitution (§262 AktG). If the unlawful conduct of an AG makes it a threat to the public good, it may be dissolved by court order on the application of the supreme state authority (§396 AktG). Once it has been dissolved, an AG is wound up by liquidators, usually members of the board of management.

Very often an AG comes to an end through fusion or merger (Verschmelzung). Merger has an extensive meaning in the law of competition (see Chapter 15 below), but in company law it designates a process whereby, without any liquidation, the assets of the companies involved are amalgamated and vested in a single juristic person, shares in which are given in exchange for shares in any AG that ceases to exist. The participating companies may all cease to exist and their property be transferred to a new AG (Fusion durch Neubildung), or one of the companies may survive and take over the property of the other (Fusion durch Aufnahme) (§339 AktG). Either method requires a decision of the general meetings, with a majority of three-quarters (§§340, 353 AktG).

272 *Partnerships and Companies: Business Organization*

VII COMBINED ENTERPRISES (KONZERNE)

1 The Approach of German Company Law

Problems have arisen in all the industrial countries of the West from the fact that businesses which are independent from the legal point of view can be linked together in various ways — by shareholding, by contracts, or by influence of other kinds. It may be dangerous for competition, and for investors and creditors, if a business is so influenced in its decision-making that it no longer acts independently, but only as a component of a larger economic unit, namely a combine. The danger to competition is taken care of by the law of competition (see Chapter 15 Section II below), but company law had never been used to deal with the dangers to minority shareholders and creditors that arise when their AG's freedom of action is impaired by its links with other companies until Germany enacted its Company Act of 1965. The statutory rules apply only to links between companies where an AG is involved on one end or the other.

In such cases the economic facts tend to be extremely various and complex. In order to classify them, the law sets out several definitions, which overlap to a certain extent (§§15–19 AktG). If any of the following descriptions is satisfied, the case will be one of 'linked enterprises' (verbundene Unternehmen):

— businesses in which another business has a majority interest (§16 AktG);
— dependent businesses that are subject to controlling influence by a 'dominant business'; a presumption that this is so arises from a majority shareholding (§17 AktG);
— a combine such as exists if several legally independent businesses are subject to the same managment: here one distinguishes between the subordinate combine (Unterordnungskonzern), where dependent businesses are managed by a controlling business, and the co-ordinate combine (Gleichordnungskonzern), where independent businesses voluntarily place themselves under common management (§18 AktG);
— businesses that have mutual interests in each other (§19 AktG).

Although the rules focus on combines in the strict sense (see

Section VII 2 below), linked enterprises of other kinds are subjected to duties of disclosure and some other constraints. If the X Co. takes up one-quarter of the capital or obtains a majority of the shares in the Y Co., it must immediately inform the Y Co., and the Y Co. is under a duty to publish the fact; until such publication the X Co. may not exercise any rights attaching to its shares (§§20–1 AktG). Even after such publication, companies that hold shares in each other may exercise their rights only in respect of one-quarter of the capital, though this restriction does not apply to the first business to fulfil its duty of publication (§328 AktG). The law imposes other limitations on the exercise of shareholders' rights as well: for example, if the X Co. has a majority shareholding in the Y Co., the Y Co. may not acquire any shares in the X Co. (§§56 II, 71 IV AktG).

2 Three Types of Combine

The law specifies three different categories of combine (§18 AktG): the *de facto* combine, the contractual combine, and the integrated enterprise.

(a) The *de facto* combine is the combine familiar to people the world over. It arises when a business which has the power to control another business, by reason of a majority shareholding or otherwise, actually exercises this power in order to promote a unitary business policy. Here the statute renders it illegal to cause the dependent business to enter a transaction disadvantageous to it, unless compensation is to be provided by the end of the business year at the latest (§311 AktG). Furthermore, the management of every dependent business must each year prepare, but need not publish, a 'dependency report', listing all the transactions it has entered with the dominant business and all the instructions it has received from it (§312 AktG). Unless compensation has been provided, the dependent business may sue the dominant business and its statutory agents for damages for any detriment suffered by following its instructions, and this claim can be brought on the dependent company's behalf by its creditors and shareholders (§§317, 309 III AktG).

(b) The novel and unfamiliar idea of 'contractual combine' is best approached through the institution of the 'enterprise contract'. The draftsmen of the 1965 Act realized that enterprises are not always linked by the mere fact of power, such as arises from majority shareholding and so on, but are often linked by contracts as well. Such contracts constitute a clear legal expression of the power relationship, so the law approves of them, and provides models for them, under the name of the 'enterprise contract' (Unternehmensvertrag) (§§291–307 AktG). The most important of its variants are the control contract, whereby one business subordinates itself to the management of another business, and the profit-channelling contract (§291 AktG); but there may also be contracts for part channelling of profits, the pooling of profits, and the leasing of the business (Betriebspacht) (§292 AktG).

Enterprise contracts need the approval of the general meeting, with a majority of three-quarters (§293 AktG), and must be registered in the commercial register (§294 AktG). A control contract and a profit-channelling contract generally go together. A control contract gives the controlling business the right to issue instructions to the business under control, even if these are detrimental to it (§308 AktG). Here the law makes it legitimate to influence the subsidiary enterprise, subject to a condition that protects the subsidiary and thus its creditors and minority shareholders: the controlling enterprise must compensate the controlled enterprise for any loss suffered in the year (§302 AktG). Where a control contract exists, the statutory representatives of the controlling business must act with care in issuing any instructions, and in case of breach will be liable to a suit for damages by the subsidiary business (§309 AktG), which may also be brought by its shareholders and creditors (§309 IV AktG). The minority shareholders of the subsidiary company are given further protection: the profit-channelling or control contract must provide for a fair annual compensatory payment to them (§304 AktG), and also fix a purchase price for their shares, should they wish to leave the dominated AG (§305 AktG). The fairness of these sums is subject to judicial control at the shareholder's suit (§§306, 304 III, 304 V AktG).

(c) The tightest form of linkage exists in the case of integrated enterprises, such integration being akin to a real merger (see Section VI 7 above). If one AG has all the shares in another AG, the integration can be achieved by resolution of the general meetings of both and by registration in the commercial register

(§219 AktG), the company that has been integrated remaining as a legal person, which may be desirable from the point of view of sales, for example, if the company's name or Firma is well-known. The principal company has an uncontrolled power of management over the integrated company and an uncontrolled power of disposition over its property (§§323, 324 AktG), but it is also liable to the creditors of the integrated company for all existing and new debts (§324 AktG). Here the legislator has thoroughly 'pierced the corporate veil'. Integration is also possible if the principal company has only 95 per cent of the shares in the other company, but then it must provide the minority shareholders with fair compensation, usually in its own shares (§320 AktG).

What has been the practical effect of the law of combines as we have described it? It is too early to say. Certainly, enterprise contracts on the legal models have been made and published, enterprises have been integrated, and lawsuits have successfully questioned the fairness of the payment for minority shareholders who wish to get out.[28] The notion of 'dependency' (§17 AktG) came up for consideration by the Bundesgerichtshof in the following case. A total of 55 per cent of the shares in an AG were held by three different GmbHs, the shares in each GmbH being held equally by two branches of a single family. Although there was no finding that in fact there had been any common management of the AG (§18 AktG), there was nevertheless the possibility of common control of the AG by the GmbHs (§17 AktG). This was enough to impose on the AG a duty to draw up a dependency report (§312 AktG). Since it had failed to do so, the resolution of the general meeting to discharge the members of the board of management was defective (§120 AktG), and the minority shareholders who sought a declaration of nullity were therefore successful.[29] No decisions have yet been handed down on the special liabilities of dominant companies, or on their duty to make compensatory payments to a subsidiary,[30] though there has been one case outside company law (see Section VII 3 below).

(d) A true combine (§18 AktG) is subject to a special duty of disclosure. If the components of a combine are under the single management of an AG or KGaA situated in Germany

---

[28] BGH in: Die Aktiengesellschaft (AG) 1976, 218; OLG Düsseldorf, AG 1977, 168.
[29] *BGHZ* 62, 193.
[30] Wiedemann, I *Gesellschaftsrecht* §8 IV 1, p. 463 (Munich 1980).

(Obergesellschaft, the parent company) the board of management of the parent company must draw up a consolidated balance sheet and profit and loss account for the whole combine and publish an annual report on the combine's business (§§329–38 AktG). The same duty is imposed on any parent company that is subject to the duty of disclosure under the 1969 Act. If the parent company is not under a duty of disclosure under either of these laws, any component companies of the combine that are AGs or KGaAs must prepare a balance sheet for their part of the combine (§330 I AktG); such a balance sheet for part of a combine must also be prepared when the parent company is situated abroad (§330 II AktG).

3 Law of Combines not Involving an AG: the ITT case

The ITT case is important because in it the Bundesgerichtshof, without relying on the provisions of the Company Act regarding combines, held not only that a dominant enterprise could be liable for issuing detrimental instructions to a subsidiary, but also that minority members of the subsidiary were entitled to claim damages on its behalf, as well as on behalf of other enterprises dependent upon it.

The action was brought by the minority partner in two GmbH & Co. KGs (see Section III 3 above); these businesses, which had subsidiaries under their control, were themselves controlled by the managing GmbH, which belonged to the ITT combine. The controlling GmbH issued instructions to the dependent enterprises to enter 'consultancy agreements' with a company that also belonged to the ITT combine. The high 'consultancy fees' so payable were really just a means of shifting money around within the combine. The plaintiff's claim, based on the disadvantage thus accruing to the businesses, was upheld in principle by the Bundesgerichtshof, though the case was actually remanded to the appellate court for findings of fact. Instead of analogizing from §317 AktG, the BGH based the liability on the duty of fair play that members of a GmbH & Co. KG owe to each other.[31]

VIII CO-DETERMINATION OF EMPLOYEES

1 Co-determination Legislation in its Political and Legal Setting

Throughout the industrialized West, the law of partnerships

[31] *BGHZ* 65, 15.

and companies that establishes the legal structure for business organizations is based on the idea that the enterprise and its managment should be under the control of the owner or investor of the equity capital. The German laws on the co-determination of employees do not abandon this viewpoint, but they add a new element to it: in large corporate enterprises employees and trade unions should be able, at least in an intermediate way through their representatives, to take a part in the managerial decision-making process.

Co-determination was first put into practice after the war in the coal, iron, and steel industry, and the practice was sanctioned by the 1951 Act on the Equal Co-determination of Employees in the Coal, Iron, and Steel Industry. Under this Act the supervisory board of the company (see Section VI 2 (b) above) is composed of five representatives of capital and five of labour, plus an additional eleven 'neutral' members; the board of management must contain a 'labour director' (Arbeitsdirektor). A Co-determination Amendment Act of 1956 extended this legislation to combines containing companies in the coal, iron, and steel industry. The Labour Management Relations Act of 1952 (see Chapter 16 below) requires companies with more than 500 employees to have supervisory boards on which one-third of the members represent the workers. In 1976 the Co-determination Act established a new scheme for companies with more than 2,000 employees: there is to be formal parity of labour and capital on the supervisory board, and one member on the board of management who is especially responsible for personnel and labour matters (Arbeitsdirektor, but in a different sense).

The new law reflects the belief that in a modern industrial society the idea of an economic partnership of labour and capital should replace the outmoded concept of the class struggle. This new social philosophy is not unchallenged, and the Co-determination Act has by no means displaced the adversary system in collective bargaining and labour conflicts, but there were several factors which favoured the new law — the comparatively high degree of industrial peace that the Bundesrepublik has always enjoyed (see Chapter 1 Section II above), the power of the trade unions, and the Germanic predilection for formalized compromises — and it must be seen in the context of the works councils and other features of Germany's highly developed labour law (see Chapter 16 below).

2 The Legal Mechanics of Co-determination

The Co-determination Act of 1976 (Mitbestimmungsgesetz, MitbG) (which leaves intact the Acts of 1951 and 1956 on the coal, iron, and steel industry) applies to companies (AG, GmbH) and certain GmbH & Co. KGs with more than 2,000 employees, the employees of the different companies in a combine being computed together (§5 MitbG). As a rule, the supervisory board of a company is to be composed of equal numbers of representatives of shareholders and employees — six, eight, or ten, on each side, depending on the size of the company (§7 MitbG).

The representatives of labour are elected by the employees of the enterprise or combine (§16 MitbG), but the law divides them into different classes. There must be at least two representatives of the trade unions, who, unlike the others, need not be employees of the enterprise (§7 MitbG). The employees are divided into three groups — workers, office staff, and senior executives, a special group within the office staff (§3 MitbG) — and each group must be proportionally represented on the board, with a minimum of one representative (§15 MitbG). The supervisory board makes its decisions on the basis of a simple majority of the votes cast (§29 MitbG), but two considerations tilt this formal parity in favour of capital: first, that any representative of the senior executives is apt, as a matter of fact, to vote in accordance with the interests of the enterprise as a whole rather than just those of the workers; and, second, that the chairman of the board, who, because of the election procedures, is normally a representative of capital (§27 MitbG), has a casting vote if two votes end in a tie (§29 MitbG). The powers and functions of the supervisory board remain as laid down in the Company Act (§30 MitbG): as we have seen (Section VI 2 (b)), it does not manage the company, but does, in its general control of the management, have the important rights of recalling and appointing the members of the board of management. Here, too, the Co-determination Act gives the chairman a casting vote in the case of a tie, if the complicated procedures laid down are unable to produce a reconciliation

(§31 MitbG). The Act finally provides for a labour representative on the board of management. The Labour Director, as he is called, has all the rights and duties of a normal managing director (§33 MitbG), and is responsible for personnel and labour matters, but his exact legal position and qualification is still a matter of debate.

3 Issues and Prospects of Co-determination

The Co-determination Act of 1976 is a political compromise between the trade unions, who are unsatisfied because they have failed to obtain full parity in all respects, and capital, which is concerned that in the long run it may be disadvantageous to the company and the economy as a whole if decisions concerning management are affected by trade unions and workers pursuing their own interests.

The new law was challenged before the Constitutional Court as being in violation of the constitutional rights of investors, relating to their private property, the exercise of their profession as entrepreneurs, and their right to form their own trade associations free from trade union influence. In March 1979 the Constitutional Court declared that the Act was consistent with the Constitution. In relation to the forebodings expressed about the effect of the decision-making process on big enterprises and the economy generally, the Court upheld the right of the legislator to legislate on the basis of some optimism about the future co-operation of industrial partners.[32]

IX TAXATION

Since any decision to invest in equity capital depends largely on tax considerations, some basic information on taxes is needed for a realistic picture of partnerships and companies as business organizations. Equally clearly, this is no place for a detailed guide to German tax laws.

1 Individual and Company Income Tax

Partners in an OHG or KG pay individual income tax on the income they derive from their share in the profits of the partner-

---

[32] *BVerfGE* 50,290

ship. The partnership is not taxed as such, but its annual profit, as evidenced in its financial statement, must be included in the return filed by the individual partners. Individual income tax starts at a rate of 22 per cent and rises to a rate of 56 per cent.

Under the Law on Taxation of Incorporated Associations (Körperschaftssteuergesetz), companies are taxed at a rate of 36 per cent on distributed income and of 56 per cent on retained income. Distributed profits, such as dividends, are taxed again in the hands of the individual shareholder who receives them. The double tax burden that resulted from this system has been eliminated by a new tax credit system, contained in the Tax Reform Act of 1976. Now, the shareholder obtains a tax credit (36 per cent) along with the net dividend (64 per cent); and while the whole of the dividend is subject to individual income tax in his hands, the tax credit is offset. This tax credit system has put an end to the long-criticized discrimination against shares as compared with other forms of investment, but it has created a new difficulty: it discriminates against the foreign shareholder, for example, foreign parent companies of German subsidiaries which are unable to take advantage of the tax credit in their home country. Double taxation treaties may be amended so as to afford some relief on this point.

Dividends are also subject to a capital proceeds tax (Kapitalertragssteuer) at a rate of 25 per cent; this is withheld from all dividends, and is likewise treated as a prepayment of personal income tax and offset against personal tax liability.

Long before the enactment of the new Company Act of 1965, the tax laws had recognized that combines called for special treatment. With respect to incorporated association income tax, a parent company and its subsidiary may be treated as a unit, and the profits and losses of both parent company and subsidiary offset in the financial statements of the parent company provided that they are in a relationship described as 'Organschaft' or integration; the subsidiary (Organgesellschaft) must be financially, economically, and organizationally integrated into the dominant company (Organträgergesellschaft). The parent must have at least a 50 per cent interest in the equity of the subsidiary (financial integration), the subsidiary must be subordinated to the decisions of the parent, and there must be a contract between parent and subsidiary (in writing, for a minimum of five years) by which the parent undertakes to assume the subsidiary's profit or loss.

2 Capital Transfer Tax

A capital transfer tax of 1 per cent (Kapitalverkehrssteuer) payable by the company is imposed when membership rights are first acquired in an AG, GmbH, other incorporated business associations, or GmbH & Co. KG. A similar tax is imposed on any other capital contributions made to the company by its shareholders or members. In addition to this so-called company tax (Gesellschaftssteuer), an exchange turnover tax (Börsenumsatzsteuer) is imposed under the same law on any transfer of certain securities (shares, debentures, investment fund certificates, shares of limited partners in a GmbH & Co. KG), whether they are listed or acquired on a stock exchange or not. The tax rate is 0.10 to 0.25 per cent of the purchase price. The real estate acquisition tax of 7 per cent which is levied on the acquisition, purchase, or transfer of land or buildings within Germany, is also levied on the acquisition or the vesting in single ownership of all the shares in a corporation or partnership that owns real property in Germany.

3 Business Tax (Gewerbesteuer)

The Gewerbesteuer or business tax is a municipal trade tax imposed by federal law (Gewerbesteuergesetz, GewStG) which allows the municipalities to determine their own tax rates, within certain limits. The tax is imposed on all business enterprises. Although it is calculated on the basis of both trading profits and trading capital, its incidence cannot be reduced by electing to operate with debt capital rather than equity capital: for taxation purposes, interest on long-term debts is included in income and the sum borrowed is included in capital.

15

The Law of Competition

I THE LAW AGAINST UNFAIR COMPETITION

1 *The Unfair Competition Act (UWG): Purpose, Scope and Sanctions*

The purpose of the Unfair Competition Act of 1909 (Gesetz gegen den unlauteren Wettbewerb; UWG; most recently amended in 1969, 1970, 1974, and 1975) is to ensure that businessmen play fair when they are competing: according to §1 UWG, 'A person who in the course of business and for the purposes of competition conducts himself in a manner which offends against good morals may be enjoined and held liable in damages.' After this general clause, the statute proceeds to prohibit specific behaviour, such as giving misleading information on commercial matters by misrepresenting the source, method of production, or calculation of the price of goods (§3 UWG), or deliberately giving false information in advertising (§4 UWG). The terms of the general clause of §1 UWG make it clear that the statute is concerned only with commercial, not private or official, acts, and only with such commercial acts as are done with a view to competition. The concept of 'good morals' denotes not simply the prevalent ethic, but the judgment of 'all right-thinking persons' in business life (see Chapter 8 above); it is thus normative rather than purely descriptive, and the real problem is to decide its effect in the individual case (for examples, see Section I 2 below). A person in breach of this provision is liable in damages only if he acted culpably, that is, unreasonably at least;[1] but the courts have gone further, and insist that he must have been aware of the facts that make for the immorality.[2] The question of fault is most acute when the

---

[1] *BGHZ* 27, 264, 273 ('Boxveranstaltung', the boxing match case); *BGHZ* 48, 12, 17. The fault is related to the immorality of the conduct.

[2] *BGHZ* 8, 387, 393; *BGHZ* 23, 184, 193f.; according to BGH *LM* §1 UWG no. 181, it is not enough to show that the defendant's negligence was the cause of his unaware-

kind of conduct in question has not yet been finally appraised in the competition context, and no judicial consensus has crystallized.[3] Proof of fault is, however, unnecessary when the suit is for an injunction.[4] The law of unfair competition is designed to protect consumers as well as competitors and, indeed, competition in general. This is clear from the sanctions that are available and from the rights of suit that the statute confers. If a competitor has suffered injury, he may bring a claim for damages as well as an injunction, but any competitor at all in the same line of business may claim an injunction, as may an association for the advancement of economic interests (§13 I UWG), such as the Central Incorporated Association for the Suppression of Unfair Competition in Frankfurt, or consumer associations (§13 Ia UWG). German law does not, however, have the 'class action' that is such a feature of the law in the United States. Certain types of conduct may attract criminal sanctions and punishment, such as, for example, intentionally putting out false and misleading information in advertising (§4 UWG), or bribing employees (§12 UWG); in such cases action is taken by the prosecuting authorities of the state, sometimes spontaneously, sometimes on application (§22 UWG).

The Unfair Competition Act must now be seen in conjunction with the more recent Law against Restraints on Competition of 1957 (Gesetz gegen Wettbewerbsbeschränkungen, GWB; see Section II below). One can distinguish them by saying that the purpose of the GWB is to protect and maintain competition as an institution, and the purpose of the UWG is to say what rules of play are to apply in the competition whose survival the GWB assures. This distinction is not, however, an absolute one, for it is admitted that both laws are needed for the proper legal protection of competition, and that the cases they cover must abut or overlap.[5] This occurs in certain forms of unfair competition, for example, when a very large enterprise

---

ness. The matter is disputed. When the claim is for an injunction, subsequent knowledge (provided by the suit itself) is sufficient, since only future behaviour is in issue.
　[3]*BGHZ* 27, 264, 273 ('Boxveranstaltung', the boxing match case): yes as to the injunction, no as to the damages claim.
　[4]BGH *LM* §1 UWG no. 181.
　[5]Rittner, *Wirtschaftsrecht* §12 C IV (Karlsruhe 1979); Emmerich, *Kartellrecht* §3 I (3rd ed. Munich 1979).

keeps offering goods at knock-down prices in order to put a smaller enterprise out of business. Such behaviour constitutes a breach of §1 UWG, but it might also be an abuse of a dominant market position under §22 GWB, and a breach of the prohibition of discrimination under §26 GWB.

2 What is Unfair?

No general formula can tell us what conduct is morally offensive and thus unfair, for a great deal depends on the cultural traditions of the country and the current problems of commercial life; besides which, views change. The practical lawyer therefore takes note of the different categories into which the cases fall and the special principles that apply to each. Sometimes the specification is effected by the draftsman himself when he individuates the statutory prohibitions, but to a large extent it has been done by the courts in their progressive application of §1 UWG. We must mention a few of these types of cases and rules.

The numerous court decisions applying §1 UWG fall into four classes: entrapment of customers, obstructing a competitor, cashing-in, and breach of the law.

Customer entrapment (Kundenfang) involves manipulating the customer's choice in different ways: by misleading him, pressuring him, bothering him, or playing on his feelings. To misinform customers is obviously immoral; it is also forbidden under §3 UWG. It may be illicit to use psychological pressure on customers, for example, where purchasers of goods are encouraged to compete for a prize or take part in a game of chance or receive some additional bonus.[6] It would be bothersome conduct under §1 UWG if a monumental mason went to a house of mourning to solicit an order for a gravestone or if a garage or car-hire company touted for business at the scene of an accident.[7] Only in special circumstances is it immoral to play on the customer's feelings, for example, if one exploits people's pity for the wretched for purposes of profit rather than for charity; otherwise appeals to the feelings are quite acceptable in competition, if only to make it more interesting.

Measures directed against other competitors, such as price wars, discrimination, boycotts, or comparative advertising, may constitute

[6] BGH *GRUR* 1973, 474, 475f. ('Preisausschreiben', the prize competition case); *BGHZ* 65, 68, 72 ('Vorspannangebot', the tied-offer case).
[7] *BGHZ* 56, 18 ('Grabsteinwerbung', the gravestone case); BGH *MDR* 1975, 381–2 ('Werbung am Unfallort', touting at the scene of accident).

'hampering and obstructing' (Behinderung). To underbid one's competitors and even to sell below one's own costs is in principle licit,[8] but there was one famous case where the Reichsgericht quite correctly held an oil cartel liable for deliberately and concertedly underselling in order to destroy a particular chain of service stations: this was immoral under §1 UWG and §826 BGB.[9] Exceptionally, too, it may be immoral to give away one's own produce when the effect is to block the market and permanently prejudice one's competitors' sales.[10] Boycotts are unlawful when used for competition, and may even be so when used to express political views.[11] On the other hand, discriminating or differentiating between those with whom one deals is immoral only under special circumstances; the prohibition of discrimination in the GWB is likewise limited (see Section II 4 below).

The law of competition in Germany differs from that in other systems by taking a rather strong stance against comparative advertising: a person who compares the services he renders with those of his competitors is making himself judge in his own cause, and in principle this is not allowed. Under §14 UWG, which prohibits the dissemination of prejudicial facts about competitors and their goods, there is no liability if the facts published are demonstrably true, but the courts have held that under certain circumstances there may be a breach of §1 UWG even where the ventilated facts are perfectly accurate. On the other hand, the courts have introduced many exceptions to the prohibition of comparative advertising, in deference to the constitutional guarantee of freedom of expression and the public interest in protecting the consumer. Today advertisers may make an objective comparison of types of goods or sales techniques, or a comparison that is necessary in order to make it clear what they are offering.[12] In certain cases the courts have allowed an accurate and unbiased comparison of prices.[13] A consumer journal is entitled to publish the results of fair and neutral comparative tests of goods, even if they are unfavourable to one producer.[14]

Cases of cashing-in (Ausbeutung) involve making unfair use of a competitor's efforts rather than obstructing and hampering him. By

[8]BGH *WuW/E BGH* 767, 771 ('Bauindustrie', building industry case); BGH *GRUR* 1979, 321, 322ff. ('Verkauf unter Einstandspreis', case of selling under cost price).
[9]*RGZ* 134, 342, 347f. ('Benrather Tankstelle', the Benrath service station case).
[10]*BGHZ* 43, 278, 280ff. ('Kleenex').
[11]*BVerfGE* 25, 256, 263ff., disapproving BGH *NJW* 1964, 29, 30ff. ('Blinkfuer').
[12]BGH *GRUR* 1964, 208, 210 ('Fernsehinterview', the television interview case).
[13]*BGHZ* 49, 325 ('40% sparen', the 'save 40%' case); the leading case is BGH *GRUR* 1952, 416 ('Dauerdose', the can that lasts).
[14]BGH *LM* §823 (Ai) BGB no. 28 ('Warentest I', the first goods-testing case); BGH *NJW* 1976, 620 ('Warentest II', the second goods-testing case).

and large, it is perfectly permissible to do what a competitor does, provided that no specially protected rights are involved, such as patents and trade-marks. But a breach of §1 UWG may nevertheless be involved under certain circumstances, for example, where a person systematically imitates another's products, latches on to his brainwaves, or exploits his good reputation by echoing his publicity or aping the get-up of his goods.[15] So, too, while there is nothing wrong in employing a competitor's ex-employees and even benefiting from their know-how, one is not entitled to woo away his staff in a systematic manner in order to exploit his efforts.[16]

Another way of infringing the rules of competition in order to gain a competitive advantage is by breaking the law, for example, by selling medicaments in a drug-store which may be lawfully sold only in a chemist's shop,[17] or by inducing a senior employee of a competitor to work for one in breach of his contract of employment or restrictive covenant.[18] §1 UWG applies only when it is for competitive purposes that the law is broken.[19]

These cases and principles have been decided and adopted under §1 UWG, but one must not forget the specific prohibitions elsewhere in the Act; they are often invoked in conjunction with the general clause of §1 UWG. Conduct specifically forbidden includes inaccurate advertising, putting on special sales, bribing employees, disclosing business secrets, and impugning commercial probity. The infringement of trade-marks is covered by other statutes as well, and there are also the enactments on bonuses and rebates (see Sections I 3 and 4 below).

If an advertiser makes statements that are factually inaccurate he may be enjoined under §3 UWG. The statements might relate to the method of production, composition, or origin of his goods or services, or perhaps to the prizes he has won with them at trade fairs and so on. Because §3 UWG does not require any showing that the defendant was aware of the facts that rendered his conduct immoral, liability is easier to establish than under §1 UWG, and in practice the courts

---

[15] *BGHZ* 5, 1, 10ff. ('Hummelfiguren', the Hummel figures case); *BGHZ* 44, 288, 300 ('Apfelmadonna', the Apple Madonna case); BGH *GRUR* 1968, 371, 377 ('Maggi'); see also Section 4 below.
[16] BGH *GRUR* 1966, 263 ('Bau-Chemie').
[17] *BGHZ* 22, 167, 179 ('Arzneifertigwaren', the packaged-drug case).
[18] BGH *GRUR* 1966, 263, 265f. ('Bau-Chemie').
[19] *BGHZ* 22, 167, 179 ('Arzneifertigwaren'); Baumbach-Hefermehl, *Wettbewerbsrecht* §1 UWG Rdn. 536, 540; a more extreme position is taken by Schricker, *Gesetzesverletzung und Sittenverstoss* 250ff. (1970).

found their judgments on it more frequently. No claim for damages arises under §3 UWG, but a person who knowingly gives misleading information in his publicity may incur criminal sanctions under §4 UWG.

In days gone by, the practice of holding special sales was much abused, so the UWG is relatively severe. It permits such sales only in very limited circumstances: on giving up business (§7 UWG), on clearing warehouse stock for a special stated reason (§7a UWG), summer and winter sales (§9 UWG), and sales on special occasions for which permission has been given by the authorities (§9a UWG).

A person who seeks to gain an advantage over a competitor by bribing one of his executives or senior employees may be fined or imprisoned under §12 UWG, and §§17–20a UWG render civilly and criminally liable those who disclose, exploit, or try to procure the disclosure of other people's business secrets. Business reputation, which is already protected in principle by §1 UWG, is further protected by §14 UWG, which specifically forbids the dissemination of prejudicial facts about competitors or their products unless the facts are demonstrably true. A claim for an injunction and damages lies, and a person who knew the truth when he so acted is also punishable criminally (§15 UWG).

3 Bonuses and Rebates

The mischief which the Ordinance on Bonuses of 1932 (Zugabeverordnung) was intended to counter was the offer of free goods or services along with the goods or services being paid for. Such bonuses and free gifts are now generally prohibited, but there are several exceptions: it is permissible to provide standard commercial concomitants, such as a tea caddy with the purchased tea, or normal ancillary services, such as guarantees, free parking, free transport, free advice, or discounts for cash and for quantity. The enormous growth in the granting of discounts for cash payment and bulk orders that ensued after the passing of the Ordinance on Bonuses led to the enactment in 1933 of the Rebate Law (Gesetz über Preisnachlässe, RabattG). While it does not forbid rebates altogether, it imposes strict limitations on the discounts that may be given to consumers: rebates for cash must not exceed 3 per cent (§3 RabattG), and rebates for quantity must not exceed what is normal in the trade (§7 RabattG). Now that competitive pricing is thought to be so desirable, this enactment is thought to be hampering it. Indeed, although the GWB has been amended so as to prohibit vertical price-fixing by producers, except publishers of books and periodicals, and to allow manufacturers at the most to recommend resale prices (§38a GWB), retailers tend to treat these recommended prices as normal prices and use the Rebate Law as a pretext for not giving a decent discount.

4 The Protection of Trade-Marks and Business Marks

Business names and marks are protected by a great many different provisions, whose scope of application overlaps to some extent. A person whose name is used by someone else without authority may sue him under §12 BGB for an injunction and the reparation of the harm caused by the infringement. This provision also covers names used in business, names that are part of the business or company name (Firma), names of legal persons, and finally even assumed names if they are generally known and unmistakable.[20] Even a mark on goods can acquire such currency in commerce that it serves as an indication of the undertaking and denotes the firm just as a name does.[21] If the infringement is culpable, liability in damages may be imposed under §823 I BGB. A Firma or business name is protected against unauthorized use by §37 II HGB (see Chapter 13 Section II 3 above).

A trade-mark enjoys protection against unauthorized use under §15 of the Law on Trade-Marks (Warenzeichengesetz, WZG; as amended in 1968) once it has been entered in the register of trade-marks that is kept by the patent office. The same enactment forbids the unauthorized use of names, business names, or trade-marks (§24 WZG) as well as the unauthorized use of a get-up of goods which is recognized in the trade as indicating that they come from a particular manufacturer (§25 WZG). An action lies for an injunction and, where the conduct is culpable, for damages as well.

The protection afforded by §16 UWG is quite comprehensive: any person who makes unauthorized use of the name, Firma, or business marks of another in such a way as to be apt to cause confusion in trade is liable to be enjoined and, where his conduct is blameworthy, held liable in damages. §16 UWG is

[20] OLG Munich *NJW* 1960, 869 ('Romy').
[21] BGH *NJW* 1956, 1713 ('Meisterbrand').

rather more extensive than §37 II HGB, for example, which applies only when the business names that are apt to be confused are in current use in the same place or area covered by the local commercial register (§30 HGB); furthermore, §16 UWG protects not only the business name but also any other indication of the business, and it allows a claim for damages which does not arise under §37 II HGB. The cover afforded by §16 UWG is also wider than that given by the Law on Trade-Marks, since §16 UWG does not require that the goods be similar, or that there be any competition between the parties, but only that there be the likelihood of confusion.

5 International Protection of Industrial Property

Like most other countries in the world, the Bundesrepublik is a signatory of the International Convention for the Protection of Industrial Property (the Paris Union Treaty) of 20 March 1883, whereby the citizens of any signatory state are to enjoy in the territory of any other signatory state the same treatment as the latter's citizens in respect of unfair competition and the protection of their industrial property such as patents, designs, and trade-marks. The owner of a protected right enjoys priority in Germany from the time it is first registered in any other member state. The Bundesrepublik is also a signatory of the Madrid Treaties of 14 April 1891 regarding False Indications of Origin on Goods, and the International Registration of Manufacturers' Marks and Trade-Marks. These treaties are all in force in Germany in their most recent version.

II THE LAW AGAINST RESTRAINTS ON COMPETITION

1 Survey: Scope and Sanctions of the Law

For over a century all the industrial countries of the West have been concerned with the ways in which cartels, big business, and combines can use their private economic power to inhibit or destroy free competition. The first counter-measures were taken in the United States, whose anti-trust legislation has had an influence on German law. Today Germany has the most comprehensive and extensive law on competition and cartels in all Europe. The Allied Occupation Powers had enacted laws to control the concentration of such economic power in Germany as had survived the war, but while its prohibitions were severe, they were not very satisfactory from the technical point of view, and they were repealed by the Law against Restraints of Competition of 1957 (Gesetz gegen Wettbewerbsbeschränkungen, GWB) which came into force on 1 January 1958, at the same time as the European law of competition contained in the Treaty of Rome (see Section II 7 below). The fact that the GWB has already been amended four times (in 1965, 1973, 1976, and 1980) shows the constant concern of the legislator that the legal protection of competition should actually work. In the GWB the following types of restraint of competition are separately regulated: cartels (§§1–14 GWB), 'other agreements' in restraint of competition (§§15–21 GWB), enterprises with dominant positions in the market, and mergers (§§22–4b GWB), anti-competitive and discriminatory conduct (§§25–7 GWB), codes of competition (§§28–33 GWB), and recommendations for competition (§§38 I no. 11, 12, II, III, 30a GWB).

In order to make for effective implementation and enforcement, the GWB provides for both suits by private individuals and action by governmental authorities. Whoever infringes a provision of the GWB or an administrative disposition based on such a provision (§35 I GWB) may be sued for damages by any person who has suffered harm thereby, provided that the plaintiff can show that the provision or disposition infringed was designed to protect a particular group of people of whom he is one. The prohibition of boycotts and discrimination (§26 GWB) are protective provisions of this kind and so, probably, is the prohibition of cartels (§1 GWB),[22] but it is more doubtful in the case of other provisions. The claim for damages requires proof of fault, but a claim for an injunction also lies, and this is independent of any fault on the part of the infringer. In general, liability under §35 GWB is analogous to liability in tort under §823 II BGB, but since in practice proof of the existence and extent of the harm often gives rise to difficulties, and since it is often unclear whether the rule infringed is protective in the sense of giving rise to a claim for damages or an injunction, as the case may be, the result is that private claims are relatively less important in practice than administrative action.

[22] *BGHZ* 64, 232, 238 ('Krankenhaus-Zusatzversicherung'); so also for §27 GWB: *BGHZ* 29, 344, 350f.

The Law Against Restraints on Competition

For the enforcement of the GWB the appropriate authorities are the Bundeskartellamt, or Federal Cartel Office, in Berlin, the cartel authorities of the Länder, and the Federal Minister for the Economy (§§44–50 GWB). In addition, a Monopolies Commission, set up in 1973, prepares reports on economic concentrations and on the application of the provisions on businesses with a dominant position in the market. The Bundeskartellamt is by far the most important law-enforcement body (§§48—50 GWB). It has very wide powers to make investigations and to impose sanctions (for its procedure, see Section II 5 below). It can exact heavy fines for breach of the law (§§38, 39, 81 GWB), and it may forbid the implementation of an agreement or resolution in restraint of competition even if no one is to blame (§37a GWB). In specific cases it can issue prohibitions, for example, forbidding businesses from going forward with a merger (§24 II GWB) or abusing a dominant position in the market (§22 V GWB). Exceptionally it can also authorize certain cartels (for example, see §4 GWB). The Bundeskartellamt is under the instructions of the Federal Minister for the Economy, who is empowered to grant special permissions: under certain circumstances he may approve an application for permission to form a cartel (§8 GWB) or a merger (§24 III GWB). Decisions of the cartel authorities (the Bundeskartellamt and the cartel authorities of the Länder) are subject to judicial review (see Section II 5 below). The numerous decisions of the Kammergericht in Berlin and the Bundesgerichtshof have contributed much to the clarification and development of the law of competition.

2 Cartels and Concerted Conduct

(a) The GWB starts out with a general prohibition of cartels. §1 GWB reads:

> Agreements concluded by enterprises or associations of enterprises for a common purpose and resolutions of associations of enterprises are invalid in so far as they are apt to restrain competition and thereby affect the production or marketing of commercial goods or services. This provision does not apply in so far as this Act provides otherwise.

This prohibition is primarily aimed at 'horizontal' arrangements, the parties to which are enterprises operating at the same market level; but under certain conditions 'vertical' arrangements between enterprises at different levels of marketing may fall within §1 GWB if the parties are thereby promoting a common purpose. §1 GWB applies to 'enterprises' in a wide sense, all the way from individual merchants up to government-controlled enterprises (§98 GWB) — even, indeed, to the government itself or other public agencies, to the extent that they are involved in market processes. The 'agreements' mentioned in §1 GWB are contracts of private law, but gentlemen's agreements, where the parties' only commitment is a moral one, have also been considered to violate §1 GWB.[23]

(b) In situations of oligopoly, firms quite often behave in an identical manner ('Parallelverhalten', parallel conduct). If there is no more than this, §1 GWB does not apply. Where the similar practices are concerted, it was a doubtful question whether the provision applied until the Bundesgerichtshof rendered its decision in the famous 'breakfast cartel' case (Teerfarben), which was very much on the borderline.

The major European producers of aniline dyestuffs arranged for regular meetings where, in the most informal way, company representatives exchanged views on the future pricing policies of their respective enterprises. They avoided any arrangements beyond small talk at breakfast, but the result was that all the companies involved subsequently raised their prices at the same time and by the same percentage. Relying on §§1, 38 GWB, the Bundeskartellamt fined companies that had participated in one such meeting in Basle. The companies sought review of this order by way of complaint (Beschwerde: Section II 6 below) to the Kammergericht in Berlin. The Kammergericht could not find that the breakfast talks in Basle constituted an agreement or a resolution within the meaning of §1 GWB, so it disaffirmed the order of the Bundeskartellamt, and this decision was upheld by the Bundesgerichtshof.[24] It should be noted that the Court of Justice of the European Communities in Luxembourg, in July 1972, affirmed the substantial fines that the European Commission had imposed in 1969 on ten aniline dyestuff producers, and held that conscious parallelism in pricing policy may, and in that case did, constitute a concerted practice in violation of art. 85 of the Treaty of Rome.[25]

[23]KG *WuW* 1976, 135, 138f. = *WuW/E* OLG 1627, 1630f. ('Mülltonnen').
[24]*BGHSt* 24, 54, 61f.
[25]Case 51/69 Farbenfabriken Bayer AG v. Commission of the European Communities (1962) Sammlung 745, 775f.; *AWD* 1972, 470, 471; *WuW/E EWG* MuV 269, 272.

The fact that §1 GWB, unlike art. 85 of the Treaty of Rome, had failed as a weapon against concerted practices stimulated a reaction from the law-maker in the 1973 amendment to the GWB: §25 I GWB now prohibits 'concerted practices... which this Act renders incapable of forming the subject matter of a binding contract'.

(c) Cartel agreements within the meaning of §1 GWB must have a 'common purpose'. Such a common purpose is typically, though not only, found in what can loosely be called partnership agreements (see also Chapter 14 above). The question how and to what extent a restraint of competition must stem from or be expressed in the agreement or resolution is highly controversial. If §1 GWB applied only to those arrangements that expressly provide for restrictive measures and practices ('subject matter theory'), the scope of application would be fairly narrow, whereas it would be quite wide if §1 GWB were satisfied by the mere fact that an agreement or resolution had a restrictive effect ('effect theory'). The Bundesgerichtshof is trying to find a middle path.

This is shown by the case of the ZVN (Cement Sales Office of Lower Saxony). The cement producers of Lower Saxony had long been organized in a cartel, and in 1945 they set up a common sales office, the ZVN–GmbH. After the GWB came into force, the producers applied for permission to continue with the cartel, but the Bundeskartellamt refused. The statutes of the ZVN–GmbH were therefore amended so as to do away with the obligation of the shareholders of the GmbH to offer their entire production to ZVN. The purpose of the company, the centralized sale of cement, remained unchanged, however, and the contracts whereby ZVN purchased cement from the producers, its shareholders, had identical prices and terms, as did its sales, which amounted to more than 80 per cent of the cement produced in the area. The Bundeskartellamt issued a prohibition under §37a GWB for violation of §1 GWB, prohibiting further adherence to the statutes of ZVN and the maintenance of the purchase and sales system. The Kammergericht in Berlin upheld the order and the Bundesgerichtshof affirmed this decision, on the ground that the statutes of the ZVN and the contracts of purchase and sale together constituted a system of organization that operated as a cartel even though the producers, the shareholders of ZVN, were not legally obliged to sell to or through ZVN.[26]

[26] *BGHZ* 65, 30, 36ff.

(d) Read literally, §1 GWB simply invalidates agreements and resolutions, and does not prohibit conduct at all; but conduct that disregards such invalidity is nevertheless illegal, and the Bundeskartellamt can sanction it with a fine (§38 I no. 1 GWB) or simply prohibit it (§37a GWB). Thanks to §1 GWB, cartels such as were common in branches of industry in Germany before the war are no longer set up; but the provision goes further than this, and seeks to put an end to other agreements in restraint of competition as well.

(e) §1 GWB has another function, too. Certain kinds of cartels are permitted. They are listed in §§2–8 GWB, which, as a matter of legislative technique, are drafted as exceptions to the general prohibition in §1 GWB. The result is that, before a cartel can be permitted, it must be established that the specified preconditions exist; this makes it more difficult to slip through loopholes in the law. Some cartels are permitted on mere notification to the cartel authorities (§9 II GWB): these include cartels for the adoption of uniform standards and classifications (§5 I GWB) and cartels for export (§6 I GWB). Other cartels are permitted only if the cartel authorities raise no objection within three months of notification: these include cartels for the uniform application of general conditions of business, supply or payment (Konditionenkartelle; §2 GWB), and cartels that seek to rationalize production and distribution either by inducing firms to specialize in particular products (Spezialisierungskartelle; §5a GWB), or by encouraging small and medium-sized businesses to co-operate (Kooperationskartelle; §5b GWB). A third group of cartels requires express authorization from the cartel authority, which can be granted for only a limited period; these include cartels to overcome a deep-seated crisis in some branch of the economy (§4 GWB), cartels that aim at rationalization of kinds other than those already mentioned (§5 II, III GWB), and export cartels which affect competition in Germany (§6 II GWB). Finally, and exceptionally, the Federal Minister for the Economy may authorize a cartel when the economy as a whole and the general good overwhelmingly require competition to be limited in this way (§8 GWB). §8 GWB is employed very rarely, but one could instance the 1972 agreement among cigarette manufacturers to limit advertising on television, and the short-lived coal–oil cartel of 1969. Energy supplies have since been assured by the Energiesicherungsgesetz of 1974.

(f) Business and professional associations may set up codes of competitive conduct in their field. These rules of competition (Wettbewerbsregeln) may either denounce specific behaviour in competition as unfair (Lauterkeitsregeln, rules of fair trading) or be designed to maintain or protect effective competition (Leistungsregeln, rules of effective competition) (§28 I, II GWB). The difficulty is that any resolution that made such a code of competition binding on the members might constitute a prohibited cartel under §1 GWB. This is why it has been made possible to apply to the Bundeskartellamt for the registration of such codes of competition conduct (§28 III GWB). The application is published and the Bundeskartellamt calls a hearing open to interested businesses and economic associations in order to determine whether the rules serve the prescribed purpose and are consistent with the law of competition (§§30–3 GWB). If the code of conduct passes this test and is registered, the association may resolve to adhere to it without infringing the prohibition of §1 GWB (§29 GWB).

Codes of conduct that seek to make competition more efficient became licit only in 1973. Since then, many specialist groups in industry and the retail trade have had such codes registered. They are often directed against preference agreements whereby major retail outlets, such as supermarkets and discount stores, obtain specially favourable terms from large suppliers in return for stocking their products.[27] The courts have since held such conduct immoral under §1 UWG.[28]

3 Vertical Agreements and Price Recommendations

In its section on 'Other Agreements' (§§15–21 GWB) the statute strikes at 'vertical' agreements, that is, agreements between firms at different levels of marketing which limit a party's freedom to contract with third parties. Thus a manufacturer might seek to control the wholesaler's contracts with his retailers, or a supplier of parts or materials might restrict a manu-

---

[27]For example, the competition rules of the Markenverband e. V. of 10 May 1976, *WuW/E* BKartA 1633.
[28]BGH *BB* 1977, 262f. ('Eintrittsgeld', the entrance money case); BGH *NJW* 1977, 631f. ('Schaufenstermiete', the window-hire case).

facturer's dealings with his customers. A distinction is drawn depending on whether the restriction relates to the terms of such 'secondary contracts' or to the range of persons with whom they may be formed. In the former case §15 GWB applies. Under §15 GWB agreements are null and void to the extent that they restrict the freedom of a party to determine at what price or on what terms to deal with third parties. §15 GWB is effective not only against resale price fixing, but also against 'most favourable treatment' clauses, whereby a seller is prevented from selling to other customers on better terms. §15 GWB also applies to restrictions that are achieved by merely economic rather than contractual means, but not to the arrangements that a firm may make with commercial agents or commission agents (see Chapter 13 Sections III 3 and 4 above) regarding its organization of sales and purchases.

Until 1973 there was an important exemption from §15 GWB: producers of branded goods were allowed to make contractual arrangements for the resale price maintenance of their products. After a heated debate, this exemption was abolished by the 1973 amendment to the GWB, with only one exception: prices of books and other publications may still be fixed (§16 GWB) in order to protect professional retail booksellers; this exception is narrowly interpreted by the courts, and does not extend to the sale of gramophone records.[29] The law does grant the possibility to producers of branded goods, that is, goods produced under a guaranty of constant or improved quality and marketed under a brand name or mark of origin, to make a recommendation as to the resale price (§38a GWB), provided they make it clear that the recommendation is not binding. The exception of price recommendations for branded goods under §38a GWB is a derogation from the general rule that recommendations are strictly prohibited if they affect competition in terms of price or result in an evasion of the prohibitions of the law (§38 I nos. 11, 12 GWB). Accordingly, the cartel authority is carefully watching for any misuse of such recommendations, and is authorized to issue prohibition orders in case of abuse, e.g., if the recommendation is likely to result in an unjustifiably high price level or in a restriction of production, or if it may

---

[29] *BGHZ* 46, 74.

mislead the consumer with respect to the resale prices actually prevailing in the market. Furthermore, if a producer of goods tries to enforce his price recommendation by refusing to furnish goods to a retailer who disregards it, the cartel authorities may intervene and the courts recognize the retailer's claim, under §26 II GWB, to continued supplies.[30]

Contracts that restrict a party's freedom to choose with whom to deal are not flatly prohibited, but are reviewable by the cartel authorities (§18 GWB). Four types of such restriction are specified: on the freedom to decide how to use the purchased goods or other goods or services; on the freedom to choose to what parties to resell the purchased goods; on the freedom to contract with third parties for the sale or purchase of other goods or services; and on the freedom to refuse to buy from the other party other goods or services unrelated to the goods or services purchased. The cartel authority may declare such agreements ineffective in so far as the restraints substantially impair competition, e.g. by imposing uniform constraints on a substantial number of enterprises or by making access to the market more difficult for other firms.

Licensing agreements for patents, trade-marks or other exclusive rights are bound, by their very nature, to restrict competition. These restrictions are tolerated by the law as a natural consequence of the legal protection of industrial property, but since the restrictive effect should be no greater than is required in order to protect the property, §20 GWB provides that agreements for the acquisition or use of patents, registered designs or protected rights in seed varieties (Sortenschutzrechte) are ineffective in so far as the restraints they impose exceed the scope of the protected right. Clauses restricting the exercise of the protected right as regards the manner and field of use, quantity, territory, or time are regarded by the law as falling within the scope of the protected right.

4 Abuse of a Dominant Position in the Market; Discriminatory Practices

Considering perfect competition to be the best of all economic worlds, the draftsman of the GWB was perfectly well aware that

---

[30] BGH *BB* 1976, 198 ('Rossignol'); Emmerich, *AG* 1976, 57, 92, 93.

domination of the market by big enterprises and combines constitutes a constant and serious threat to this ideal. Accordingly, he established a system of control over such enterprises in order to achieve a situation as if competition were unhampered (§22 GWB). The aim is to prevent any abuse of dominant positions in the market, whether the victims are competitors (abuse by restriction) or contractors (abuse by exploitation). The cartel authority can prohibit any abusive conduct by enterprises that dominate the market, and can declare agreements invalid (§22 V GWB). In the latter case, since to invalidate the contract as a whole might cause detriment to the weaker party, the cartel authority can declare that the other contractual arrangements remain valid (§19 II GWB). If an enterprise continues its abusive conduct after a prohibition order has been issued by the Cartel Office, the Cartel Office can impound any extra profits attributable thereto (§37b GWB (1980 amendment)).

Economic power is a highly complex phenomenon, and the statutory definition of the criteria for controlling it poses one of the most difficult tasks of modern business law. Under the original formulation an enterprise is dominant in the market 'if as a seller or buyer of a particular type of goods or commercial services it has no competitors or is not exposed to any substantial competition'. Since a dominant position is defined in relation to a particular market for specific goods or services, the court must first determine the 'relevant market' before it can apply the statute.[31] This approach has sometimes proved too narrow, so writers have proposed that the enterprise in question should be viewed as a whole, taking into account its position with regard to both potential and actual competitors in other markets as well.[32] For like reasons the law was amended in 1973 so as to include an alternative description of dominance in the market: this is now satisfied by a business that has a superior market position in relation to competitors, particular regard being paid not only to its share of the market, but also to its financial strength, its access to supply and sales markets, its involvement with other businesses, and also any legal or factual

[31] *BGHZ* 68, 23, 27 ('Valium I'); BGH *WuW* 1978, 375ff. = *WuW/E BGH* 1501, 1502f. ('Kfz.-Kupplungen', the truck-clutch case).
[32] Rittner, *Wirtschaftsrecht* §23 B I. 1 (Karlsruhe 1979).

barriers that prevent other firms having access to the market. There are thus several different ways of determining whether a business has a dominant market position.

Since true monopolies are quite rare, the real problem of market power arises in situations of oligopoly: the law therefore takes two or more businesses together and treats them as having a dominant market position when in any particular market there is in fact no real competition between them and when, taken together, they satisfy the criteria of having a dominant position in the market (§22 II GWB). The application of the law is facilitated by a presumption that a dominant position in the market exists if any of the following conditions is satisfied (§22 III GWB): in the case of one business, if it has at least one-third of a particular market; in the case of two or three businesses, if between them they have at least half the market; in the case of up to five businesses, if together they have a two-thirds or greater share of the market. The presumptions do not apply if, in the first case, the annual turnover of the business is less than DM250 million, and in the other cases if the businesses have an annual turnover of less than DM100 million.

Although it is not defined in the statute, 'abuse' is a legal concept; certainly §22 GWB does not authorize the cartel authorities to implement pure economic policy, as may be done in England under the law of restrictive trade practices and monopolies.[33] The writers and courts have sought to define 'abusive conduct' by comparing the conduct in question with the conduct to be expected under competitive market conditions. To do this one needs to discover an actual market, or invent a hypothetical one, as a unit of comparison, and this is naturally very difficult, since all markets are different. Other attempts to define abuse more closely take market behaviour, market structure or market results as the criterion. Since there are special provisions which prohibit unfair obstruction and discrimination by big businesses that dominate the market (§§26 II, 18 GWB), the principal area to which §22 GWB applies is abuse by overpricing. But here arises a new difficulty. The statute was clearly not intended to introduce price control, which is anathema to believers in the market economy, and the

---

[33] For a comparative viewpoint see Baur, *Der Missbrauch im deutschen Kartellrecht* 7ff., 66ff. (1972).

Bundeskartellamt is not supposed to be a Commissar of Prices. How, then, is one to determine when prices are abusively excessive and therefore to be restrained?

The 1980 amendment did not resolve these difficulties, but it did specify three types of abuse (§22 IV 2 GWB), based on past experience with the law. An enterprise abuses a position of dominance in the market if it (1) unjustifiably obstructs the possibilities of other competitors to compete in the market; (2) quotes prices and conditions different from those that would probably result if the competition were efficient (here the behaviour of enterprises in comparable markets with effective competition is to be taken into account); (3) quotes prices or conditions less favourable than those which the same enterprise quotes for comparable customers in comparable markets ('market splitting').

Only a few cases arose for decision before the 1980 amendment, but they illustrate all the difficulties.

Two cases relating to the market in pharmaceutical products demonstrate the problems. The first case concerned a Vitamin B12 preparation, which the producer sold for DM1.66 on the general market and to hospitals at a much lower price. In 1974 the Bundeskartellamt issued an order prohibiting the producer from selling the product on the general market for more than DM0.70, on the ground that any higher price was an exploitative abuse of a dominant market position in the sense of §22 IV, V GWB. In the ensuing court proceedings, the producer pointed to his high research and development costs for new pharmaceutical products, and contested the possibility of establishing a hypothetical market price. The Kammergericht upheld the decision of the Bundeskartellamt but raised the permitted price to DM1.00 in order to take account of the research and development costs. The Kammergericht accepted that there was a position of dominance in the market because of the producer's share of the market, his financial strength, and his very high visibility, which made it more difficult for competitors to enter the market. The Kammergericht also found that an abuse with regard to price was proved, after it had made comparisons with the Swiss market and the market with chemists in German hospitals. Further proceedings took place before the Bundesgerichtshof, and here the producer was successful. The Bundesgerichtshof held that, although the producer's position in the market was strong, it was not a dominant one, for his accounts showed that higher prices caused him a reduction of turnover.[34]

[34] *BGHZ* 67, 104, 118.

The second case concerned the pricing policy of another producer, who also had to face proceedings in England and before the authorities of the European Community, in respect of the drug Valium. The Bundesgerichtshof left it an open question whether the producer had a pre-eminent position in the market, but disapproved of the method employed by the Kammergericht to determine a hypothetical competitive price: there were so many differences between the German pharmaceutical market and the Dutch market, which the Kammergericht had used as a basis for comparison, that the many adjustments to the price which the Kammergericht had had to make in order to take account of them had rendered its conclusion too doubtful.[35]

The statute also combats other ways in which economic power can be used to restrain competition, mostly in its section on 'Anti-competitive and Discriminatory Practices' (§§25–7 GWB). We have already mentioned that concerted practices are prohibited (see Section II 2 above). The statute also makes it illegal to use promises or threats of economic consequences so as to cause a firm to engage in conduct that would be prohibited if it were the object of a cartel agreement; nor may such pressures be used to persuade a firm to join a permitted cartel (§§25 II, III GWB). Of course, boycotts are prohibited: boycotts, that is, requiring other firms to stop dealing with a particular firm (§26 GWB), are one of the cardinal sins of competition law, and naturally also an infringement of §1 UWG (see Section I above). It is more difficult when one comes to obstructive and discriminatory conduct, since, unless one has the right to discriminate between business partners, competition and economic intercourse on the basis of contracts freely entered into becomes impossible. Even under §1 UWG special circumstances must exist before such conduct can be discountenanced. Accordingly, discriminatory market behaviour is forbidden by law only when it is practised by firms with a dominant position in the market, cartels, firms that fix prices, and enterprises occupying a 'relatively strong market position' (§26 II GWB).

A firm may have a 'relatively strong market position' for this purpose simply by producing such a famous brand of a commodity, such as

---

[35]*BGHZ* 68, 23, 34ff. with the remand ('Valium I'); BGH *BB* 1980, 543, 545 ('Valium II').

skis, that certain customers, such as sports shops, would have an incomplete stock if it did not include them.[36] In order to decide whether or not a firm with a relatively strong market position has unfairly obstructed another firm contrary to the prohibition of §26 II GWB or treated it in a discriminatory manner without sufficient reason, the courts have to weigh the respective interests of the firms involved. If a manufacturer can persuade the court that he needed to discriminate in order to protect the good name of his branded articles, he has a good defence.[37] The application of §26 II GWB is facilitated in certain cases by the fourth amendment to the law in 1980: a strong market position is presumed if a firm regularly obtains special advantages from suppliers which they do not grant to other firms. This presumption can be employed to justify an injunction under §37a III GWB.

Certain kinds of recommendation in restraint of competition can be punished by statute. These include proposals to co-operate in contravening rules of competition law or to evade prohibitions of cartels by behaving alike, and also vertical price recommendations (§38 I nos. 10–12 GWB), except that, as we have seen, the law permits recommended prices for branded goods under certain conditions (see Section III above). Permission is also granted to other specified recommendations that the legislator supposes will be useful to the economy without endangering competition; these include the so-called 'medium-class' recommendations, whereby associations of small and medium-sized businesses try to help their members compete with larger firms. Recommendations for the use of technical standards and classifications, and for the adoption of contractual conditions, are also included here (§38 II nos. 1–3 GWB), recommendations on contractual conditions having grown in importance since the passing of the Law on General Conditions of Business (AGBG; see Chapter 5 above). By permitting and encouraging businesses of small and medium size to co-operate in the manner indicated, the legislator hopes to maintain and strengthen workable competition in markets where large enterprises are active.[38]

[36]BGH *BB* 1976, 198 ('Rossignol'); BGH *NJW* 1979, 2152; BGH *NJW* 1979, 2154 is more restrictive.
[37]BGH *WuW/E BGH* 509, 514 ('Originalersatzteile'); a different attitude to §22V GWB is taken in KG *WuW/E* OLG 995 ('Meto-Handpreisauszeichner').
[38]See Emmerich, *Die Zulässigkeit der Kooperation von Unternehmen aufgrund des reformierten GWB* 57f. (1974).

5 Control of Mergers

One of the principal aims of the statutory amendment of 1973 was to prevent the creation of dominant positions in the market by means of mergers. Whereas previously the law has imposed no sanctions on mergers, but only a duty to report them, the Bundeskartellamt was now given the power to prohibit them. Mergers in the publishing industry were subjected to tighter control in 1976 in order to counter the increase in the power of large publishers and the disappearance of small ones. The statutory amendment of 1980 tightened merger control still further by imposing general limitations on the purchase of small businesses by large ones.

The rules for the control of mergers are complex, but in summary form turn on the following elements: (a) mergers are widely defined; (b) mergers must be reported to the cartel authorities; (c) advance notice of a forthcoming merger may, and in some cases must, be given; (d) the cartel authority has power to prohibit and unscramble mergers; (e) the Federal Minister for the Economy has power to grant special permission for a merger.

(a) A merger exists in the following situations: the acquisition of the whole or a substantial part of the property of another business; the acquisition of shares in another business whereby the acquirer's interest amounts to 25 per cent or 50 per cent or a majority holding in the sense of §16 AktG; contracts with another business that create or extend a combine, or constitute an enterprise contract within §§291ff. AktG (see Chapter 14 Section VII 2 above); interlocking directorates; any other link between firms which could subject one of them to the dominant influence of another or others, or confer on the other, by any contractual arrangement, a position comparable to that of a shareholder with 25 per cent of the voting capital.

(b) If such a merger is of significance to competition it must be notified forthwith to the Bundeskartellamt. The requisite significance, according to the statute, exists in three cases: if the merger creates or increases a share of 20 per cent or more in a particular market; if a participating firm has a share of 20 per cent or more in another market; or if any of the participating firms had as many as 100,000 employees or a turnover of

DM500 million in the previous business year (§23 I GWB). The Bundeskartellamt publishes the mergers notified to it.

(c) The statute also provides for advance notification of a merger which is in view (§24a GWB). Such notification is advantageous, because the Bundeskartellamt then has only one month in which to make it known that it has started its enquiry into the merger process, and only four months in which to decide whether or not to forbid the merger. Advance notification is generally voluntary, but in certain cases it is mandatory, for example, when each of two participating firms has an annual turnover of DM1 milliard or any single enterprise has an annual turnover of DM2 milliard.

(d) If a merger may well create or reinforce a position of dominance in the market, the Bundeskartellamt has the power under §24 GWB to prohibit it, if planned, or to undo it, if consummated, unless the participants can show that the merger will result in improvements in the conditions of competition sufficient to outweigh the disadvantages of their control of the market. The concept of market-dominant position was originally identical with the concept used in §22 GWB to control abuse of a market-dominant position (see Section II 4 above), but while §22 GWB remains the starting-point, the 1980 amendment had modified the concept for the purposes of merger control. The Bundeskartellamt may base its prohibition decree on prognosis of future developments;[39] it does not need to prove that as of its effective date the merger created or reinforced a market-controlling position. To facilitate such a prognosis, the 1980 amendment introduces legal presumptions, using turnover figures as indications of resources. Under §23a GWB it is presumed that a merger creates or reinforces a market-dominant position (1) if an enterprise with an annual turnover of DM2 milliard merges with a small enterprise and (a) the market is one in which small or medium-sized competitors have a two-thirds share of the total market (i.e., the market invaded by the large enterprise is still polypolistic) or (b) the (small) enterprise to be acquired is itself dominant in a market with a total turnover of at least DM150 million; finally,

---

[39] According to §23a GWB (added by the 1980 amendment) there is a presumption that a dominant position in the market is being created or reinforced if the participating firms or the relevant markets are of a stipulated size.

(2) if the total annual turnover of all participating enterprises is DM12 milliard, and two of them have DM1 milliard each.

The 1980 amendment furthermore removes one obstacle to merger control of oligopolies. Under §22 GWB, several enterprises can be held jointly market-dominant only if there is no substantial competition between or among them. As this criterion is difficult to establish, the burden of proof for purposes of merger control is now shifted to the enterprises: they have to show that competition between them can be expected to be effective after the merger, or that the participating enterprises will not have a pre-eminent position in the market (§23a II GWB).

De minimis situations, such as small enterprises with only DM50 million annual turnover, or mergers affecting only a regional or a small market, were originally exempted from merger control (§24 VIII GWB; 'Bagatellklausel'). Big enterprises, however, abused these exemptions by buying out the owners of small enterprises by the dozen and by invading many markets hitherto polypolistic. This was first felt in the press industry: many small and regional newspapers disappeared, and the diversity of the press and of public opinion suffered. The 1976 amendment therefore rendered most of the exemptions inapplicable to the press industry. The 1980 amendment went further; in order to stop big enterprises swallowing up small ones, it tightened the exemption requirements by making them inapplicable where an enterprise with an annual turnover of more than DM1 milliard acquires an enterprise with an annual turnover of more than DM4 million. The exemption for mergers affecting only regional markets was abolished.

In the case of a planned merger, the prohibition decree invalidates any legal transaction in violation of the decree, except for certain agreements that have been entered in the commercial register (§24 II GWB). Where the merger has already been carried out, the prohibition decree obliges the participants to dissolve it (§24 II GWB), though it may not be necessary to revert to the *status quo ante* if the restraints on competition can be removed otherwise (§24 VI GWB). The decision of the Bundeskartellamt is subject to review by the Kammergericht in Berlin at the instance of the parties (§62 GWB).

(e) The parties may also apply to the Federal Minister for the Economy for special permission to effect a merger (§24 III, IV GWB). Unless the restraint on competition is so extensive as to endanger the system of market economy, such permission may be granted if the restraint on competition is outweighed by the overall economic advantages of the merger, or if the merger is justified by an overriding public interest. Mergers that involve smaller firms were originally exempted from control (§24 III GWB), but large enterprises made such great use of this exemption in order to acquire smaller firms that it was severely limited by the 1980 amendment (§24 VIII no. 2 GWB).

Merger control ran into practical difficulties at the outset, but it has since proved itself to be an effective weapon against business concentration.

Both the difficulties and the successes of merger control are illustrated by the cases. We take first the Veba–Gelsenberg case in 1974. In an attempt to form a powerful German oil company in order to secure the country's oil supply, which was dominated by seven foreign multi-national oil companies, the Bundesrepublik acquired a 51 per cent majority holding in Gelsenberg AG (48 per cent from the main shareholder RWE), with a view to transferring it to Veba AG, which was state-owned. Both Veba and Gelsenberg had annual turnovers of over DM1 milliard.

The Bundeskartellamt prohibited the merger under §24 GWB, but the Minister for the Economy subsequently granted special permission on the ground of overriding public interest.[40]

In the GKN case, the British-based multinational combine GKN intended in 1976 to acquire 75 per cent of the capital, less two shares, of Sachs-AG, a holding company with major interests in a number of companies including Fichtel & Sachs AG. While the products of the Sachs group included ballbearings and various types of clutches, and GKN produced a wide range of products including clutches and other parts for cars, the acquisition of Sachs did not enlarge the existing markets for GKN products; the Bundeskartellamt nevertheless prohibited the acquisition under §24 GWB because it expected the position of dominance in the German market for two types of clutches to be reinforced, and the participating enterprises could not prove that the merger would result in improvements of competitive condi-

---

[40]BKA *WuW* 1974, 263 = *WuW/E BKartA* 1457; BMW, Decree of 1 February 1974, *WuW/E BMW* 147. The same decision has been reached most recently on the acquisition of parts of Veba-Gelsenberg by BP, BKA *WuW/E BKartA* 1719 and BMW, Decree of 5 March 1979, *WuW/E BMW* 165.

tions that outweighed the detrimental effect of their domination of the market.⁴¹ The firms appealed to the Kammergericht in Berlin and the prohibition decree was reversed. The court held that there was no sufficient evidence that Fichtel & Sachs AG would use the additional financial strength of GKN in its market. On further appeal by the Bundeskartellamt on point of law (§73 GWB), however, the Bundesgerichtshof reversed the decision of the Kammergericht and affirmed the Bundeskartellamt's prohibition decree,⁴² holding that the increased financial strength of Fichtel & Sachs would deter other potential competitors and make access to the market more difficult for them. GKN subsequently withdrew its application to the Minister for special permission for the merger.

6 Procedures

The Bundeskartellamt reaches its decisions much like a court. For this purpose it has eight divisions, each with a chairman and two assessors, permanent officials who must be qualified for judicial or high administrative office (§48 II–IV GWB). As a superior federal authority, the Bundeskartellamt comes under the orders of the Federal Minister for the Economy, but it is generally accepted that in view of the Bundeskartellamt's judicial features the Minister must issue no instructions regarding the decision to be reached in any pending case.⁴³ In the Länder the cartel authorities are specialist divisions of the economic ministries.

Investigations of possible breaches of the GWB often start out quite informally, but even here the cartel authorities have extensive powers to obtain information from the firms involved and their representatives; they may conduct their own inquiry into business records (§46 GWB), and indeed they may even make spot searches, with court consent (§46 IV GWB). Many cases are settled through informal procedures and discussion between the parties, but if this fails, formal administrative proceedings take place before the authorities (§§51–8 GWB). The parties are entitled to be heard (§53 I GWB), but an oral hearing is not always held, though one may be ordered, even if

---

⁴¹Bundeskartellamt, Decision of 12 May 1976, *WuW/E* 1625.
⁴²BGH *BB* 1978, 674, 676ff.
⁴³Rittner, *Wirtschaftsrecht* §25 B II 2 (Karlsruhe 1979); Emmerich, *Kartellrecht* §31 b (3rd ed. Munich 1979).

no application has been made. The proceedings are usually not public (§53 III GWB), except in cases under §22 GWB. The cartel authorities have the powers of a court with regard to obtaining evidence (§54 GWB).

A party aggrieved by a decision of the cartel authorities has one month in which to enter a Beschwerde or complaint, and thereby apply to the appropriate Oberlandesgericht or court of appeal for review (§§62, 65 GWB). The appropriate court to review decisions of the Bundeskartellamt is the Kammergericht in Berlin, and since such cases are numerous and important, the judgments of the Kammergericht have made a special contribution to cartel law. The court reviews the entire legal basis of the order made by the cartel authority, which shows that the cartel authorities must apply positive law rather than adopt a competition policy of their own. In order to operate properly, after all, a market economy needs clearly established guidelines, and one of the most important of these is an unambiguous law of competition. Application for further review of a decision of an Oberlandesgericht lies to the Bundesgerichtshof, if there is some procedural defect (§73 GWB) or if the Oberlandesgericht gives its consent.

If the cartel authorities impose a fine, they must act in accordance with the Law on Statutory Offences (Gesetz über Ordnungswidrigkeiten; OWiG), but they need not give any special justification for their decree. Here, too, the party affected can apply for review to the Oberlandesgericht (§82 GWB), and subsequently on to the Bundesgerichtshof (§83 GWB). In civil suits arising under the GWB (§87 GWB), such as suits for damages (§35 GWB; see Section I above) or claims based on permitted cartel agreements, the Landgericht has jurisdiction. Access to the ordinary courts in cartel matters cannot be excluded by arbitration clauses (§91 GWB) for fear that permitted cartels might use such clauses in order to escape judicial scrutiny of their methods of controlling their members.

7 The GWB and the Cartel Law of the EEC

As a rule, EEC cartel law (art. 66, Treaty Establishing the European Coal and Steel Community, 1951; art. 85 and 86, Treaty of Rome Establishing the European Economic Community, 1958) and national cartel law are both applicable in a given case. In case of conflict, community law prevails.⁴⁴ It is true that in the GKN case the parties had obtained permission from the EEC Commission under art. 66 of the Treaty on the European Coal and Steel Community, but both the Commission and the German authorities were of the opinion that a prohibition of the merger under German law would not conflict with the permission granted by the Commission because it covered only a very limited aspect of the whole transaction. This is in accordance with the principle that a no-action letter under Regulation no. 17, s. 2, does not exclude a prohibition decree by the national authority.⁴⁵ The conflict may be more serious in cases where the national authorities issue a prohibition decree after the Commission has granted an express exemption. Here it seems appropriate to allow the national authorities to exercise merger control under national law notwithstanding community law.

---

⁴⁴Case 14/68 Walt Wilhelm and others *v.* Bundeskartellamt (1969) *ECR* 1, 16ff. (dyestuff case); Mestmäcker, *Europäisches Kartellrecht* §10 I.
⁴⁵Markert, *BB* 1978, 678.

16

Labour Law*

I INTRODUCTION

The labour law of a country tends to be much more deeply marked by national history and social mores than its law of contract or tort. Certainly there are peculiar difficulties in presenting German labour law to a foreigner; and for all that Great Britain and the Bundesrepublik are at much the same level of industrialization and are fairly comparable in economic and demographic terms, the English lawyer in particular will feel that he is entering very unfamiliar territory when he comes to German labour law. We will find differences on many points, but perhaps the most striking is that in Britain the system of industrial relations is basically voluntary, in the sense that both sides of industry have always preferred to settle their disputes in their own way without assistance or guidance from anyone else, including the state and the courts. In Germany, too, of course, employers and trade unions have the right — indeed, the constitutional right — to make, and indeed to use pressure on each other to procure, collective agreements on matters of importance such as the conditions of work and the formation

---

*In addition to general works on labour law, the following are especially valuable: Aaron and Wedderburn (eds), *Industrial Conflict: A Comparative Legal Survey* (London 1972) (with comparative essays including German law); Blanpain, 'Influence of Labour on Management Decision-Making: A Survey of Belgium, France, Germany, the Netherlands and the Societas Europaea', 15 *Riv.dir.int.lav.* 200 (1975); Farnsworth, *Productivity and Law* (Farnborough and Lexington, Mass. 1975) (with a discussion of the German Labour Management Relations Act); Gamillscheg, 'Outlines of Collective Labour Law in the Federal Republic of Germany', in *Western European Labour and the American Corporation* (ed. Kamin) 253 (Washington 1970); Kahn-Freund, '*Pacta Sunt Servanda*: A Principle and Its Limits, Some Thoughts Prompted by Comparative Labour Law', 48 *Tul.L.Rev.* 894 (1974); Seyfarth *et al.* (eds), *Labor Relations and the Law in West Germany and the United States* (III *Michigan International Labor Studies*) (Ann Arbor 1969); Simitis, 'Workers' Participation in the Enterprise: Transcending Company Law?' 38 *Mod.L.Rev.* 1 (1975); Summers, 'Worker Participation in the U.S. and West Germany: A Comparative Study from an American Perspective', 28 *Am.J.Comp.L.* 367 (1980).

and termination of employment relationships; and it is accepted that it would be unconstitutional for the state directly to trench on this freedom to bargain, for example, by setting up and operating a system of compulsory arbitration or forced conciliation of industrial conflicts. On the other hand, it is equally accepted that a collective agreement, just like any other contract, is legally binding on the parties to it, and that either of them may go to court and seek an injunction or damages if the other has broken or threatened to break it. The principle that direct state intervention in labour disputes is unconstitutional does not mean that the state may not limit the scope of industrial conflict and seek to contain its harmful effects by laying down the rules of the game and empowering the courts, as a last resort, to see that they are respected.

One reason that German labour law seems very complicated is that its rules are at many different levels and are to be found in many different enactments. Right at the top of the hierarchy, the Basic Law provides that 'the right to form associations to safeguard and improve working and economic conditions is guaranteed to everyone and to all trades, occupations and professions' (art. 9 par. 3 GG). Far from being merely hortatory, this is a rule of directly applicable law which invalidates agreements, contracts, and even statutes to the extent that they infringe this basic right of association. Many statutes deal with questions of industrial life. One of the most important is the Collective Agreements Act of 1949 (Tarifvertragsgesetz). This law specifies who can make collective agreements, what kind of matters they may cover, and what their legal consequences are. Mention must also be made of the Labour Management Relations Act 1972 (Betriebsverfassungsgesetz), which contains rules for the creation, composition, and powers of works councils. A great many special statutes deal with different aspects of the individual contract of employment such as notice, protection against unfair dismissal, sick pay, and holiday entitlements. Other enactments have been passed to protect special groups of workers such as children and minors (Jugendarbeitsschutzgesetz, 1960), expectant and recent mothers (Mutterschutzgesetz, 1968), and the severely disabled (Schwerbeschädigtengesetz, 1961). And we must remember that the Bundesrepublik has a full complement of courts to deal with labour law questions — Labour Courts, Labour Appeal Courts, and the Federal Labour Court (see p. 33 above); the procedure these courts are to apply is contained in the Labour Courts Act of 1953 (Arbeitsgerichtsgesetz).

The BGB itself has rules that relate to labour law. §§611–30 BGB concern Dienstverträge or contracts of service(s), that is, contracts whereby one party 'undertakes to provide the promised service, and the other party to provide the agreed remuneration' (§611 BGB). This definition covers the agreements between doctor and patient, tennis coach and pupil, and lawyer and client, but it also applies where the service is subordinate, where the party who works is subject to the instructions of the party who pays. Such contracts are called Arbeitsverträge, or employment contracts. But while the rules in §§611ff. BGB apply to contracts of employment in principle, in practice they are almost completely overshadowed by the rules contained in the special enactments we have mentioned. It is generally admitted that labour law has long since decamped from the BGB, and it is treated as an independent area of law both in the universities and in the legal literature. It is therefore not surprising that the federal government has been at work for many years drafting a labour code that would consolidate the whole of labour law. It is, however, very doubtful whether this will come to fruition in the foreseeable future.

II COLLECTIVE AGREEMENTS AND INDUSTRIAL ACTION

1 Trade Unions (Gewerkschaften)

Another important difference between Great Britain and Germany in the field of industrial relations is that the unions in Germany are organized as industrial unions: this means that the trade union a worker belongs to depends not on what his own trade or skill may be but on what branch of industry his employer is engaged in. Of these industrial unions there are at present sixteen, all members of a parent organization, the Deutscher Gewerkschaftsbund (DGB). The two biggest unions are the Metal Workers Union, with 2.6 million members, and the Union of Public Service and Transport Workers, which has 1.1 million. The Deutsche Angestelltengewerkschaft (DAG), which is not allied to the DGB, has a special position because its

473,000 members, predominantly white-collar workers, or Angestellte, are employed in all branches of industry and commerce. About 31 per cent of the 25 million employees in the Bundesrepublik belong to a union, that is about 8 million workers in all.

Union activity in the factory is guaranteed by German law to a considerable extent. The employer must allow unions to distribute promotional and informational material in the factory, either outside working hours or during breaks,[1] and to canvass there for union members who are standing for election to the works council.[2] These rights are not conferred in terms by statutory provisions: they have been derived by the courts from art. 9 par. 3 GG on the argument that 'the right to form associations to safeguard and improve working conditions' would be futile if the unions could not do any canvassing for members within the factory itself.

On the other hand, the courts have interpreted art. 9 par. 3 GG to mean that an individual has not only the right to join a union but also the right not to join one. It follows that a collective agreement whereby an employer undertook not to take on non-union members (closed-shop agreement) would be unconstitutional, and that a strike would be unlawful if it was designed to achieve an agreement that contained such an undertaking.

Given that union membership cannot help one to obtain employment, the decision not to join a union involves hardly any disadvantages at all. It is true that an employer who has entered a collective agreement is free, so far as the law goes, to pay lower wages to non-union members than the agreement provides for, but in fact this never happens: the employer treats all his employees alike, especially as this deprives them of an incentive to join the union. Since the closed-shop agreement is unlawful, many employees see no point in joining a union, for they can save the union dues and still benefit, as free riders, from any increases in pay and so on that the union fights for and obtains. It is true that certain facilities offered by the unions, such as legal advice, insurance cover, and the like, are available only to members, and that a non-member will get no support

---

[1] BAG *NJW* 1967, 843.   [2] BVerfG *NJW* 1966, 491.

from union funds if a strike closes the factory down and stops his pay, but the risk of having to put up with the rather lower welfare payments provided by the state is one that many employees are quite content to accept, especially in branches of industry where strikes are uncommon. Because of this the unions have occasionally tried to force the employer to agree to give union members a preference over non-union members with regard to certain benefits.

In a case decided by the Federal Labour Court, the Textile Workers Union wanted to force the employer to sign a collective agreement to create a holiday benefit fund from which unionized workers would receive higher payments than their non-unionized colleagues. The union also wanted the employer to agree not to make voluntary equalizing payments to non-unionized staff. It called a strike for this purpose. The employer claimed that the strike was unlawful in that what it sought to achieve was unlawful, and his action for an injunction succeeded in every court. The union argued that it must be possible to use collective agreements to establish some difference between members and non-members and thus create some incentive to join a union, for otherwise the state would be in breach of its duty under art. 9 of the Basic Law to guarantee the existence of the unions and to see that they can perform their functions effectively. The Federal Labour Court rejected this argument and held that the use of even mild pressure on workers constituted an infraction of art. 9 of the Basic Law.[3] This decision has occasioned much dispute: many authors can see nothing unconstitutional in a collective agreement that discriminates against non-union members, at any rate if the relative disadvantage no more than equals the cost of union membership.

2 *The Parties to a Collective Agreement*

Another important feature of German industrial relations is that collective bargaining takes place above the plant level. A trade union very rarely negotiates with a single enterprise, unless it be a very large private enterprise such as Volkswagen or a public employer such as the Post Office or the Railways. What usually happens is that in a whole region, such as Schleswig-Holstein or Northern Baden-Württemberg, the employers in the metal industry, construction industry, textile

---

[3] BAG *NJW* 1968, 1903.

industry, or whatever form an employers' association, which negotiates with the local section of the appropriate industrial union. The bargain struck in the first such regional negotiation to be concluded in, say, the construction industry often sets the tone for similar negotiations in other regions, with the result that they can be concluded more quickly.

The principle that unions are industry-wide means that these negotiations often involve only one trade union, but it is not infrequent for some union other than the appropriate industrial union to be involved as well, normally the Deutsche Angestelltengewerkschaft, whose members come from all branches of industry. In these cases the unions tend to develop a common front and speak at the conference table with one voice, but on several occasions in recent years it has happened that the unions have been unable to agree on a common policy and have pursued different objectives throughout the negotiations. To find a compromise that suits both unions in such cases is clearly very difficult.

3 Collective Agreements

Collective agreements between unions and employers' associations have effects at two levels. First, there is the so-called 'normative effect'. In regulating 'the content, formation, and termination of employment relationships' (Collective Agreements Act, §1), they apply directly, just like a law, to every individual contract of employment between any employer who is a member of the employers' association that is bound by the agreement, and any employee who is a member of the union that is similarly bound.

Apart from the 'normative effect', a collective agreement also has 'obligational effects'; that is, it generates duties, just like any other contract, for the parties themselves — the employers' association on the one hand and the union on the other. The most important of these 'obligational effects' is the duty to keep the peace (Friedenspflicht). This means that during the currency of the agreement both parties must avoid taking any step that would be apt to provoke or promote industrial conflict on a point covered by the collective agreement. This duty to keep the peace bears especially on the unions: they must maintain industrial peace until the collective agreement has expired, and indeed, in the case of a wildcat strike, that is, a strike called without their consent, the duty to keep the peace is held to require the union to do its best to make the members involved return to work. When there has been a breach of the duty to keep the peace, the affected employer may sue the union for an injunction and damages; of course, it is the employers' association rather than the individual employer that is party to the collective agreement, but the courts treat the collective agreement as a contract for the benefit of third parties (§328 BGB) and on that basis allow the individual employer to sue in his own right.

The Federal Labour Court has given a very wide interpretation to the duty to keep the peace. The leading case arose out of the Metal Workers Strike in Schleswig-Holstein in the winter of 1956–7. This strike lasted nearly four months and is one of the longest in the history of German industrial relations. At a time when it was still subject to the duty to keep the peace, the union made and published a resolution not only to poll its members, in accordance with its rules, on the question whether to strike or not, but also to advise them to vote for the strike. When the strike was over, the employers sued the union for damages for the harm that the strike had caused them. The Federal Labour Court granted the claim on the view that any premature step that puts pressure on the employer and reduces his freedom of decision constitutes a breach of the union's duty to keep the peace.[4] Writers have suggested that this is to carry the duty to keep the peace too far, and they also doubt whether the Federal Court was right to hold the union liable for the totality of the harm caused by the strike: if the union had waited until the duty to keep the peace had come to an end before taking the steps in question, it would have conducted itself quite lawfully and there would still have been a strike, just a day or two later. Be this as it may, the employers did not take the matter any further after their victory in the Federal Labour Court, since the union was then ready to meet their wishes and enter an agreement to lay any matter on which negotiations should founder in the future before a neutral arbitrator, whose decision would be binding only if acceptable to both parties.

4 *Industrial Conflict*

If collective negotiations collapse, the union may call a strike and the employer can react by ordering a lockout.

[4] BAG *NJW* 1959, 356.

(i) *Strikes* The strike is a legally recognized weapon of conflict which unions use in order to force employers to improve the workers' conditions of employment. Since strikes are in principle lawful, the employer can neither seek to enjoin a strike as such nor claim damages for the harm it causes him; but matters are different if the strike is unlawful, for then the employer may seek an injunction and damages on the basis that his right to the undisturbed exercise of his business, an interest protected by §823 par. 1 BGB, has been unlawfully infringed (see p. 150 above). The critical question is therefore when a strike is unlawful and when it is not. Surprisingly enough, the relevant rules in Germany are all judge-made. Writers have accused the legislature of abdicating their responsibilities by leaving it to the courts to resolve a problem of such massive political and social importance. Its very importance, however, makes it easy to understand the legislature's finding it too hot a potato to handle.

A strike can be lawful only if it is called and conducted by a trade union. This is because the function of strikes is to lead to the conclusion of a collective bargain, and only unions can be party to such bargains. It follows from this that wildcat strikes are unlawful,[5] and so are strikes called by works councils. A strike called by a union is unlawful if it is called in breach of the union's duty to keep the peace (see p. 315 above), or if it is called to further an aim other than those that may be achieved by a collective agreement, viz., 'the improvement of economic and working conditions' of employees (Collective Agreements Act, §1). Thus it would be unlawful for a union to call a strike in order to make the employer take on or dismiss particular individuals. There is some debate whether this applies to strikes by way of warning, protest, or demonstration; such strikes are normally quite short and constitute a reaction by the union to specific conduct of the employer, such as offering too low an increase in wages.[6] Sympathetic strikes, on the other hand, are held to be lawful provided that the strike in support of which they are called is itself lawful. A strike called to put pressure on Parliament or the government rather than the employer cannot

---

[5] BAG *NJW* 1964, 883.
[6] The Bundesarbeitsgericht has held that a short warning strike called by the trade union was not illicit: see BAG *NJW* 1977, 1079.

be justified in terms of industrial conflict, but it remains an open question whether it might not under certain circumstances be lawful on *constitutional* grounds, as where it is called to thwart a threat to the democratic system of government.

The Federal Labour Court has also said that a strike is unlawful if it offends against the 'principle of proportionality', that is, if the harm the strike may do to the general public, by reason of intensity or mismanagement, is out of all proportion to what the union stands to achieve thereby.[7] In the absence of any leading decision it is a matter of doubt and debate how this principle of proportionality would apply to particular facts.

(ii) *Lockouts* The principal weapon available to employers in industrial disputes is the lockout. This is normally used defensively, as a reaction to a strike. A lockout may apply only to those workers who are on strike already, or it may apply also in factories where no strike has yet been called. If the union has called 'pressure-point' strikes (Schwerpunktstreiks), that is, strikes in only a few chosen factories, the employers may on their side react with a lockout in all the factories in the area; the union may then have to pay very high strike benefits.

While the courts accept the legality of the lockout on the ground that there must be 'parity of punch' (Kampfparität) between the parties, there has been much public discussion of late on the question whether the lockout should not be banned. The unions argue that the employers start off in a strong position because they own the means of production; this is balanced by the unions' right to call strikes; if the employers are given the right of lockout, this balance is converted back into a preponderance of power for the employers. The employers answer this by saying that it costs the unions almost nothing to strike in selected factories one after another, and that they need the right of lockout in order to counter this enormous pressure.

In a recent decision the Bundesarbeitsgericht confirmed the legality of defensive lockouts as a reaction to strikes affecting only a few, or even just one, of the plants subject to the collective bargain. Even in such a case, however, the lockout would be unlawful if it infringed the 'principle of proportionality', as would be the case, for example, if the lockout affected all or most

---

[7]BAG (Grosser Senat) *NJW* 1971, 1668

of the workers in the field covered by the collective bargain when the strike involved only a quarter of them.[8]

III WORKS COUNCILS

Another leading feature of German labour law is the Betriebsrat, or works council. This is a body elected by the workers in a plant, whether union members or not, which participates in decisions at the *plant level* on social, personal, and economic matters.

To be distinguished from this is joint decision-making at the *company level*. This has existed in the mining industry since 1951, and since 1977 in undertakings, about 650 of them, with more than 2,000 workers. Workers in these undertakings are entitled to choose half the members of the supervisory board (see p. 277 above).

Detailed rules for the election of the works councils, their legal position, and their powers are contained in the Labour Management Relations Act of 1972. The works council, it is important to note, is elected by all the workers in the plant and represents the interests of them all. The number of members depends on the size of the plant, and they need not be members of the union, though in practice they very often are. In 1975 there were about 191,000 members of works councils in about 34,000 plants: about 80 per cent of their members were members of unions. Since members of the works council who do their job may be something of a nuisance to the employer, they are given special protection against dismissal. The employer must give members of the works council such time off work, with pay, as is required for the proper performance of their functions, given the size and nature of the plant. In plants with more than 300 employees at least one member of the council must be freed from work entirely, as must, roughly speaking, one more member for every 1,000 employees over that figure.

Just as unions and employers' associations make collective bargains with normative effects on employment relationships, so the works council and the employer can make plant agreements (Betriebsvereinbarungen) in the narrower domain of

---

[8] BAG *NJW* 1980, 1642

working conditions in the plant. But here the parallel with the collective bargaining system ends, for the role of the works council is to co-operate with the employer rather than to conflict with him. The law expressly requires the works council and the employer 'to work together in good faith for the welfare of the workers and the factory'. It follows from this that the works council can never go in for confrontation; to put it another way, the works council is subject to a permanent and absolute duty to keep the peace.

Apart from its power to conclude plant agreements, the works council has many statutory rights and duties. Sometimes it has a true right of co-determination; sometimes it is entitled only to be informed, consulted, or involved in some other way in decisions taken by the employer. Where the works council enjoys a statutory right of co-determination the employer cannot take any steps against its will. This applies, for example, to fixing the hours of the working day, including breaks, and to formulating general rules regarding behaviour in the plant, the timing of holidays and the provison of sick pay (§87, Labour Management Relations Act). In addition, the works council has important rights to be heard and to be involved in decisions to hire or to dismiss (see p. 322 below).

IV THE CONTRACT OF EMPLOYMENT

Individualarbeitsrecht, or the law of the individual contract of employment, is the name given to the rules which apply to disputes between an individual employee and his employer. There are applicable provisions in the BGB and a large number of special laws (see p. 311 above), but since they are often patchy or unclear, the rules developed by the courts have an enormous importance in this area as well: many decisions of the Bundesarbeitsgericht are virtually legislative in effect as well as in style. Only a few of the manifold legal problems to which the individual contract of employment gives rise can be discussed here.

1 Continuation of Pay during Business Interruptions

If an employee is able and willing to work as promised, must the employer continue to pay him wages even if there is no work for him to do? The business may have been brought to a halt by a fire in the factory, by adverse market conditions, by an energy crisis, or by non-delivery of raw materials owing to strikes against the suppliers. The general rules of private law provide that the employer must continue to pay wages if he is to blame for the fact which prevents him from having work for his staff to do. This would be the case where the factory burnt down because the employer failed to provide proper fire alarms or fire extinguishers. It would follow logically from this general principle that wages are not payable if the employer is not to blame for the interruption of his business, for example, where deliveries of raw materials cease because of an export ban or the blockage of the Suez Canal. The courts however have not accepted this logical consequence, but hold that, even if he is entirely innocent, the employer must continue to pay wages if the cause of the interruption falls within his 'sphere' or, what comes to the same thing, his 'business risk'. This applies not only to market difficulties and recessions, but also to shortages of energy or raw materials, and even to natural events and other acts of God that prevent the employer carrying on with his business. The only interruptions of business that the courts do not attribute to the 'sphere' of the employer are those resulting from strikes, and this is true even of strikes in some quite different branch of industry, and whether the strike is lawful or not. Clearly, the underlying thought is that there is a kind of class solidarity between workers throughout industry.

2 Industrial Accidents

§618 BGB lays down that a person who is entitled to demand a service pursuant to a contract of service(s) must ensure the safety of the premises, installations and equipment with which the other party will come in contact in performing the service; damages are payable in case of breach. For contracts of employment this provision has been, otiose for some time, for all employees are covered by the statutory scheme of insurance against industrial accidents, which forms part of the social insurance system. This means that any employee who suffers personal injury in an accident arising in the course of his employment obtains compensation for the ensuing loss of income, though not for pain and suffering, and obtains it from the current statutory insurer. His entitlement does not depend on anyone's being at fault, whether his employer, a fellow-workman, or a third party; and it is unaffected by his being at fault himself, unless indeed he caused the accident intentionally. The same rules apply to claims by his survivors where the accident is fatal.

The employers bear the full cost of this accident insurance, so in return the law exempts the employer and his employees from all the claims arising from the accident that the victim or his survivors might have brought under the general rules of contract and tort.[9] This means that the worker may sometimes receive less from the insurance carrier than he would have obtained from the employer under the general rules of liability. Such cases are not very frequent, for where the injury is slight, or only moderately serious, and has no permanent effect on the victim's earnings, the sums provided by social insurance are normally higher than those that would be granted under the general rules. By contrast, however, it is precisely where the injuries are so serious as to render the victim incapable of earning his living that he would be much better off if he could bring a normal claim for damages against his employer; for then, instead of just receiving a pension, he would obtain damages for pain and suffering as well as an indemnity for all his lost earnings.

Only the claims that an industrial accident victim has against his employer or his fellow-employees are abolished. A third party remains liable under the general rules if he is responsible for an industrial accident. This may arise where the accident is due to a defective tool supplied by the third party, or where the third party is a careless motorist who injures the workman on his way to or from work, such an event qualifying as an industrial accident in Germany. To the extent of the benefits the insurance carrier is bound to provide, it is subrogated to the workman's and his survivors' claims against the third party,[10] but they can sue the third party for financial loss not covered by the insurance and for any pain and suffering.

3 Dismissal

Another important problem for which there are detailed statu-

---

[9]See §§636–7 Code of Social Security (Reichsversicherungsordnung; RVO).
[10]See §1542 RVO.

tory rules is in how to protect the worker against dismissal. At the very least he must be given due notice, but as this is not thought to be enough in Germany, additional job security is provided in various ways. Here one must first distinguish between 'ordinary' dismissals and 'extraordinary' or 'instantaneous' dismissals.

The employer (and, of course, the employee) may terminate the employment relationship without giving any notice by way of an extraordinary dismissal under §626 BGB if there is an 'important reason' for so doing. An 'important reason' exists, according to §626 BGB, where there are

'facts such that in all the circumstances of the individual case and after giving due weight to the interests of both parties the party effecting the termination could not be expected to continue with the employment relationship until the expiry of the period of notice.'

Here, as so often, the legislature is virtually leaving the decision to the courts. According to the case law, an 'important reason' exists only where the worker has deliberately and seriously prejudiced his employer, for example, by giving away business secrets, or by repeated malingering, or by criminal conduct on the job, such as stealing from his employer or his colleagues. The Federal Labour Court has found an 'important reason' in the fact that a worker had taken part in an illegal wildcat strike and refused to return to work despite repeated demands.[11] On the other hand, matters that fall within the employer's 'business risk', such as a collapse of the market or natural events which bring his business to a standstill, do not constitute an 'important reason' for dismissing an employee.

In 'ordinary' dismissals the employer must give the proper period of notice. Here §622 BGB makes a distinction between white-collar and blue-collar workers.

A white-collar worker is entitled to six weeks' notice, and his employment may be terminated only at the end of a calendar quarter. If the white-collar worker has been employed by that employer for five years or more, the period of notice is extended to three months, and it goes up to four, five, or six months if the employment has lasted eight, ten or twelve years. The period of notice for blue-collar workers is much less generous, normally two weeks, and this period is not greatly increased even if they are employees of long standing in the

[11] BAG *NJW* 1970, 487.

business: employment of five, ten, and twenty years can be terminated by notice of one, two, and three months respectively.[12] Of course, these periods are statutory minima and may be extended in the workers' interest by collective agreement; in practice this is quite often done.

If the employer is minded to dismiss an employee, he must inform the works council of his reasons and hear its views, which they have one week to express. If he fails to inform the works council or makes a dismissal to which it is opposed before hearing its views, the dismissal is void. The works council must also be consulted in cases of 'extraordinary' dismissals, but here its opposition must be communicated to the employer within three days or it will be deemed to have agreed to the dismissal.[13] The works council's expressed opposition to a dismissal does not in law prevent the employer from effecting it, and a dismissal is never invalid on the mere ground that the works council was opposed to it; but where such opposition exists many employers will think twice before going on with the dismissal and facing a conflict with the works council.

If the employer, in effecting an ordinary dismissal, has consulted the works council and given due notice, the Protection against Dismissals Act, 1951 (as amended) offers the employee under notice a further means of self-defence.[14] He has three weeks in which to ask the Arbeitsgericht to hold that the dismissal is 'socially unjustified' and that the employment is consequently still on foot. A dismissal is 'socially unjustified' unless it is based on grounds relating to the person or conduct of the employee or to 'imperative business necessity'. It is for the employer to prove the existence of such grounds.

Reasons relating to the conduct of the employee are mainly breaches of duty such as would, if they were very grave, justify an 'extraordinary dismissal'. Reasons relating to the person of the worker exist when it appears that he is not now up to the demands of the workplace, and no suitable occupation can be found for him anywhere else

---

[12]For details, see §622 par. 2 BGB.
[13]For details, see §102 Labour Management Relations Act (Betriebsverfassungsgesetz, BVG).
[14]See the Protection against Dismissal Act (Kündigungsschutzgesetz) of 10 August 1951 (as amended).

in the business.[15] Dismissals that turn on 'imperative business necessity' are usually based on a collapse of the market, a change of product, a closure of part of the business, or a rationalization of the production process which entails some redundancy. But a dismissal may still be 'socially unjustified' in such cases unless the employer can prove that, even after retraining, the worker could not be employed anywhere else in the same factory or in any other factory forming part of the same enterprise.

Even if all these conditions are satisfied, an employer who is dismissing some workers and retaining others must take 'social considerations' into account in deciding who should be dismissed. He cannot simply give notice to those with whose services his business can most easily dispense, but must take into account each person's age, family circumstances, and length of service in the business. The Arbeitsgericht will hold a dismissal 'socially unjustified' if he has not done this conscientiously enough and has given notice to the 'wrong' employee. Since the Protection against Dismissals Act clearly gives the worker very little protection when rationalization and automation cause a loss of jobs, collective agreements today often contain provision for longer periods of notice, exemption of older employees from dismissal, and the payment of severance pay.

If, but only if, the works council opposed the dismissal (see p. 324 above), a worker who has raised a claim under the Protection against Dismissals Act can require the employer to continue to employ him at the old rate of pay until the court proceedings are over. If the employer refuses to do so, the workman may obtain an interim court order. The only way the employer can avoid such an order is to convince the Arbeitsgericht that the works council's opposition was manifestly groundless, that the employee's claim has no chance of success, or that it would constitute an unfair economic burden on him to be required to continue with the employment.

---

[15]See BAG *NJW* 1977, 2132, where the question was raised whether prolonged illness of a worker could constitute a reason related to him personally so as to render his dismissal not socially unjustified.

Appendix I

Abbreviations

ABGB	Allgemeines Bürgerliches Gesetzbuch (Austrian Civil Code of 1811)
AbzG	Gesetz betreffend die Abzahlungsgeschäfte of 16 May 1894 (Instalment Sales Act)
ADHGB	Allgemeines Deutsches Handelsgesetzbuch (previous German Commercial Code of 1861)
AG	Aktiengesellschaft (company)
AG	*Die Aktiengesellschaft (periodical)*
AGBG	Gesetz zur Regelung des Rechts der Allgemeinen Geschäftsbedingungen of 9 December 1976 (Law on General Conditions of Business)
AGG	Arbeitsgerichtsgesetz of 3 September 1953 (Labour Court Act)
AktG	Aktiengesetz of 6 September 1965 (Companies Act)
ALR	Allgemeines Recht der Preussischen Staaten (General Prussian Land Law of 1794)
Am.J.Comp.L.	*American Journal of Comparative Law*
AO	Abgabeordnung of 16 March 1976 (Revenue Code)
AWD	*Aussenwirtschaftsdienst des Betriebsberaters (periodical)*
BAG	Bundesarbeitsgericht (Federal Labour Court)
BayObLG	Bayrisches Oberlandesgericht (Bavarian Court of Appeal)
BayObLGZ	*Entscheidungen des BayObLG in Zivilsachen* (Reports of Civil Cases Decided by the BayOLG)
BB	*Betriebsberater (periodical)*
BbauG	Bundesbaugesetz of 23 June 1960 (Federal Building Law)
Betrieb	*Betriebsberater (periodical)*
BetrVG	Betriebsverfassungsgesetz of 15 January 1972 (Labour–Management Relations Act)
BeurkG	Beurkundungsgesetz of 28 August 1969 (Notarial Documents Act)

BGB	Bürgerliches Gesetzbuch of 18 January 1896 (Civil Code)
BGBl	*Bundesgesetzblatt* (official gazette)
BGH	Bundesgerichtshof (Supreme Court, Ordinary Jurisdiction)
BGHSt	*Entscheidungen des Bundesgerichtshofs in Strafsachen* (Reports of Criminal Cases decided by the BGH)
BGHZ	*Entscheidungen des Bundesgerichtshofes in Zivilsachen* (Reports of Civil Cases decided by the BGH)
BKA, BKartA	Bundeskartellamt (Federal Cartel Office)
BMW	Bundesminister(ium) für Wirtschaft (Federal Ministry of the Economy)
BRAO	Bundesrechtsanwaltsordnung of 1 August 1959 (Federal Ordinance on Attorneys)
BSHG	Bundessozialhilfegesetz of 30 June 1961 (Federal Social Welfare Act)
BundesbahnG	Bundesbahngesetz (Federal Railway Act)
BVerfG	Bundesverfassungsgericht (Federal Constitutional Court)
BVerfGE	*Entscheidungen des Bundesverfassungsgerichts* (Reports of Cases decided by BVerfG)
BVerfGG	Gesetz über das Bundesverfassungsgericht of 12 March 1951 (Federal Constitutional Court Act)
BVG	Betriebsverfassungsgesetz of 15 January 1972 (Labour–Management Relations Act)
BWahlG	Bundeswahlgesetz of 7 May 1956 (Federal Election Law)
DAG	Deutsche Angestelltengewerkschaft (Trade Union of Upper Echelon Employees)
DB	*Der Betrieb* (periodical)
DGB	Deutscher Gewerkschaftsbund (Trade Union Congress)
DDR	Deutsche Demokratische Republik (East Germany)
DepotG	Gesetz über die Verwahrung und Anschaffung von Wertpapieren of 4 February 1937 (Law on Deposit and Acquisition of Securities)
DM	Deutsche Mark
DöV	*Die öffentliche Verwaltung* (periodical)
DRiG	Deutsches Richtergesetz of 8 September 1961 (Law on the Judiciary)
ECR	*European Court Reports*

EEC	European Economic Community
EGBGB	Einführungsgesetz zum Bürgerlichen Gesetzbuch of 18 August 1896 (Introductory Law to BGB)
EheG	Ehegesetz of 20 February 1946 (Marriage Law)
EKG	Einheitliches Gesetz über den internationalen Kauf beweglicher Sachen of 17 July 1973 (Uniform Law on International Sales of Goods)
EMRK	Europäische Konvention zum Schutze der Menschenrechte und Grundfreiheiten of 4 November 1950 (European Convention on Human Rights)
ErbbauVO	Verordnung über das Erbbaurecht of 15 January 1919 (Heritable Building Right Ordinance)
EWG	Europäische Wirtschaftsgemeinschaft (European Economic Community)
EWGV	Vertrag zur Grundung der Europäischen Wirtschaftsgemeinsschaft of 25 March 1957 (Treaty of Rome)
GBO	Grundbuchordnung of 24 March 1897 (Land Register Ordinance)
GenG	Gesetz betreffend die Erwerbs- und Wirtschaftsgenossenschaften of 1 May 1889 (Law on Co-operatives)
GewO	Gewerbeordnung of 21 June 1889 (Trade Regulation Ordinance)
GewStG	Gewerbesteuergesetz, now of 25 May 1965 (Trading Tax Law)
GG	Grundgesetz of 23 May 1949 (Basic Law)
GKN	Guest, Keen, and Nettlefolds
GmbH	Gesellschaft mit beschränkter Haftung (limited liability company)
GmbHG	Gesetz über die Gesellschaft mit beschränkter Haftung of 20 April 1892 (Law on Limited Libility Companies)
GRUR	*Gewerblicher Rechtsschutz und Urheberrecht* (periodical)
GVG	Gerichtsverfassungsgesetz of 27 January 1877, now 9 May 1975 (Constitution of Courts Act)
GWB	Gesetz gegen Wettbewerbsbeschränkungen of 27 July 1957 (Law Against Restraints on Competition)

HaftpflG	Haftpflichtgesetz of 7 June 1871, now 4 January 1978 (Law of Civil Liability)
HGB	Handelsgesetzbuch of 10 May 1897 (Commercial Code)
I.C.L.Q.	*International and Comparative Law Quarterly*
i.v.	in Vertretung (on behalf of)
JGG	Jugendgerichtsgesetz of 4 August 1953 (Juvenile Courts Act)
JW	*Juristiche Wochenschrift* (periodical)
JWG	Gesetz für die Jugendwohlfahrt of 7 January 1975 (Juvenile Welfare Act)
Kfz	Kraftfahrzeug (lorry)
KG	Kammergericht (Berlin Court of Appeal)
KG	Kommanditgesellschaft (limited partnership)
KGaA	Kommanditgesellschaft auf Aktien (private company with partners)
KSchG	Kündigungsschutzgesetz of 10 August 1951 (Protection against Dismissal Act)
LG	Landgericht (district court)
LM	*Lindenmaier-Möhring, Nachschlagewerk des Bundesgerichtshofes* (Digest of cases decided by the BGH)
MDR	*Monatschrift des Deutschen Rechts* (periodical)
MitbG	Mitbestimmungsgesetz of 4 May 1976 (Co-Determination Act)
MuV	Montanunionsvertrag of 18 April 1951 (Treaty Establishing the European Coal and Steel Community)
NJW	*Neue Juristische Wochenschrift* (periodical)
OHG	Offene Handelsgesellschaft (general commercial partnership)
OLG	Oberlandesgericht (court of appeal)
OWiG	Ordnungswidrigkeitengesetz of 24 May 1968 (Law on Minor Offences)
PostG	Postgesetz (Post Office Act)

Q.B.	*Queen's Bench Reports*
RabattG	Gesetz über Preisnachlässe of 25 November 1933 (Law on Rebates)
RabelsZ	*Zeitschrift für ausländisches und internationales Privatrecht* (periodical)
Rdnr.	Randnummer (marginal number)
Rev.tri.dr.civ.	*Revue trimestrielle de droit civil* (French periodical)
RG	Reichsgericht (Imperial Supreme Court)
RGBl	*Reichsgesetzblatt* (official gazette, 1871–1945)
RGZ	*Entscheidungen des Reichsgerichts in Zivilsachen* (Reports of Civil Cases decided by the RG)
Riv.int.dir.lav.	*Rivista di diritto internazionale e comparato del lavoro* (Italian periodical)
RuStAG	Reichs- und Staatsangehörigkeitsgesetz of 22 July 1913 (Law of Nationality)
RVO	Reichsversicherungsordnung of 19 July 1911 (Insurance Ordinance)
ScheckG	Scheckgesetz of 14 August 1933 (Cheques Act)
SeuffArch	*Seufferts Archiv für Entscheidungen der obersten Gerichte in den deutschen Staaten* (Reports of decisions till 1944)
SGB	Sozialgesetzbuch of 11 December 1975 and 23 December 1976 (Social Security Code)
SGG	Sozialgerichtsgesetz of 3 September 1953 (Social Court Act)
StGB	Strafgesetzbuch of 15 May 1871, now 2 January 1975 (Criminal Code)
StPO	Strafprozessordnung of 17 September 1965, now 7 January 1975 (Code of Criminal Procedure)
StVG	Strassenverkehrsgesetz of 19 December 1952, now 16 November 1970 (Road Traffic Act)
StVollG	Gesetz über den Vollzug der Freiheitsstrafe und der freiheitsentziehenden Massregeln zur Besserung und Sicherung of 16 March 1976 (Law on Imprisonment)
Tul.L.Rev.	*Tulane Law Review* (American periodical)
TVG	Tarifvertragsgesetz of 9 April 1949 (Collective Agreements Act)

UWG	Gesetz gegen den unlauteren Wettbewerb of 7 June 1909 (Unfair Competition Act)
VAG	Versicherungsaufsichtsgesetz of 6 June 1931 (Insurance Supervision Act)
VerschG	Gesetz über die Verschollenheit, Todeserklärung und Feststellung der Todeszeit of 4 July 1939 (Missing Persons, etc., Act)
VersR	*Versicherungsrecht* (periodical)
VVaG	Versicherungsverein auf Gegenseitigkeit (mutual insurance association)
VVG	Gesetz über den Versicherungsvertrag of 30 May 1908 (Insurance Contract Act)
VerwG	Verwaltungsgericht (Administrative Court)
VwGO	Verwaltungsgerichtsordnung of 21 January 1960 (Administrative Court Ordinance)
VwVfG	Verwaltungsverfahrensgesetz of 25 May 1970 (Law on Procedure in Administrative Matters)
WEG	Wohnungseigentumsgesetz of 15 March 1951 (Home Ownership Act)
WG	Wechselgesetz of 21 June 1933 (Law on Bills of Exchange)
WM	*Wertpapiermitteilungen* (periodical)
WPM	*Wertpapiermitteilungen* (periodical)
WuW	*Wirtschaft und Wettbewerb* (periodical)
WZG	Warenzeichengesetz, now of 2 January 1968 (Law on Trade-marks)
ZHR	*Zeitschrift für das gesamte Handelsrecht* (periodical)
ZPO	Zivilprozessordnung, now of 12 September 1950 (Code of Civil Procedure)
ZVG	Gesetz über die Zwangsversteigerung und die Zwangsverwaltung of 24 March 1897, now 20 May 1898 (Law on Foreclosure and Sequestration)
ZVN	Cement Sales Office of Lower Saxony

Appendix II

Laws and Publications on German Law in English

(a) *Bibliography*

Bibliography of German Law in English and German (edited by German Association of Comparative Law). (Karlsruhe 1964); Supplement 1964–8 (1969); Supplement 1969–73 (1975).

Cohn, E.J., *Manual of German Law*, 2 vols. (British Institute of International and Comparative Law, Comparative Law Series, nos. 14 and 15) (2d ed. London 1968, 1971).

Lansky, R., *Books in English on the Law of the Federal Republic of Germany (Arbeitshefte der Arbeitsgemeinschaft für juristisches Bibliotheks- und Dokumentationswesen*, no. 4, Hamburg 1979). A selection of recent titles with bibliographical introduction.

Medicus, D., 'Federal Republic of Germany', in I *International Encyclopedia of Comparative Law: National Reports* F24–7 (Bibliography) (Tübingen, The Hague, Paris 1972).

Price, A.H., *The Federal Republic of Germany: A Selected Bibliography of English-language Publications* (2d ed. Washington. DC, 1978).

Szladits, C., *Guide to Foreign Legal Materials: France, Germany, Switzerland* (New York 1959).

(b) *Enactments in English Translation*

Basic Law of the Federal Republic of Germany. English translation with a short introduction (Press and Information Office of the Government, Bonn, 1977).

Civil Code of the German Empire. Translated by W. Loewy. 2 vols. (Boston 1919).

Co-determination in the Federal Republic of Germany. (Federal Ministry of Labour and Social Affairs, Bonn 1978).

German Banking Law. Text of the Gesetz über das Kreditwesen, with translation into English by M. Peltzer and J. Brooks and an English Introduction (Cologne 1976).

German Civil Code. Translated by Chung Hui Wang (London 1907).

German Civil Code (as amended to 1 January 1975). Translated with an introduction by J. S. Forrester, S. L. Goren, and H.-M. Ilgen (Amsterdam, Oxford 1975).

German Code of Criminal Procedure. Translated by H. Niebler, with an introduction by E. Schmidt (South Hackensack, NJ, London 1976).
German Co-Determination Act of 1976. Translated by D. Hoffman (Deventer, Frankfurt 1976).
German Commercial Code. Translated by S. L. Goren and J. S. Forrester (South Hackensack, NJ 1979).
German Law against Restraints of Competition. Bilingual edition with introduction. Translated and edited by R. Mueller and H. Schneider (Frankfurt 1973).
German Law governing Standard Business Conditions. Translated by J. Gres and D. J. Gerber (Cologne 1977).
German Penal Code of 1871. Translated by G. O. W. Mueller and T. Buergenthal, with an introduction by H. Schröder *(American Series of Foreign Penal Codes,* no. 4; South Hackensack, NJ, London, 1961).
German Standard Contracts Act. German text with English translation by H. Silberberg, and detailed introduction (Frankfurt 1979).
German Stock Corporation Law. The German Law on Accounting by Major Enterprises other than Stock Corporations. Text, translation and introduction; *Aktiengesetz, Publizitätsgesetz* (text and translation) (2d ed. edited by R. Mueller and E. G. Galbraith; Frankfurt 1976).
GmbH. German Law Concerning Companies with Limited Liability. English and German text. Translation and introduction by R. Mueller (3rd ed. Frankfurt 1977).
Labour Management Relations Act (Betriebsverfassungsgesetz). Text and translation with commentary by M. Peltzer (2d ed. Frankfurt 1977).
Law against Restraints of Competition. Text and commentary in German and English by A. Riesenkampf and J. Gres (Cologne 1977).
The Private Company In Germany. A translation and commentary by Mary C. Oliver (Plymouth 1976).

(c) *Works in English on German Law*

Baur, F., 'Introduction to German Law' in *Bibliography of German Law in English and German* (edited by German Association of Comparative Law) (Karlsruhe 1964).
Esser, K., Rüster, B. and Zahn, J., 'The Commercial Laws of the Federal Republic of Germany and West Berlin', in *Digest of Commercial Laws of the World* (New York 1975).
Heidenheimer, A. J. and Kommers, D. P., *The Governments of Germany* (4th ed. New York 1975).
Hesse, K., *Basic Principles of the Constitutional Law of the Federal Republic of Western Germany* (Deventer, in preparation).

Heyde, W., *The Administration of Justice in the Federal Republic of Germany* (Bonn 1971).

Ilgen, H. M. and Forrester, J. S., *German Legal System* (South Hackensack, NJ 1972).

Langbein, J. H., *Comparative Criminal Procedure: Germany (American Casebook* series; St Paul, Minnesota 1977).

Medicus, D., 'Federal Republic of Germany', in I *International Encyclopedia of Comparative Law: National Reports* F1–27 (Tübingen, The Hague, Paris 1972).

Rabel, E., 'Private Laws of Western Civilization', reprinted in Rabel (ed. H. G. Leser), *Gesammelte Aufsätze*, vol. 3: *Arbeiten zur Rechtsvergleichung und zur Rechtsvereinheitlichung, 1919–54* p. 276 (Tübingen 1967).

Schewe, D., Nordhorn, K. and Schenke, K., *Survey of Social Security in the Federal Republic of Germany*. Translated by F. Kenny (Bonn 1972).

Treumann, W. and Peltzer, M., *U.S. Business Law. A bilingual Guide for the German businessman and investor* (Cologne 1978).

Würdinger, H., *German Company Law (European Commercial Law Library*, no. 3; London 1975).

Zweigert, K. and Kötz, H., *An Introduction to Comparative Law*. Translated by J. A. Weir. Vol. 1: *The Framework;* vol.2: *The Institutions of Private Law* (Amsterdam, New York, Oxford 1977).

Addendum to Chapter 3, page 50

On 18 June 1980 the Law on Legal Advice in Non-Contentious Matters (Beratungshilfegesetz) was enacted. This provides that a poor person is also entitled to legal aid if all he wants is legal advice rather than assistance with litigation. The idea that legal advice should be provided at the state's expense by public citizen's advice bureaux and similar publicly-funded institutions gave rise to heated debate in Germany, but the Bar was strongly opposed to such schemes and in the end its opposition was successful. The statute now provides that the poor person may make a direct approach to an attorney who will, before or after giving advice, file a petition for legal aid with the Amtsgericht: the petition contains information on the petitioner's financial means and the nature of the business on which the advice is sought. Legal aid is available in matters of civil, criminal, administrative and constitutional law, but not in labour law matters. Here it was believed that the services provided by the advice bureaux of the trade unions were adequate — a doubtful assumption in view of the fact that only unionized workers are entitled to such advice. Legal aid is not available in Hamburg or Bremen, for these states have for many years funded a well-established system of citizen's advice bureaux.

Index of German Terms

Abandon, 255
Abschlussfreiheit, 84
Abwicklung, 94
Abzahlungsgeschäfte, 133
Aktiengesellschaft, 71, 240, 257
Akzessorietät, 236
allgemeine Geschäftsbedingungen, 88
Amtsgericht, 29
Anfechtungsgrund, 83
Anstalt, 214
Anscheinsvollmacht, 227
Anwartschaftsrecht, 170, 183
Arbeitsdirektor, 277
Arbeitsgerichte, 33
Arbeitsvertrag, 312
Armenrecht, 49
Auflassung, 75, 179
Aufsichtsrat, 258
Ausbeutung, 285
Auslegung, 75
Ausschlagung, 196
Aussendienst, 235
Aussenverhältnis, 225
Austauschtheorie, 110
Avalkredit, 236

Beglaubigung, 44
Begriffsjurisprudenz, 60
Behinderung, 285
Beschwerde, 308
Betriebspacht, 274
Betriebsverfassungsgesetz, 8
Beurkundung, 44
Beweis des ersten Anscheins, 152
Beweissicherung, 230
Börsenumsatzsteuer, 281
Bundesgerichtshof, 31
Bundeskartellamt, 291
Bundesrat, 17
Bundestag, 16
Bundesverfassungsgericht, 20
Bürgschaft, 216, 236

Darlehen, 117
Deutsche Angestelltengewerkschaft, 7
Duetscher Gewerkschaftsbund, 7

Dienstvertrag, 117, 312
Differenztheorie, 110
Drittwirkung, 137

Eigengeschäft, 234
Eigentum verpflichtet, 172
Eigentumsvorbehalt, 183
Einbürgerung, 201
Einführungsgesetz, 66
Eingetragener Verein, 240
Entlastungsbeweis, 158
Entmündigte, 73
Entmündigung, 205
Entschädigungsanspruch, 7
Erbbaurecht, 178
Erbbauzins, 178
Erbschein, 197
Erbvertrag, 197
Erfüllungsanspruch, 109, 123
Erfüllungsgehilfen, 98, 108, 113

faktischer Vertrag, 79
Festschriften, 57
Finanzgerichte, 34
Firma, 213, 219
Formkaufmann, 212
freibleibend, 76
Freirechtschule, 60
Fremdorganschaft, 242
Friedenspflicht, 315
Fusion durch Aufnahme, 271
Fusion durch Neubildung, 271

Gattungsshuld, 91
Gefährdungshaftung, 160
gemeinschaftliches Testament, 197
genehmigtes Kapital, 269
Genossenschaft, 240
Gerichtsverfassungsgesetz, 29
Gesamthand, 222, 241, 243
Gesamtprokura, 228
Gesamtrechtsnachfolge, 195
Gesamtschuldner, 241, 246
Gesamtvertretung, 245
Geschäftsanteile, 251
Geschäftsfähigkeit, 73

Index of German Terms

Geschäftsführer, 252
Gesellschaft mit beschränkter Haftung, 71, 240, 251
Gesellschaftssteuer, 281
Gesellschaftsvertrag, 257
Gesetzesvorbehalt, 172
Gestaltungserklärungen, 83
Gewerbesteuer, 281
Gewerkschaft, 312
Gläubigerverzug, 92, 121
Gleichordnungskonzern, 272
grosse Strafkammer, 30
Grundbuch, 75, 119, 180
Grundgesetz, 14
Grundhandelsgewerbe, 212
Grundpfandrechte, 184
Grundrechte, 21
Grundschuld, 169, 184, 236

Habilitationsschriften, 57
Halter, 161
Hand wahre Hand, 176
Handelsgeschäft, 212, 215
Handelsstand, 212
Handelsvertreter, 229
Handlungsfähigkeit, 71
Handlungsgehilfe, 229
Handlungsvollmacht, 228
Hauptversammlung, 258
Herrschaftsrechte, 170
Hochschulschriften, 57
höhere Gewalt, 161
Hypothek, 169, 184

Immissionen, 171
Inhaltsfreiheit, 84
Innenverhältnis, 225
IPR, 203

Jahresabschluss, 266

Kaduzierung, 255
Kampfparität, 318
Kapitalertragsteuer, 280
Kapitalverkehrssteuer, 281
Kaufmann, 212
Kaufmännisches Bestätigungsschreiben, 216
kleine Strafkammer, 30
Koalitionsverbot, 6
Kommanditgesellschaft, 239, 248
Kommanditgesellschaft auf Aktien, 240
Kommanditist, 241

Kommissionär, 234
Kommission mit Selbsteintritt, 232
Kommittent, 232
Komplementär, 248
Konditionenkartelle, 294
Konkretisierung, 91
Konzernrecht, 242, 272
Kooperationskartelle, 294
Körperschaftssteuergesetz, 280
Kundenfang, 284

Landgericht, 29
Lauterkeitsregeln, 295
Leistungsregeln, 295
Leistungsstörungen, 91, 93, 121
Lombardkredit, 237

Maklervertrag, 117
Mangelfolgeschaden, 129
Mangelschaden, 130
Miete, 117
Minderkaufleute, 220
Minderung, 127
Mitbestimmung, 8, 278

Nachfrist, 102, 123
Nachschusspflicht, 255
NebenleistungsAG, 265
Nebenpflichte, 119
negative Publizität, 225
Niessbrauch, 181
Notar, 44
Nottestament, 197

Obergesellschaft, 276
Oberlandesgericht, 31
offene Handelsgesellschaft, 239, 243
offene Stellvertretung, 225
ordentliche Gerichtsbarkeit, 27
Organe, 242
Organgesellschaft, 280
Organträgergesellschaft, 280

Parallelverhalten, 292
Partenreederei, 240
Pfandrechte, 185
Pflichtteil, 199
positive Vertragsverletzung, 94
Preisnachlässe, 287
Prokura, 228
Publikums-KG, 239

Realkredit, 184

Index of German Terms

Rechtsanwalt, 41
Rechtsfähigkeit, 71
Rechtsgeschäft, 74
Rechtskraft, 63
Rechtsmangel, 124
Referendar, 37
Reichsjustizgesetze, 11
Reisevertrag, 117
Richtergesetz, 36

Sachsenspiegel, 9
Sammelverwahrung, 264
Satzung, 257
Scheinkaufmann, 214
Schmerzensgeld, 154
Schöffen, 29
Schuldrecht, 67
Schuldverhältnisse, 68, 147
Schwerpunktstreik, 318
Schwurgericht, 30
Selbstorganschaft, 242
Sicherheit, 252
Sicherungseigentum, 170, 186
Sicherungsübereignung, 237
Sicherungszession, 237
Sonderopfer, 173
Sortenschutzrechte, 297
Sozialgerichte, 34
Spezialisierungskartelle, 294
Stammanteil, 255
Stellvertretung, 225
Stiftung, 71
stückeloser Wertpapierverkehr, 265
Stückschuld, 91
subjektives Recht, 68

Tarifvertrag, 311
Tatbestand, 58
Testamentsvollstrecker, 197
Treu und Glauben, 86, 97

Überraschungsentscheidungen, 48
unerlaubte Handlungen, 147
Unmöglichkeit, 94
Unternehmensvertrag, 258, 274
Unterordnungskonzern, 272

verbundene Unternehmen, 272
Verein, 71
Verfassung, 14
Verfassungsbeschwerde, 22
verlängerter Eigentumsvorbehalt, 238
Verrichtungsgehilfe, 157
Verschmelzung, 271
Versendungskauf, 122
Versicherungsverein auf Gegenseitigkeit, 241
Vertragshändler, 235
Vertretener, 225
Vertretenmüssen, 113
Vertretungsmacht, 225
Verwahrung, 117
Verwaltungsgerichte, 33
Verwirkung, 138
Verzug, 94
vinkulierte Namensaktie, 264
Volljurist, 36
Vollmacht, 225
Vonselbsterwerb, 195
Vormerkung, 119
Vorstand, 8, 226, 258

Wandlung, 89, 127
Warenzeichen, 288
Werkvertrag, 117
Wettbewerbsregeln, 295
widerruflich, 76
Willenserklärung, 74
Willensmangel, 75, 80

Zugabeverordnung, 287
zugehen, 77
Zuständigkeit, 208

Index

Abortion, constitutionality, 25
Abstraction:
  agency, in, 225
  Pandectism, in, 74
  principle of, 70, 179
  rescission and, 84
Abuse:
  dominant position, of, 297
  rights, of, 138
Acceptance:
  letters of confirmation, 78, 216
  on own terms, 77
  silence and, 77, 216
Accidents:
  contract and tort, 146
  industrial, 321
Acquisition:
  community of, 191
  finder, by, 177
  good faith, in, 125, 171, 176, 217
  land, of, 179
  land register, 182
  maker, by, 177
Administration:
  anti-trust cases, 291
  lawyers in, 35
  mergers, 304
Administrative law:
  courts, 33
  judicial review, 34
  Landgericht, 30
  lay judges, 33
  right to complain, 34
  unconstitutional conduct, 22
Adoption, 207
AG:
  accounting, 266
  balance sheet, 267
  constitution, 257
  dissolution, 271
  employees on board, 260
  financial basis, 268
  formation, 257
  general meeting, 262
  management and shareholders, 258
  management board, 259
  management, control of, 261
  merger, 271
  organization, 258
  raising capital, 268
  reserves, 268
  shares in, 264
  supervisory board, 259
Agency:
  applicable law, 205
  Ausgleichszahlung, 232
  authority, lack of, 227
  commercial, forms of, 227
  commercial law, 225
  commercial manager, 229
  commission agent, 232
  creation of power, 225
  delegation, 228
  duties of agent, 230
  duties of principal, 231
  estoppel to deny, 227
  falsus procurator, 73, 227
  fees, 231
  independent contractor, 229
  internal and external, 225
  'organs', 73, 243
  ostensible authority, 230
  parent and child, 225
  Prokura, 228
  ratification, 227
  restrictive covenant, 231
  termination, 232
  transfer of, 228
  underlying contract, 230
  undisclosed, 232
Airplanes:
  security in, 169
  tort liability, 162
Amtsgericht:
  jurisdiction, 29
  land register, 180
  successions, in, 196
Animals:
  sale of, 94, 106
  tort liability, 163
Anti-trust law, 289 ff.
Anwartschaftsrecht, 183

Appeal:
  anti-trust cases, 308
  civil cases, 31
  constitutional attack, 23
  constitutional right, 31
  criminal cases, 31
  final, 31
  labour cases, 33
Arbitration:
  compulsory, 311
  foreign, 209
Arbeitsgerichte, 33
Armenrecht, 50
Assignment:
  agent's claim, 233
  banker, to, 238
  in bulk, 238
  method of, 120
  security, for, 187, 237
  vendor, to, 238
Associations:
  forms of, 240
  personality, 71
Attorneys:
  admission of, 41
  civil service, in, 35
  contingent fee, 43
  contractual duties, 42
  courts, attribution to, 42
  cross-examination, 48
  fees, 43
  legal aid, 50, 335
  litigation, role in, 46
  need for, 41
  preparation of cases, 40
  professional examination, 36
  sale of practice, 120
  scope of activity, 41
  training of, 35
  trial, role in, 46
  witnesses, and, 46

Bailment:
  chattel mortgage, 185, 237
  risk in, 121
  tort and restitution, 174
  transfer by, 233
  transfer of title, 175
Banks:
  agency of, 234
  cashier as agent, 229
  forms of, 239
  general conditions of business, 185
  legislation, 241
  merchants, as, 213
  new shares, 269
  proxies, as, 263
  suppliers, and, 188, 238
Bankruptcy:
  buyer's rights, 119
  director's liability, 156
Basic law:
  association, 311
  basic rights, 21
  co-determination, 279
  conflict of laws, and, 204
  construction of, 26
  entrenched clauses, 15
  equal rights, 23, 189
  governmental institutions, 16
  nationality, 201
  ordre public, 204
  origin, 14
  ownership in, 172
  right of personality, 166
  strike, political, 318
  trade unions, 313
Basic Rights:
  appeal, to, 32
  collective bargaining, 311
  divorce, 204
  examples, 21
  freedom of press, 24
Begriffsjurisprudenz, 60, 66
Berlin:
  Bundeskartellamt, 291
  Kammergericht, 291, 308
  special position, 15
  will, 197
BGB:
  §242, 135
  characteristics, 65
  dispositive rules, 85
  employment, 312
  family law, 189
  fundamental concepts, 68
  gaps in, 65
  general clauses, 66, 136
  General Part, 66
  HGB, and, 217
  invalid clauses, 85
  land in, 177
  limits to contractual freedom, 85
  origins, 64
  ownership in, 172
  partnership in, 239

pledge, in, 185
property law, 169
social rules, 65
sources of, 64
strict liability, 163
structure, 66
style of, 66
succession in, 195
tort, 147
vicarious liability, 157 ff.
wills in, 197
Boundaries, 180
Boycott, 301
Bundesgerichtshof, 31
Bundeskartellamt, 291
Bundesrat, 18
Bundesrepublik:
  basic rights in, 21
  Chancellor, 16
  constitution, 1, 14
  courts, system of, 27
  DDR, and, 2, 173
  elections in, 16
  European Communities, 6
  federal government, 16
  federal nature, 3
  foundation, 1
  Head of State, 17
  Länder in, 15
  President, 17
  rule of law, 4
Bundestag, 17
Bundesverfassungsgericht:
  basic rights, 22
  courts, control of, 23
  dissenting opinions, 40
  judges, appointment, 39
  legislation, control of, 25
  self-restraint, 26
Building contract, 101
Business:
  forms of organization, 239
  general conditions, 87, 143, 294, 302

Cadaster, 180
Capacity:
  conflict of laws, 205
  different types, 73
  inception of, 71
Cartels:
  control of, 291 ff.
  permitted, 294

Chancellor:
  dismissal of, 17
  nomination of, 16
  powers of, 17
Chattel mortgage, 185, 237
Children:
  BGB, in, 189
  capacity, 73
  care of, 208
  pre-natal rights, 72
Citizenship:
  acquisition of, 200 ff.
  East Germans, 2
Codes:
  citation of, 58
  commentaries on, 56
  draftsmanship, 52
  gap in the law, 12, 61
  general clauses, 136
  general principles, 62
  HGB and BGB, 217
  incompleteness, 72
  interpretation, 12, 61
Co-determination, 8, 277
Collective agreement:
  effects, 315
  history, 6
  norm and obligation, 315
  parties, 314
Combines:
  different types, 273
  disclosure, 275
Commercial law:
  acceptance by silence, 78, 216
  agency, 225
  agency, forms of, 227
  BGB and HGB, 218
  business activity, 212
  contract, modification of, 216
  custom, 219
  good faith acquisition, 217
  HGB, 211
  history, 5
  Landgericht, 30
  letters of confirmation, 78, 216
  merchant's lien, 217, 237
  nemo dat . . ., 217
  offers, 216
  penalties, 216
  Prokura, 228
  rescission, 95
  sources, 218
  transactions, 212
  unification, 5

Commercial register:
  accuracy presumed, 224
  AG, 258
  agency, 228
  duty to apply, 223
  duty to register, 223
  exclusion of liability, 222
  firm name, 219
  form and function, 223
  limited partnership, 248
  OHG, representation, 245
  Prokura, 228
  third parties, 224
Company:(see also AG, GmbH)
  accounting, 266
  anti-trust law, 272
  applicable law, 205
  books, 266
  co-determination, 276
  derivative suit, 270
  disclosure by, 266
  duty to liquidate, 156, 260
  GmbH, formation, 251
  land, on liquidation, 181
  liability to, 242
  'lifting the veil', 242
  linked enterprises, 272
  management and representation, 242
  manager, as, 252
  merchant, as, 212
  merger, 275
  one-man, 256
  organs, 242, 252
  partner, as, 250
  personality, 71, 241
  protection of public, 240
  registration, 251
  representation, 226, 252
  resolutions voidable, 264
  shareholder's liability, 242, 252, 256
  shares, 251
  shares in others, 273
  statutes, 6
  subsidiary, 273
  subsidiary, liability for, 242
  supervisory board, 254
  taxation, 280
  ultra vires, 259
  vicarious liability, 240
  workers on board, 254
Comparative law:
  legislation and, 60
  Max-Planck-Institut, 56

  similarity, assumption of, 45
Competition:
  anticompetitive conduct, 301
  boycotts, 284
  breach of law, 286
  codes of, 295
  cutthroat prices, 284
  dominant position, 298
  fairness and efficiency, 295
  false advertising, 286
  fault requirement, 283
  general conditions of business, 302
  good faith in, 139
  industrial property, and, 297
  mergers, 303
  misrepresentation, 286
  price-fixing, 288
  rebates, 287
  recommended prices, 296
  restraints on, 289
  unfair, 282 ff.
  UWG and GWB, 283
  vertical agreements, 295
Conflict of laws:
  agency, 205
  Basic Law, and, 204
  companies, 205
  contracts, 205
  EGBGB, in, 67, 202
  German and European, 309
  marriage, 206
  proof of law, 209
  property, 205
  renvoi, 204
  sales, 205
  succession, 208
  theories, 203
  tort, 205
Constitution; see Basic Law
Constitutum possessorium, 175, 233
Consumer:
  associations, 283
  comparative advertising, 285
  entrapment, 284
  hire-purchase, 183
  instalment sales, 133
  journal, 285
  misrepresentation, 284
Contract:
  acceptance by silence, 78, 216
  adaptation of, 143
  aftereffects, 93
  agency in commerce, 225

344 Index

analysis, 100
anticipatory breach, 104
applicable law, 205
attorney's duties, 42
avoidance of, 80
bailment, 171
bilateral, 99
breach, types of, 93, 107
building contract, 101
capacity, 73
capacity of foreigner, 205
closed shop, 313
collective bargaining, 311
combination, of, 273
commercial modifications, 216
commercial transaction, 215
commission contract, 218
composite nature, 95
conduct, and, 294
contributory negligence, 153
control contract, 274
conveyance, and, 70, 75, 170
conveyance, as, 119
deceit, 81
defective performance 107
delay, 94, 101
delay, risk in, 103
dispatch of goods, 78
disproportional expense, 96
dominant position, 298
duress, 81
duties in, 90, 106
duty of disclosure, 81
duty to, 84
effect of breach, 98
employment, BGB, 312
employment, of, 320
enterprise contract, 303
exemption clauses, 87 ff.
factual, 79
fault in, 98, 112
formalities, 87, 144
formation of, 76
foundation of, 142
fraud, 81
freedom of, 84
freedom, limits to, 85
freedom to create, 69
frustration, 141
function of, 84
general conditions of business, 87, 143
gentlemen's agreements, 292
good faith, 86, 97, 135

illegal, 85
immoral, 86
implied terms, 90, 138
impossibility, 96
impossibility, initial and subsequent, 97
index clauses, 141
inflation, and, 140
information, for, 157
inheritance, 197
intention to, 80
interpretation, 92
invalid clauses, 85
Leistungsstörungen, 93
management of affairs, 230
minors, of, 74
mistake, and, 82, 129
mixed, 120
notarization, 44
offer and acceptance, 76
oppressive, 86
parking-place, for, 80
partnership, 239
penalities, 216
performance, 90
performance and damages, 109
performance, place of, 92
performance, time of, 91
positive breach, 65, 94, 106
precontractual liability, 108
putting in default, 102
rebus sic stantibus, 141
reciprocity, 99
remedies, 109
remoteness of damage, 113
repudiation, 104
rescission and restitution, 112
restitution on avoidance, 70
restraint of trade, 291
restraints on, 296
restrictive covenant, 231
sale and other types, 117
specific performance, 103
standard terms, incorporation, 219
strict liability, 113
third parties, and, 115, 160, 316
tort, and, 146
tort, replacing, 159
unitary nature, 65
vicarious liability, 98, 114, 159
Conveyance:
    abstraction, principle of, 70
    choses in action, 120

contract, and, 70, 75, 119, 170
land, formalities, 179
land register, 181
minors, by, 74
moveables, 175
rescission, 83
unauthorized, 176
Cooperatives, 240
Copyright, 166
Courts:
 administrative, 33
 admission of attorneys, 42
 assignments, and, 187
 cartel authorities, and, 307
 cooperation, 209
 costs, 49
 decisions commented, 57
 foreign jurisdiction, 208
 good faith, and, 137
 general conditions of business, and, 88
 judgments, validity of, 209
 jurisdiction of, 27
 labour law, 312, 320
 legal aid, 50
 ordinary, 27
 ordinary, reform of, 32
 separate opinions, 40
 strikes, law on, 317
 system of, 27
 tax, 34
 unfair dismissal 325
 unfairness in competition, 284
 venue, 208
Crime:
 capacity, 73
 competition, in, 283
Criminal law:
 jurisdiction, 30
 jurors, 30
 Schwurgericht, 30
Culpa in contrahendo, 108, 159

Damage:
 extent of, 316
 kinds of, 152
 remoteness of, 151
Damages:
 anti-trust, 290
 breach of contract, 101
 causality, 111
 claim by nasciturus, 72
 computation of, 100
 consequential harm, 129

culpa in contrahendo, 108
delay, for, 102
flexibility, 101
industrial accidents, 322
invasion of personality, 168
negative interest, 109
pain and suffering, 154, 168
payment by instalments, 154
performance, additional to, 109
remoteness in contract, 113
repudiation, for, 105
rescission, on, 83
sales, in, 128
Schmerzensgeld, 154
subsidiary company, 273, 276
tort, quantification, 154
trade marks, 288
trade unions, 311
unfair competition, 282
Death:
 agency, effect on, 228
 agent's fees, 232
 gifts, effect on, 199
 land register, 181
 matrimonial property, 191
 merchant, of, 222
 moment of, 73
 partner, of, 246
Deceit:
 competition, in, 287
 contract, and, 81
Declaration:
 mistake in, 82
 rescission, of, 83, 104
 theory, 75
 time of efficacy, 77
Disclosure, duty of, 81
Discovery, 45
Dismissal:
 agent, of, 231 f.
 employee, of, 322 ff.
 manager, of, 253
Divorce:
 applicable law, 207
 BGB, in, 189
 fault in, 193
 foreign judgments, 209
 foreigners, of, 204
 jurisdiction, 29
 Law of 1976, 192
 pension rights, 195
 property rights, 191
 right to, 204

Drugs:
  contract to supply, 85
  tort liability, 163
Duress, 81

Easements, 181
East Germany:
  citizenship, 2, 201
  ownership in, 173
  treaty with, 201
EEC:
  cartel law, 308
Eigentumsvorbehalt, 183, 238
Elections, 16
Employment:
  BGB, in, 312
  collective agreement, 314
  co-determination, 276
  contract of, 320
  dismissal, 323
  enticing employee, 286
  industrial accidents, 321
  job security, 323
  'keeping the peace', 315
  lockouts, 318
  manager, of, 253
  minority groups, 311
  notice, 323
  redundancy, 325
  shareholding employees, 269
  strikes, 318
  wages for no work, 320
  works councils, 319
Enlightenment, 10
Enrichment:
  commixtion, 177
  contract, avoidance of, 70
  land register, 182
  minors, of, 74
  proceeds of sale, 176
  rescission, on, 84
  security, extinction of, 185
Environment, water pollution, 163
Equal rights:
  man and woman, 23, 190
  shareholders, 266
  union members and non-members, 314
Equality, basic right, 23
Equity, §242 BGB, 135 ff.
Erbschein, 197
Erfüllungsgehilfen:
  contract, 98, 113
  culpa in contrahendo, 108

Estoppel:
  §242 BGB, 144
  agency, 227
Europe:
  anti-trust law, 290, 292
  Germany in, 6
  human rights, 21
Evidence:
  expert, 48
  foreign law, 209
  rules of, 46
Exequatur, 209
Expropriation, 173

Familiengericht, 29
Family law:
  BGB, in, 67
  changes in, 67
  conflicts rules, 206
  intestate succession, 198
  present law, 189 ff.
Fixtures, 237
Foreclosure, 184
Formality:
  company foundation, 257
  guarantee, 216
  lack of, 144
  legal transactions, 120, 179
Freedoms, see Basic Rights

General clauses:
  Codes, in, 66, 136
  unfair competition, 282
General conditions of business:
  acceptance on, 77
  banks, 185
  commerce, in, 219
  competition, and, 302
  contract, and, 87
  incorporation, 219
  law on, 89, 302
  unfairness of, 88
  utility of, 88
German Law:
  bibliography, 52
  citation of, 57
  codes, 64
  divisions of, 51
  English law, and, 12
  English procedure, and, 45
  judge-made, 63, 135 ff., 147, 190, 284
  unification, 11
Germanic law, 9, 116

Germany:
  civilian law, 8
  codes in, 10, 64
  democracy in, 3
  Germanic law in, 9
  history, 1
  industrial development, 5
  nationality, 200
  ownership in, 173
  post-war division, 14
  post-war economics, 289
  Roman law in, 9
  unification, 2
Geschäftsgrundlage, 142
Gifts:
  notarization, 44
  revocable on death, 199
GmbH:
  creditors, liability to, 255
  dissolution, 256
  liability to company, 255
  partner, as, 250
  shareholder's liability, 255
  shares in, 255
Good faith:
  §242 BGB, 135
  acquisition in, 125, 171, 176, 217
  contract, and, 86
  limited partnership, 249, 276
  role of courts, 137
  statutory requirements, 143
Government, see Basic Law.
Grundbuch, see Land Register.
Grundgesetz, see Basic Law.
Grundrechte, see Basic Rights.
Grundschuld, 184, 236
Guarantee, 182, 236
  civil and commercial, 215
  shareholder, by, 256

Hand wahre Hand, 176
Handelsvertreter, 229
Harm:
  breach of statute, 155
  different kinds, 146
  economic loss, 149
  intentionally caused, 156
  personality, to, 165
  radiation, 162
  strike, due to, 316
Human Rights, see Basic Rights
Hypothek, 184, 236

Immorality:
  awareness of, 282
  company resolutions, 263
  contracts, in, 86
Index clauses, 141
Inheritance, see Succession
Injunction:
  anti-trust law, 291
  absolute rights, 174
  fault unnecessary, 283
  mergers, 304
  trade marks, 288
  trade union, 311
Infants, capacity, 73
Instalment sales:
  consumer protection, 133
  not in BGB, 65
Insurance:
  damages, and, 164
  duty to effect, 164
  forms of business, 241
  subrogation, 322
Intention, see Declaration, Will
Interpretation:
  §242 BGB, 138
  Basic Law, 26
  codes, 12
  contracts, 92
  declaration theory, 75
  declarations of will 75
  gap in the law, 61
  legislation, 61
  objective and subjective, 61
  will theory, 75

Jhering, 11, 62
Judges:
  administration, review of, 34
  Amtsgericht, 29
  appointment of, 38
  basic rights, and, 21
  cartels, control of, 291
  commercial, 30
  constitutional court, 21
  control of conduct, 144
  creativity, 26, 63, 135 ff., 147, 190, 284
  development of law, 61
  dissenting opinions, 40
  German and English, 40
  German and English view, 12
  independence, 39
  Landgericht, 30

lay judges, 30
lay, labour cases, 33
lay administrative, 33
  legislation, and, 61
  number of, 29
  precedent, 63
  promotion, 39
  revalorization, 140
  system of courts, 27
  training of, 36
  trial, role in, 46
  witnesses, questioning of, 48
  witnesses, summon, 46
Judgments:
  citation of, 58
  commented, 57
  constitutionality, 23
  dissenting opinions, 40
  execution, 209
  foreign, 209
  precedent, 63
  validity, 208
Jurisdiction:
  foreign cases, 208
  service abroad, 209
Jury, 45

KG:
  company as partner, 250
  liability, 249
  limited liability, 248
  management, 250
  modern uses, 249
  representation, 248
  taxation, 279

Kommissionär, 232

Labour law:
  closed shop, 313
  code?, 312
  co-determination, 8
  courts, 33
  dismissal, 323
  diverse sources, 311
  history, 7
  job security, 323
  'keeping the peace', 315
  minority groups, 311
  minors, 311
  Weimar's contribution, 7
  works councils, 311, 319

Länder:
  attorneys, training of, 37
  Bundesrepublik, in, 15
  cartel authorities, 307
  courts in, 27
  de facto powers, 19
  disputes with federation, 20
  finances, 20
  law libraries, 56
  legislation by, 18
  Oberlandesgerichte, 31
  role in legislation, 19
Land, (see also Property, Ownership)
  building, 178
  condominium, 178
  encumbrances, 180
  moveables and, 169
  priority, 180
  transfer of, 179
Land register:
  accuracy, 181
  Amtsgericht, 29, 180
  constitutive effect, 181
  securities, 237
  succession, 197
Landgericht, 29
Landlord and tenant:
  §242 BGB, 138
  BGB, in, 66
  jurisdiction, 29
  long leases, 178
Law books:
  approach to, 56
  different types, 57
  libraries, 56
  periodicals, 57
Law reform:
  §242 BGB, through, 138
  anti-trust, 293
  BGB, in, 66
  courts, ordinary, 32
  death, time of, 73
  dominant position, 298
  drugs, 163
  equal rights, 190
  family law, 67, 189
  general conditions of business, 89
  illegitimacy, 192
  instalment sales, 133
  legal aid, 50, 335
  legal training, 37
  ownership, 178

Index 349

security rights, 188
strict liability, 165
vicarious liability, 160
Law report, 58
Lawyers, see Attorneys, Judges, Notaries
Legal act, validity of, 73
Legal Aid, 50, 335
Legal advice, 41, 355
Legal method, 58
Legislation:
  application of, 58
  basic rights, 21
  Bundesrat's role, 18
  Bundestag, 16
  company law, 240
  competition, in, 287
  constitutionality, 22
  equal rights, 189
  excess of, 51
  federal competence, 18
  interpretation of, 61
  main instances, 52
  nationality, 200
  Occupying Powers, 290
  ordre public, 204
  'protective laws', 155, 290
  provincial powers, 18
  Roman law, and, 59
  strict liability, 160
  strikes, on, 317
  supremacy of, 58
  various types, 59
Life-tenancy, 181
Limitation, see Prescription, Time-Bar
Litigation:
  attorney and judge in, 47
  class action, 283
  contingent fee, 43
  costs, payment of, 49
  derivative suit, 270
  duty to inform opponent, 139
  expenses of, 49
  fees, 43
Loan:
  company, to, 256
  security for, 238
Lockouts, 318

Marriage, see Divorce
  notarization, 44
  property regime, 191
Mergers:
  notification, 303 f.

small businesses, 305
Merchant:
  appearance, by, 214
  duties, 213
  law, by, 213
  lien, 217, 237
  registration, by, 213
  silent partner, 250
  usage, 219
  who is, 212
Minderung, 127
Minors:
  capacity, 73
  international protection, 207
Mistake:
  content and expression, in, 82
  contract, in, 82
  deceit, induced by, 81
  principal and agent, 227
Mommsen, 95, 105
Money:
  nominalism, 140
  non-payment of, 110, 131
Morality:
  contract, and, 86
  effect of §242 BGB, 135
  tort, in, 156
  unfair competition, 282
Mortgage:
  chattels, of, 185, 237
  inflation, and, 140
  payments on, 182
  priority, 180
Motor-car:
  custodian's liability, 161
  parking of, 80
  sale of, 81

Nachfrist, 102
Name:
  agent's signature, 228 f.
  applicable law, 206
  business, 219
  HGB and UWG, 289
  letterhead of AG, 259
  letterhead of GmbH, 252
  right to, 165
  trade protection, 288
  transfer of, 221
  unauthorized use, 223
Nationality:
  conflicts law, and, 206
  German, 200

loss of, 201
naturalization, 201
succession, and, 208
Natural law, 10
Naturalization, 201
Neighbours:
  boundaries, 180
  conduct, 175
Negligence:
  attorneys, by, 42
  burden of proof, 152, 156
  company board, 260
  contract, in, 98
  economic loss, 149, 156
  'established business', 150
  master, required of, 157
  objective standard, 113
  rescission, and, 83
  standard of care, 148
  victim's, 153
Negotiable instruments:
  security, as, 236
  special laws, 218
  uniform laws, 218
Nemo dat . . .:
  in commerce, 217
  not followed, 176
Notaries, 44 ff.
Notarization:
  company contract, 258
  function of, 44
  lack of, 144
  land transactions, 120
  transfer of land, 179
  wills, 197
Nuclear damage, 162
Nuisance, 171, 174

Oberlandesgerichte, 31
Obligations:
  assignment, 187
  BGB, in, 67
  culpa in contrahendo, 108
  Gattungsschuld, 91
  inheritance, 196
  performance and, 170
  property, and, 68, 169
  security, 182
  Stückschuld, 91
Offer:
  binding nature, 76
  commercial law, 216
  duty to reject, 78

OHG:
  competition with, 244
  death of partner, 246
  dissolution, 246
  liability of, 246
  management of, 244
  partners, capacity, 244
  partners, powers of, 244
  representation of, 245
  requirements for, 243
  taxation, 279
  voting, 244
Ordre public, 203
Ownership:
  companies, 242
  condominium, 178
  expropriation, 173
  firms, of, 221
  injunction, 174
  joint, 177
  new things, 177
  OHG, by, 243
  OHG, in, 243
  partnership, 241
  real action, 174
  security, as, 185
  variants, 178

Pandectism:
  contract, and, 93
  effect of, 10
Parent and child:
  applicable law, 207
  BGB, in, 189
  law reform, 190
  legitimacy, 192
  nationality, 201
  representation, 225
  succession, 198
Parking-place, contract for, 80
Partnership, see also KG, OHG
  BGB and HGB, 239
  business name, 220
  civil and commercial, 218
  company control of, 276
  company partner, 250
  dissolution, 246
  exclusion of partner, 247
  general and commercial, 243
  labour and capital, 277
  liability of, 241
  limited partnership, 248
  'organs', 242

ownership, 241
personality, 241
restraint of trade, 293
silent, 250
Patents:
　competition and, 297
　improper claim, 150
　libraries, 56
Penalties, contractual, 216
Personal injury, 148
Personal security, 182
Personality:
　basic right, 25
　company, 241
　partnership, 241
Persons:
　capacity, 71
　death, 73
　majority, age of, 73
　natural and legal, 71
　pre-natal rights, 72
Pflichtteil, 199
Picture, publication of, 165
Pleadings, 45
Pledge:
　applicability of rules, 237
　moveables, 185
Prescription:
　property, 177
　sale, 131
　substantive law, 144
　tort, in, 131
Price-fixing, 296, 302
Possession:
　direct and indirect, 171, 237
　disturbance of, 171
　ownership, and, 171
　pledge, needed for, 185
　prescription by, 177
　publicity function, 171, 175, 238
　transfer by possessor, 175
Press:
　freedom of, 24
　invasions of privacy, 166
Privacy:
　basic right, 21
　threats to, 166
Private international law, see Conflict of laws
Procedure:
　anti-trust cases, 307
　civil, 45 ff.

expert witnesses, 48
foreign cases, 209
right to be heard, 48
Products liability, 153
Principal and agent, see Agency
Prokura, 228, 245
Proof:
　mergers, 305
　products liability, 153
Property:
　acquisition in good faith, 125
　applicable law, 205
　bankruptcy of vendor, 119
　BGB, in, 67
　commixtion, 176
　condominium 178
　contract and conveyance, 70, 75, 170
　divisibility, 170
　division of, 169
　easements, 181
　expropriation, 173
　form of conveyance, 179
　good faith acquisition, 176, 217
　inchoate title, 183
　industrial, 289
　industrial, and competition, 297
　joint ownership, 177
　kinds of, 169
　long leases, 178
　lost and found, 177
　matrimonial, 191
　neighbours, 175
　nemo dat . . ., 124
　new things, 177
　numerus clausus, 69, 170, 184
　obligations and, 68
　ownership, 169, 172
　possession, and, 171
　prescription, 177
　repossession, 174
　reservation of title, 183
　sale, notarization, 44
　security in, 184
　stolen goods, 171, 176
　tort, protection by, 148
Provinces, see *Länder*
Public law:
　libraries, 56
　private, and, 51

Quia timet, 174

352 *Index*

Rabel, 151
Railways:
 merchant, not, 214
 strict liability, 161
Real action, 174
Real security, 184
Rechtsgeschäft, 74
Referendar, 37
Representation: (see Agency)
 AG, 259
 company, of, 252
 unformed company, 258
Rescission:
 agents, 226
 commercial law, 95
 company resolutions, 264
 contract, of, 112
 conveyance, of, 84
 credit sale, 184
 damages, and, 83
 deceit, for, 81
 defective goods, 127
 delay, for, 103
 duress, for, 81
 irrevocability, 83
 mistake, for, 82
 need for declaration, 83
 repudiation, on, 105
 restitution, required for, 112
 risk, and, 112
Restitution: (and see Enrichment)
 abuse of position, 298
 breach of contract, on, 99
 contract, avoidance of, 70
 land register, 182
 proceeds of sale, 176
 property, and, 174
 rescission required, 112
Rights:
 absolute and relative, 68
 abuse of, 138
 assignment of, 187
 basic, see Basic Rights
 individual, 68
 injunctive protection, 174
 legal personality, 71
 personal and real, 169 f.
 pre-natal, 72
 protected by tort, 149
 security rights, 184
 use in competition, 297
Roman Law:
 effect of, 9

exceptio doli, 137
German law, in 51
legislation and, 59
Pandectism, 11
sale in, 116
vindicatio, 174

Sale:
 abatement of price, 127
 analysis, 69
 applicable law, 117, 205
 assignment of claim, 233
 bankruptcy, 119
 business, of, 221
 cancellation, 133
 carrier, liability for, 115
 collateral duties, 119
 commission agent, through, 232
 comparative advertising, 285
 consequential harm, 129
 contract and conveyance, 70, 118
 contract law, source of, 116
 credit, on, 183
 damages, 111, 124
 damages as surrogate, 95
 defect of title, 124
 defective goods, 125
 delivery, 122
 delivery date, 102
 duties in, 118
 duty to inspect, 126, 217
 duty to accept, 132
 exemption clauses, 87 ff.
 express warranties, 128
 fitness for purpose, 126
 forced, 184
 furniture, used, 88
 general conditions of business, 89
 generic goods, 91
 good faith acquisition, 125
 guaranteed attributes, 127
 HGB, and, 218
 history, 116
 impossibility and delay, 123
 infectious animal, 94, 106
 instalment sales, 65, 133
 manufacturer, liability for, 115
 minors, to, 74
 mistake and defect, 129
 motor-car, of, 81
 non-owner, by, 124
 other contracts, and, 117
 overpricing, 299

performance, claim to, 123
positive breach and defect, 130
possible objects, 120
post-contractual duties, 139
precontractual liability, 108
price payable, 131
rebates, 287
remedies, 122 f.
rescission for mistake, 83
reservation of title, 121, 183, 238
revaluation, 141
right to supplies, 297
risk, 92
Romalpa-clause, 238
self-service store, 75
seller's duties, 125
selling twice, 119
short delivery, 124
special rules, 117
special sales, 287
time-bar, 131
title of credit buyer, 183
title, reservation of, 121, 183, 238
transfer of risk, 120
unfair methods, 284
uniform law, 105, 118
vendor's security, 188, 237
ways of effecting, 234
Savigny, 11, 203
Schmerzensgeld, 154, 168
Security:
 assignment, by, 187
 commerce, in, 235
 conflicts, 188
 extinction of, 184
 foreclosure, 186
 multiple debtors, 236
 negotiable instrument, as, 236
 personal, 182, 236
 publicity, 186, 237
 real, 184, 237
 registration of, 184
 rights, in, 185, 237
 specification, 236
 stock-in-trade, 238
Servitudes, 181
Shareholders:
 AG, in, 258
 banks vote for, 263
 company, in another, 273
 company, liability to, 242
 companies as, 265
 derivative suit, 270

different types, 265
disclosure to, 267
duty to company, 266
employees as, 269
exclusion, 256
general meeting, 262
GmbH, in, 254
guarantee by, 256
minority rights, 270
loan by, 256
rights issues, 269
sale of shares, 255
taxation, 280
Sicherungseigentum, 186
Silence:
 acceptance, as, 78, 216
 deceit, as, 81
Social security:
 courts, 34
 history, 7
 industrial accidents, 322
 subrogation, 154
Spouses: (and see Divorce)
 applicable law, 206
 equality, 23, 190
 matrimonial property, 191 f.
 succession, 198
Strike:
 damages, 316
 dismissal, 323
 law, source of, 317
 legitimate purposes, 314
 liability, 316
 proportionality, 318
 sympathetic, 317
 union benefits, 314
 wages, effect on, 321
 wild-cat, 316
Succession:
 applicable law, 207 f.
 automatic, 196
 BGB, in, 67
 business name, to, 221
 disclaimer, 196
 executor, 197
 heir's liability, 196
 indefeasible rights, 199
 intestate, 198
 land register, 181
 matrimonial property, 191
 partner, to, 247
 wills, 197
Suretyship, 182, 236
Swiss Civil Code, 66

Taxation:
    capital transfer, 281
    companies, 270
    company and partnership, 250
    company law, 279
    courts, 34
    land sales, 179
    limited partnership, 249
    revenues, distribution of, 20
    share acquisition, 281
    share transfer, 281
Theft, stolen goods, 171, 176
Time-bar:
    conflict of laws, 209
    improper invocation, 144
    sale, in, 131
    substantive law, 144, 209
    tort, in, 131
Tort:
    airplanes, 162
    breach of statute, 155
    burden of proof, 153, 156
    capacity, 73
    causation, 151, 316
    company creditors, 242
    company, liability to, 242
    course of employment, 157
    damages for outrage, 168
    damages, quantum, 154
    dependants' claims, 155
    drugs, 163
    economic loss, 149, 156
    'established business', 222
    exculpation of employer, 158
    fault requirement, 147, 157
    honour, protection of, 165
    impairing credit, 165
    improper conduct, 156
    industrial accidents, 321
    intentional harm, 156
    fatal accidents, 155
    limited liability, 163
    manufacturer's liability, 153
    master and servant, 157
    motor-cars, 161
    nuclear damage, 162
    pain and suffering, 154, 163 f.
    personality, protection of, 166
    privacy, 165
    property, and, 174
    protected interests, 147
    remoteness, 151
    rules or principle?, 146
    strict liability, 160
    unborn child, 72
    unfair competition, 282 f.
    unfair criticism, 151
    unlawfulness, 147
    vicarious liability, 157
    water pollution, 162
    wrongful death, 148
Trademarks:
    competition and, 297
    protection, 286, 288
Trade unions:
    co-determination, 277
    damages, 311
    factory activity, 313
    history, 7
    industry-wide, 312
    injunction, 311
    'keeping the peace', 315
    legal advice, 335
    members' benefits, 313
    membership, 313
    nomination of judges, 33
    purposes, 317
    statistics, 313
    strikes, call, 317
Traffic accidents:
    limited liability, 164
    pain and suffering, 164
    passengers, 162
    strict liability, 161
    vicarious liability, 158
Trespass, 174
Trial:
    attorney and judge in, 47
    class action, 283
    costs in, 49
    costs, liability for, 152
    England and Germany, 45
    evidence, rules of, 46
    experts in, 48
    legal aid, 50
    orality, 46
    preliminary proceedings, 46
    proof of foreign law, 209
    witnesses, 46
    witnesses, questioning of, 48

Unification:
    commercial law, 5
    German law, 11
    German law, codes, 64
    sales law, 105

Universities:
  law libraries, 56
  law teaching in, 36
  role of, 9
Usufruct, 181

Venue, 208
Verfassungsbeschwerde, 22
Vertragshändler, 235
Vicarious liability:
  associations, 240
  contract and tort, 114
  culpa in contrahendo, 108
  tort, in, 157 ff.

Wandlung, 127
Weimar:
  constitution, 14
  labour law, 8
Will:
  Berlin, 197
  contract, required for, 79
  declarations of, 74
  formalities, 44, 197
  internal and external, 75
  limits to, 199
  theory, 75
  witnesses, 197
Willenserklärung, 74
Willensmangel, 75
Witnesses:
  expert, 48
  questioning, 46, 48
  summoned by judge, 46
  wills, 197
Works council:
  composition, 319
  dismissals, and, 324
  functions, 319
  history, 7